PROGRESS IN PHARMACEUTICAL AND BIOMEDICAL ANALYSIS
Series Editors: C. M. Riley, W. J. Lough & I. W. Wainer

VOLUME 1

Pharmaceutical and Biomedical Applications of Liquid Chromatography

Related Pergamon Titles of Interest

BOOKS

LUNTE & RADZIK: Capillary Electrophoresis*
WONG & WHITESIDES: Enzymes in Synthetic Organic Chemistry
PELLITIER: Alkaloids, Chemical & Biological Perspectives Volume 9*
HANSCH: Comprehensive Medicinal Chemistry

* In preparation

JOURNALS

BIOORGANIC & MEDICINAL CHEMISTRY
BIOORGANIC & MEDICINAL CHEMISTRY LETTERS
JOURNAL OF PHARMACEUTICAL AND BIOMEDICAL ANALYSIS
TALANTA

Full details of all Elsevier Science publications/free specimen copy of any Elsevier Science journal are available on request from your nearest Elsevier Science office

Pharmaceutical and Biomedical Applications of Liquid Chromatography

Edited by

C. M. RILEY

University of Kansas, USA

W. J. LOUGH

University of Sunderland, UK

I. W. WAINER

McGill University, Canada

PERGAMON

U.K.   Elsevier Science Ltd, The Boulevard, Langford Lane,
              Kidlington, Oxford OX5 1GB, U.K.
U.S.A.   Elsevier Science Inc., 660 White Plains Road,
              Tarrytown, New York 10591-5153, U.S.A.
JAPAN   Elsevier Science Japan, Tsunashima Building Annex,
              3-20-12 Yushima, Bunkyo-ku, Tokyo 113, Japan

First Edition 1994

Library of Congress Cataloging in Publication Data

Pharmaceutical and biomedical applications of liquid chromatography / edited by Christopher M. Riley, W. John Lough, and Irving W. Wainer.
-- 1st ed.
p. cm. -- (Progress in pharmaceutical and biomedical analysis; v. 1)
Includes bibliographical references and index.
1. Drugs--Analysis. 2. Liquid chromatography. 3. Biomolecules--Analysis.
I. Riley, Christopher M. II Lough, W. J. (W. John) III. Wainer, Irving W.
IV. Series
[DNLM: 1. Chromatography, Liquid. QD 79. C454 P536 1994]
RS 189. P444   1994
615'.1901--dc20
DNLM/DLC
for Library of Congress   94-20192

British Library Cataloguing in Publication Data

A catalogue record for this book is available from the British Library

ISBN 0 08 041009 X

Printed and Bound in Great Britain by Galliard (Printers) Ltd, Great Yarmouth

CONTENTS

LIST OF CONTRIBUTORS

Anliker, Sally L., Eli Lilly and Company, Indianapolis, IN 46285, U.S.A.

Ault, Joseph M., DuPont Merck Pharmaceutical Company, Wilmington, DE 19880, U.S.A.

Bolton, Sanford M., College of Pharmacy and Allied Health Professions, St. John's University, Jamaica, NY 11439, U.S.A.

Bopp, Ronald, J., Eli Lilly and Company, Indianapolis, IN 46285, U.S.A.

Kakodkar, Sunil V., Research Laboratory, J.T. Baker, Inc., 222 Red School Lane, Phillipsburg, NJ 08865, U.S.A.

Lang, J. Ronald, Glaxo Inc., Research Triangle Park, NC 27709, U.S.A.

Lloyd, David K., Department of Oncology, McGill University, 3655 Drummond, Room 717, Montreal, Quebec, Canada, H3G 1Y6

Lough, W. John, School of Health Sciences, University of Sunderland, Fleming Building, Chester Road, Sunderland, SR1 3SD, U.K.

Lunte, Craig E., Center for BioAnalytical Research and the Department of Chemistry, University of Kansas, Lawrence, KS 66047, U.S.A.

Lunte, Susan M., Center for BioAnalytical Research, University of Kansas, Lawrence, KS 66047, U.S.A.

Mical, Alfred J., DuPont Merck Pharmaceutical Company, Wilmington, DE 19880, U.S.A.

Narayanan, Sunanda R., Drug Metabolism, Oread Laboratories, Inc., 1501 Wakarusa Drive, Lawrence, KS 66047, U.S.A.

Noctor, Terrence A.G., School of Health Sciences, University of Sunderland Fleming Building, Chester Road, Sunderland, SR1 3SD, U.K.

Palmer John, MAC-MOD Analytical, Inc., Chadds Ford, PA 19317, U.S.A.

Perrin, Scott R., Regis Technologies, Inc., 8210 Austin Avenue, Morton, IL 60053, U.S.A.

Perry, John A., Regis Technologies, Inc., 8210 Austin Avenue, Morton, IL 60053, U.S.A.

Riley, Christopher M., Center for BioAnalytical Research and the Department of Pharmaceutical Chemistry, University of Kansas, Lawrence, KS 66047, U.S.A.

Stobaugh, John F., Center for BioAnalytical Research and the Department of Pharmaceutical Chemistry, University of Kansas, Lawrence, KS 66047, U.S.A.

Wainer, Irving W., Department of Oncology, McGill University, Montreal General Hospital, 1650 Cedar, Montreal, Quebec, Canada, H3G 1Y6

Wozniak, Timothy J., Eli Lilly and Company, Indianapolis, IN 46285, U.S.A.

Wuonola, Mark A., DuPont Merck Pharmaceutical Company, Wilmington, DE 19880, U.S.A.

Zief, Morris, Research Laboratory, J.T. Baker, Inc., 222 Red School Lane, Phillipsburg, NJ 08865, U.S.A.

INTRODUCTION

Liquid chromatography is used in every phase of pharmaceutical development, from initial synthesis or isolation of the active substance, through pharmacological and pharmacokinetic evaluation to the quality control of the finished product. This has been especially true since the introduction of modern high performance liquid chromatography in the late 1960s. Conversely, the tremendous growth of liquid chromatography over the last 25 years can be attributed in large part to its widespread acceptance by the pharmaceutical industry.

The continued popularity of liquid chromatography may be explained by the versatility of the approach, which can be used to separate and quantify large or small, polar, non-polar or ionic, and chiral or achiral molecules. The methodology can also be adapted to the preparative scale and is especially useful for the isolation of small samples of biologically active substances including proteins. In addition, methods are easily automated, increasing the number of analyses per unit time as well as improving the accuracy and precision and reducing the costs.

This volume follows a previous publication on the subject, published by one of the present authors in 1985[*] and is intended to provide an update on recent developments in the field. Consequently, this volume reflects the changes that have taken place in the pharmaceutical industry over the last ten years, most notably the increased importance attached to the question of chirality, the growing influence of biotechnology and the need for more rigorous documentation and validation of analytical methods and procedures.

The first part of this book deals with the application of new technology to pharmaceutical and biomedical analysis, reflecting the present needs for increased speed, sensitivity and selectivity in the analysis of drugs. The first chapter provides an overview of capillary electrophoresis, which represents one of the most important analytical developments to impact directly on pharmaceutical development in recent years. Although not a chromatographic technique, capillary electrophoresis was considered too important to be ignored. Moreover, Volume 2 of this book series on Progress in Pharmaceutical and Biomedical Analysis will be devoted entirely to the application of capillary electrophoresis to pharmaceutical and biomedical analysis.

---

[*]I.W. Wainer (ed.) "Liquid Chromatography in Pharmaceutical Development: An Introduction", Aster, Springfield, OR, 1985.

Over the last 25 years, liquid chromatography has grown into a mature analytical technique and many of the fundamental issues concerned with retention and separation are well defined. The practitioners of modern liquid chromatography spend as much time in the development of techniques for sampling, sample preparation and automation as they do in the development of the separation. Therefore, Part Two of this book describes recent advances in the areas of sample handling and the isolation of compounds from biological samples, including solid phase extraction, restricted access media for direct injection, coupled column technology and micro-dialysis. Similarly, Part Three contains two chapters concerned with liquid chromatographic methods for the isolation of drug substances, peptides and proteins from other complex media.

The pharmaceutical industry and the process of drug development are highly regulated and the increased importance that the regulatory authorities attach to validation has had a significant impact on the analytical techniques used for the analysis of drug. Although this has increased the workload of analysts in the pharmaceutical industry, it has also improved the quality of analytical methods used in the support of investigational and new drug applications as well the quality of methods published more recently in the literature. Consequently, Part Four of this volume describes approaches to the optimization and validation of liquid chromatography methods for the analysis of drugs in the bulk form, in pharmaceutical formulations and biological fluids. It is anticipated that Volume 3 of this Book Series will be devoted to a comprehensive treatment of the validation of analytical methods used in drug development.

This volume could not have been completed without the excellent efforts of the contributors who are named elsewhere or without the help of Nancy Harmony, who prepared all the manuscripts and final copies for publication. The assistance of the publisher's staff is also appreciated, especially the efforts of David Claridge, Peter Shepherd and Helen McPherson.

Christopher M. Riley
W. John Lough
Irving W. Wainer
1994

Part One: **Application of New Technology to Pharmaceutical and Biomedical Analysis**

CHAPTER 1

Pharmaceutical and Biomedical Applications

of Capillary Electrophoresis

DAVID K. LLOYD

Department of Oncology, McGill University,
3655 Drummond, Room 717,
Montreal, Quebec, Canada, H3G 1Y6

1. Introduction

Capillary electrophoresis (CE) separation techniques have become popular over the past decade for an extremely wide range of analytes. Their use has flourished with the introduction of commercial CE instrumentation. Many of the applications that have been reported are separations of pharmaceuticals, ranging from inorganic ions such as lithium to proteins such as human growth hormone. At this time much of the emphasis is on the analysis of bulk drugs, but there are an increasing number of papers reporting analyses of xenobiotics and endogenous compounds in biological matrices.

Modern CE methods are a refinement of open-tube zone electrophoresis techniques introduced by Hjerten (1967). Much early work concerned isotachophoretic methods (Everaerts *et al.*, 1976); however, zonal separation methods have attracted more attention of late. The recent growth of interest in CE methods was sparked by publications from Mikkers *et al.* (1979), who reported the use of PTFE capillaries with diameters of hundreds of microns, and from Jorgenson and Lukacs (1981a,b), who used glass capillaries less than 100 μm in diameter. It is interesting to note that one of these publications appeared in the journal *Clinical Chemistry* (Jorgenson and Lukacs, 1981b), revealing an early appreciation of the potentially wide applicability of CE methods. The advantages demonstrated by Jorgenson and Lukacs of using sub-100 μm capillaries to contain the separation medium are due to elimination of convection and of eddy

3

migration, due to frictional effects of the walls; to reduced current flow and, thus, reduced Joule heating in the capillary; and to highly effective heat dissipation due to the favorable surface area-to-volume ratio. These factors reduce heating-related zone-broadening processes to minimal levels, allowing the use of high separation potentials to give rapid, high-efficiency separations.

A variety of different CE separation modes are available. In capillary zone electrophoresis (CZE), separations are based solely on differences in electrophoretic mobility (Jorgenson and Lukacs, 1981a). In electrokinetic capillary chromatography (EKC) additives that complex or include the analytes are used. The most frequently reported EKC method is micellar electrokinetic capillary chromatography (MECC or MEKC), where separations of uncharged compounds are based on differences in partitioning between the hydrophobic interior of the micelle and the aqueous buffer outside the micelle (Terabe *et al.*, 1984). Most of the analyses described here will be either by CZE or MECC. However, many applications also exist for isotachophoresis (Everaerts *et al.*, 1976; Thormann, 1990), isoelectric focusing (Hjerten *et al.*, 1987; Mazzeo and Krull, 1991), gel-electrophoresis (Cohen *et al.*, 1988) and packed capillary (Knox and Grant, 1991) and open-tubular electrically driven capillary chromatography (Bruin *et al.*, 1990).

2. Instrumentation and Theoretical Aspects

2.1. Instrumentation

A diagram of a typical CE instrument is shown in Fig. 1. The heart of the system is the liquid-filled capillary tube in which the separation takes place. This is usually made of fused silica coated on the outside with polyamide to give greater resistance to breakage. Other capillary materials may be used, for example, teflon. The capillary is filled with a buffer, and electrical contact is made via buffer-containing vials at each end of the capillary. Application of an electric field across the capillary may give rise to a bulk flow of the liquid inside the capillary, the electroosmotic flow. The use of a number of types of inner-wall-coated capillaries has been described (Turner, 1991). These may be used to alter the electroosmotic flow within the capillary, or to reduce problems of adhesion of analytes to the capillary walls. Typically the capillary may be 20 to 100 cm in length, with an internal diameter of 25–100 μm. The liquid levels in the inlet and outlet buffer vials are generally kept at the same height, to avoid a gravity-flow of liquid through the capillary. A high voltage power supply capable of producing an electric field strength across the capil-

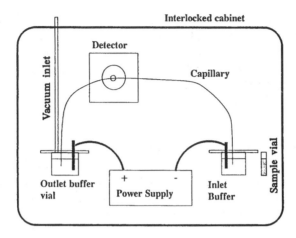

Fig. 1. Schematic diagram of a CE system.

lary of up to several hundred V cm^{-1} is connected to platinum electrodes in the buffer vials. For most reported CE separations, current flows within the capillary range from 10–200 µA. A screening and interlocking system is installed to protect the operator from contact with high voltages.

A variety of injection mechanisms are used in CE, but none are quite as accurate or reliable as a filled-loop injection in HPLC—the nanolitre injection volumes in CE make an efficient loop injector a considerable challenge in microplumbing. The sample is usually injected onto the capillary either hydrodynamically (differential pressure, gravity) or electrokinetically. All commercial instruments offer the possibility of performing electrokinetic injection. In this mode, one end of the capillary is dipped into a sample vial along with an electrode, and a voltage is applied across the capillary. Depending on the polarity of the applied field and the magnitude of electroosmotic flow in the capillary, a portion of the sample, which may or may not be representative of the total contents, is injected onto the capillary (Huang *et al.*, 1988). Provided that all conditions remain the same between analyses, this selectivity should remain constant. However, with complex and sometimes variable sample matrices such as urine, reproducibility may be difficult to achieve. A variety of hydrodynamic injection mechanisms are represented in current commercial instruments. Some use gravity flow, where the capillary end immersed in sample is raised approximately 10 cm above the outlet end. Liquid is allowed to siphon through the capillary for a given period of time, and then the ends are returned to the same level. Alternatively, a vacuum applied to the outlet end or overpressure at the sample end of the capillary may be used to inject some sample material. Unlike electrokinetic injection, hydrodynamic injection does not give a selective sampling from the matrix. Care must be taken when comparing injections between different reported analyses. The volume of liquid, v, injected onto the capillary is given by

$$v = \Delta P \pi d^4 t / 128 \eta L \qquad (1)$$

where d is the capillary diameter, L the length, η the viscosity of the buffer, t the time for which a pressure difference is applied and ΔP the magnitude of the pressure difference. Note the dependence on L^{-1} and d^4. A small change in injection volume could occur due to the common practice of cutting a few millimetres from the end of the capillary in case of blockage. Obviously, the change from a capillary of one diameter to another will drastically effect the injected quantity.

When preparing capillaries, care should be taken to ensure that the injection end of the capillary presents a squarely cut flat end, and that the polyamide external coating is burned back a few millimetres from the end. The polyamide can first be burned away with a lighter, and then the bare fused silica can be scored with a stone or glass-cutting tool before being snapped off. Inspection of the end with a magnifying glass will reveal whether a clean cut has been achieved. If not, considerable peak tailing may be seen, due to holdup of sample material at the capillary end. Burning the polyamide away from the capillary end is useful, since rough handling can cause the fracture of the silica within the polyamide and may also lead to formation of a dead volume at the capillary end. The deleterious effects of a poorly prepared capillary can be quite severe, as illustrated in Fig. 2. During development of an MECC assay for the anti-cancer agent doxorubicin, peak broadening was seen, and with some "optimisation" two peaks were resolved in a supposedly pure aqueous standard sample. These electropherograms are shown in the upper traces in Fig. 2. After some investigation, the capillary end was found to be fractured inside the polyamide. Upon re-cutting the end, the result was the third electropherogram in Fig. 2, the previous peak doublets being artifacts from the dead volume in the broken capillary end.

For optical detection, a small window is made at some point along the capillary by removing the polyamide, usually by burning or etching away the coating. Detection is then achieved on-capillary. UV absorbance is the most frequently used technique because of its wide applicability and its availability in commercial instruments. Fluorescence detection offers greater sensitivity, especially with laser excitation (Roach *et al.*, 1988). Electrochemical detectors offer good sensitivity for electroactive compounds (Wallingford and Ewing, 1987). CE has also been coupled with mass-spectrometry (CE-MS) (Edmonds *et al.*, 1989; Muck and Henion, 1989).

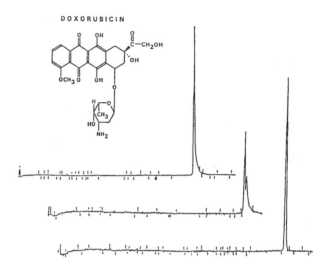

Fig. 2.   Effect of a badly cut capillary end. Upper trace, badly
tailing peak in an MECC analysis of doxorubicin; middle
trace, tailing peak resolved into a doublet after some
"optimisation"; lower trace, symmetrical doxorubicin
peak obtained after cutting capillary to give a flat end.

    The main problem with UV absorbance detection is that concen-
tration sensitivity is generally rather poor in comparison with HPLC.
This comes about because of the short detection pathlength when
looking across the width of the capillary. A variety of optical designs
have been developed to increase sensitivity, including z-shaped
capillary flowcells, bubble cells, end-illumination and rectangular
capillaries (Albin *et al.*, 1993); some of these are commercially
available. The use of rectangular capillaries has also been reported
(Tsuda *et al.*, 1990), but at the moment such devices are not widely
used.   One simple way in which reasonable sensitivity may be
achieved for many compounds is the use of very low UV wavelengths
for detection. Wavelengths of 190 and even 185 nm are offered on
many commercial instruments, and so it is possible to take advantage
of the higher absorption coefficients of many compounds at these
wavelengths. Because of the short capillary pathlength there is often
adequate transmission of light for absorption measurements, despite
absorbance by the buffer. Typically, detection limits in the range of
100 ng ml^{-1} to 10 µg ml^{-1} may be achieved with UV absorbance

detection, although in many cases on-capillary peak concentration may be used to improve limits of detection by up to an order of magnitude.

2.2. *Theoretical background*

Only a brief resumé of the theory behind CE separations will be given here. A detailed theoretical introduction to electrophoretic separations can be found in a number of texts (Shaw, 1969; Wieme, 1975). Wallingford and Ewing (1989) gave a general introduction to CE, covering both theoretical aspects and selected applications. Two articles that give a detailed background to capillary electroseparation methods have been published by Foret and Bocek (1990) and Klepárník and Bocek (1991).

2.2.1. *Ion movement and mobility*

A charged particle subject to an electric field will be accelerated by a force F_e

$$F_e = qE \qquad (2)$$

where E is the electric field strength, and q is the charge on the particle. Moving in a liquid medium, the motion of a spherical particle will be opposed by a frictional force F_f, given by Stokes law,

$$F_f = 6\pi\eta rv \qquad (3)$$

where η is the coefficient of viscosity, v is the particle velocity, and r is the particle radius. From Equations 2 and 3, the velocity of an ion moving under the influence of an electric field can be expressed as

$$v = qE/6\pi\eta r \qquad (4)$$

This simple relation is not directly useful for the calculation of ionic velocities. It is assumed that the ion is spherical, which may not be the case. In practice, the particle radius r will not be equal to the radius of the ion alone, but to the radius of the solvation shell of the ion. Also, in an electrolyte, the counterion shell around the central ion will be accelerated in a direction opposite to that of the central ion, resulting in an increased frictional force on the central ion. This results in a reduction in v, and this effect becomes greater as the ionic strength of the electrolyte is increased. Despite these caveats, equation 3 serves to show the dependence of electrophoretic motion on charge, and an inverse dependence on radius.

Instead of measuring only the velocity of the ion, another parameter, the electrophoretic mobility, u, is often useful, with

$$u = v/E = q/6\pi\eta r \qquad (5)$$

which accounts for differences in field strengths. Absolute values of mobility may be measured under specified conditions, and ionic mobilities are calculated at zero concentration of background electrolyte, with other conditions being specified.

The above equations assume that the analyte is completely dissociated. However, for weak electrolytes in solution this is not generally the case. The charge on a molecule with a single ionisable group is easily determined if the pK_a is known, from the Henderson-Hasselbalch equation

$$pH = pK_a + \log(\frac{1}{a} - 1) \qquad (6)$$

where a is the degree of dissociation of the ionisable group. The effective mobility, u_{eff}, is related to the mobility of the completely ionised species, u_i, by

$$u_{eff} = u_i a \qquad (7)$$

The overall velocity of an analyte molecule traveling through a capillary is due to its own motion resulting from the electric field, and also to a bulk flow of the liquid in the capillary, the electroosmotic flow. Electroosmotic flow arises in a fused silica capillary because of ionisation of the surface silanol groups, which gives rise to counterion condensation towards the silica surface. Upon application of an electric field there is a consequent movement of the counterions, along with their solvation atmospheres. Because the flow is driven at the walls of the capillary and there is no frictional resistance to flow in the centre of the capillary, the resulting flow profile is essentially flat. This is quite unlike the laminar flow profile seen with hydrodynamic flow. The magnitude of the electroosmotic flow is a strong function of pH, which mirrors the degree of ionisation of the capillary surface silanol groups. Thus, at pH values below about 2.5 there is almost no electroosmotic flow, whereas at pH values above about 6, electroosmotic flow is large and relatively stable. Between these values electroosmotic flow is a strong function of pH. Experimental data on the magnitude of electroosmotic flow as a function of pH is shown in Fig. 3 (Lambert and Middleton, 1990). This illustrates that the capillary history also has an effect on electroosmotic flow, and that an acid-washed capillary gives lower electroosmotic flows than an alkali-washed capillary in the intermediate pH region. Also notable is the fact that a greater variability in electroosmotic flow is seen at intermediate pH values, perhaps reflecting the strong dependence of electroosmosis on pH in this region. The electroosmotic flow has a similar dependence on ionic strength and valency of the electrolyte as the motion of a particle within the electrolyte, *i.e.*, increasing the ionic strength will decrease the electroosmotic flow velocity. An electroosmotic flow mobility may be defined as $u_{eo} = v_{eo}/E$, where v_{eo} is the velocity of the electroosmotic flow. The term v_{eo} is usually determined by injecting an uncharged compound, which migrates through the capillary with the electroosmotic flow.

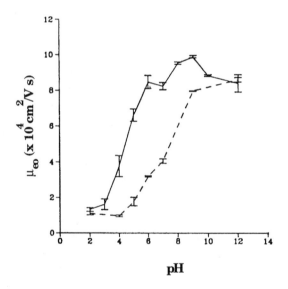

pH

Fig. 3. Effect of pH on electroosmosis. The effect of capillary history is shown, with the solid line representing data obtained on going from alkaline to acidic conditions, and the dashed line going from acid to alkaline conditions (Reproduced with permission from Lambest and Middleton 1990).

The total analyte mobility, u_t, can be measured, knowing the migration time and distance, and the electric field strength. The field strength is given by $E = V/L$, where V is the applied voltage, and L is the total length of the capillary. The analyte velocity is given by $v = l/t$, where t is the migration time, and l is the length from the injection end of the capillary to the detector. With on-capillary detection l and L are not usually equal. Thus,

$$u_t = Ll/Vt \qquad (8)$$

The electroosmotic flow velocity and thus mobility can be measured if a neutral marker compound is injected along with the analyte, and so the effective mobility of the analyte can easily be calculated, since

$$u_t = u_{eff} + u_{eo} \qquad (9)$$

It is useful to calculate the analyte mobility since variations in electroosmotic flow may occur between separations due to modifications of the capillary surface (*e.g.*, adsorption of proteins to the capillary wall). This will be evidenced by a variation in the migration time of the analyte and neutral marker, with the analyte mobility remaining constant. This can be particularly pronounced in the analysis of unextracted biological fluids. Calculation of the total mobility and thus expected migration time may also be useful in identifying a peak in a complex mixture if there is some variability in the electroosmotic flow (Beckers *et al.*, 1991). Facilities for mobilities calculations are written into at least some commercial software for data handling and control of CE instruments.

2.2.2. Band-broadening and separation efficiency

As mentioned above, the advantages of performing electrophoretic separations in capillaries were highlighted in the work of Jorgenson and Lukacs (1981a,b). The rationale for using small diameter capillary tubes is based on the anticonvective frictional effects of the capillary walls. This eliminates macroscopic mixing effects in the capillary, which would otherwise cause the merging of zones. Another important consequence of the use of small-diameter capillaries is that Joule heating in the capillary is reduced—for example, a 100 mM sodium borate solution in a 72 cm long, 50 µm internal diameter capillary passes a current of approximately 65 µA when a voltage of 20 kV is applied. This gives rise to 1.3 W of Joule heating, or approximately 18 mW cm^{-1}. Doubling the capillary diameter to 100 µm would be expected to result in a current four times higher with the same buffer system, in proportion to the increase in cross-sectional area of the capillary, and thus a fourfold increase in heat production. In practice this would probably lead to boiling of the electrolyte within the capillary. The favourable ratio of surface area to volume of small capillaries helps ensure excellent heat removal out of the capillary bore into the fused silica walls. However, there may be a considerable rise in temperature of the whole capillary if adequate measures are not taken to dissipate the heat generated into the surrounding medium. In the limit, the temperature rise of an inadequately cooled capillary may lead to boiling of the electrolyte and breakdown of the separation.

There are a number of factors that lead to band-broadening in CE. These include contributions from diffusion, effects related to Joule-heating, electromigration dispersion, injection and detection effects, and analyte interactions with the capillary surface.

Excessive Joule heating within the capillary affects the viscosity of the buffer and thus increases the diffusion coefficients of the solutes, leading to increased band-broadening. Heat transfer between the center of the capillary through the liquid and through the walls is usually excellent compared to the heat transfer from the capillary walls into the surrounding air. If band-broadening due to Joule heating is a significant problem, a solution may be to use the more effective method of liquid rather than forced-air cooling. However, this is likely to be critical only in a limited number of cases where high concentrations of high-conductivity buffers are being used. Another solution may be to reduce the buffer concentration or to investigate the use of a lower-conductivity buffer system.

Having just stated that reduction of the buffer concentration might be a good strategy for reduction of band-broadening associated with Joule-heating, it should be mentioned that at too low a buffer concentration another band-broadening mechanism may become significant. This is electromigration dispersion (Mikkers *et al.*, 1979; Sustácek *et al.*, 1991), which occurs if the sample zone conductivity is sufficiently great that there is a significant modification of the electric field within that zone. In this case, different parts of the zone will move with different velocities, leading to a spreading of the zone. This type of broadening mechanism is evidenced by peak asymmetry, with either tailing or fronting peaks. In practice, if the buffer concentration is at least 100 times greater than the sample concentration, electromigration dispersion effects are minimal. Because of the practical limits to buffer concentration due to Joule heating, electromigration dispersion may lead to a limitation on the linear range of analyte concentration that can be assayed (Weinberger and Albin, 1991). Experimental results on the analysis of naproxen over a concentration range of 9.8 μg ml^{-1} to 10 mg ml^{-1} are shown in Fig. 4 (Weinberger and Albin, 1991). Significant peak distortion is obvious at higher analyte concentrations.

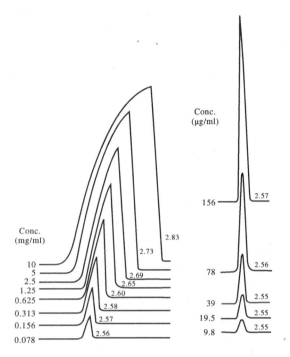

Fig. 4.   Peak broadening due to sample overloading. Symmetrical peaks are obtained at the lowest concentrations of naproxen (9.8 μg ml^{-1}) while increasingly distorted and broadened peaks are seen up to a maximum solute concentration of 10 mg ml^{-1}. Buffer is 20 mM borate, pH 9.2. (Reproduced with permission from Weinberger and Albin, 1991.)

Huang *et al.* (1989) suggested that the finite width of the injection zone is quite often the main contribution to the width of the detected zone. They noted that when the injection width is the most significant contribution to plate height, N is constant for all analytes in the system, with a value related to the capillary length (ℓ) and the injection width (w_{inj}^2):

$$N = \frac{12\ell^2}{w_{inj}^2} \tag{10}$$

For example, in a capillary where 1% of the total length is filled with sample at the beginning of an analysis, the maximum expected plate number would be 120,000. Similar reductions in efficiency may result from the width of the detection region, although usually this is negligible in comparison with injection and other effects. Another factor to be aware of concerning detection is the effect of the detector time constant. In HPLC, it is suggested as a rule of thumb that the detector time-constant should be approximately 1/10th of the peak width. In CE, temporal peak widths are often as low as 1–2 s, thus a short detection time constant should be used so as not to introduce any band-broadening.

2.3. Capillary zone electrophoresis

In CZE analyses, separation takes place with a small starting zone of sample introduced into a capillary filled with a homogeneous buffer system. At neutral to high buffer pH with uncoated fused silica capillaries there is a strong electroosmotic flow, and most analytes of both positive and negative charge are swept past the detector. Resolution is achieved on the basis of differences in electrophoretic mobility—i.e., differences in analyte charge and mass or size. Since the total analyte mobility is a result of its own effective mobility plus the electroosmotic flow mobility, the electroosmotic flow can have a strong effect on the resolution of two compounds (Eq. 9). When the electroosmotic mobility is in the same direction as the mobility of the analyte, resolution is reduced compared to the situation with no electroosmotic flow. However, if the electroosmotic mobility opposes that of the analytes an increase in resolution may be obtained (Fig. 5). In the extreme case, if electroosmotic flow exactly balances the motion of one of a pair of analyte molecules, infinite resolution can be achieved, assuming infinite patience of the analyst. Resolution can be manipulated by changing the pH of the buffer system to optimise the mobility differences between the analytes. Because electroosmosis is also a strong function of pH over a range around 3–6, the dependence of resolution on pH may not be simple at these pH values.

If the pK_a values of the analytes are known, the choice of separation conditions can be a relatively simple matter. Wren (1991) described optimisation of the CZE analysis of 2-, 3- and 4-methylpyridine, starting with knowledge of the pK_a values of the analytes. The choice of analytical pH then involves calculating the pH value that maximises the charge difference between the analytes. For peptides it is possible to make fairly accurate estimates of the overall charge at a given pH, knowing the amino acid constituents and their respective pK_a values. van de Goor et al. (1991) described the optimisation of the separation of a variety of peptide fragments derived from adrenocorticotropic hormone by calculation of their charge and relative mobility starting from the sequence data.

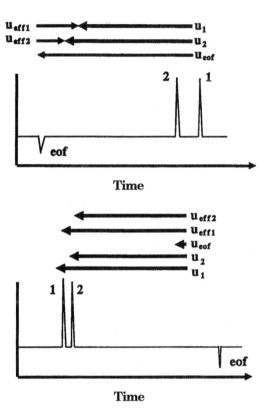

Fig. 5. Effect of electroosmosis on resolution and migration time. The overall analyte velocity is the
sum of the analyte's own electrophoretic motion
and the electroosmotic flow velocity. At high pH
anionic solutes migrate against the electroosmotic flow, while at low pH cationic analytes
migrate in the same direction as a much reduced
electroosmotic flow.

As well as variation of the buffer pH, a number of other parameters may be varied to effect a CZE separation. If there is evidence of
a partial separation with the analytes migrating rapidly along the
capillary, it is often possible to trade off resolution against analysis
time. One way to do this is simply to increase the length of the
separation capillary while maintaining the electric field strength.
Alternatively, if excessive electroosmotic flow is a problem, some
method of modifying electroosmosis may be used. Several

approaches are possible. One is to use a capillary with a modified inner surface. Polyacrylamide coatings (Hjerten, 1985; Cobb *et al.*, 1990) can almost completely eliminate electroosmosis. If this is too drastic, a coating which gives a somewhat reduced electroosmotic flow may offer a suitable alternative. Materials such as short-chain alkyl coatings have been used for this purpose (Turner, 1991). Instead of using a permanently modified capillary, variation in the separation buffer constituents may be used to achieve a desired separation. Organic modifiers such as acetonitrile and alcohols have been widely used for this purpose (Schwer and Kenndler, 1991), their effects being due to modification of the zeta-potential at the capillary surface or modifications to the buffer viscosity. A simple change in the buffer concentration may also bring about the desired result. Increasing ionic strength usually leads to an increase in separation efficiency (until heating effects become significant) and a reduction in electroosmotic flow.

2.4. Micellar electrokinetic capillary chromatography (MECC)

Micellar electrokinetic capillary chromatography (MECC) is an elegant countermigration separation technique that was first introduced for the CE analysis of uncharged solutes (Terabe *et al.*, 1984) and has since found application for a very wide variety of separations of both ions and neutrals. The analytical instrumentation is identical to that for CE, but a micelle-forming surfactant is added to the separation buffer. The surfactant is added at a concentration above the critical micelle concentration (cmc), the concentration at which micelles begin to form. The first MECC separations were demonstrated using sodium dodecyl sulfate (SDS), an anionic surfactant. Fig. 6 shows a representation of an MECC separation. Analysis takes place in a fused-silica capillary at intermediate to high pH, resulting in a strong electroosmotic flow. The micelles are themselves negatively charged due to the charge on each individual SDS molecule, so they have a strong anodic migration. Usually, the separation buffer is such that the electroosmotic flow mobility is greater than the micellar mobility; thus the net motion of the micelles is toward the cathode. The interior of the micelles is highly hydrophobic, formed from the alkyl chains of the surfactant molecules. The exterior surface is a region of high charge-density, with an associated atmosphere of condensed counterions. Outside the micelle is the aqueous buffer medium.

The mechanism by which separation of neutrals takes place in an MECC system is partitioning of the solute between the buffer phase and the interior of the micelle. However, for ionisable species, the

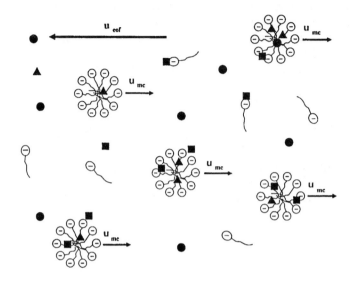

Fig. 6.   Schematic representation of an MECC separation. The micelles have a mobility u_{mc}, against
the electroosmotic flow direction.   Neutral
solutes (▲) move at the micelle velocity when
included, and with the electroosmotic flow when
outside the micelle.   Anionic solutes (●) are
generally repelled from the anionic micelles,
whereas cationic solutes (■) may be included into
the micelles or ion-pair with the micellar surface
or free surfactant monomers.

charge of the solute will lead also to electrophoretic migration and
separation, and ion-pairing rather than partitioning effects may also
be observed. Since there are several possible mechanisms by which
separation may occur, it can be quite difficult to predict the effect of
changing one parameter in an MECC separation. However, for the
simplest case, the MECC separation of uncharged solutes, analysis of
cause and effect is reasonably simple. Analogous to HPLC, the
capacity factor, k', is given in MECC by (Terabe *et al.*, 1984)

$$k' = (t_r - t_0)/t_0(1 - t_r/t_{mc}) \qquad (11)$$

where t_0 is the migration time of a neutral marker which is unretained in the micelle and moves only with the electroosmotic flow,

t_r is the migration time of the solute and t_{mc} is the migration time of a marker that is completely retained in the micelle, giving a measure of the micelle mobility. k' may be related to the partition coefficient between the buffer and micelle phases, P_{wm}, by:

$$k' = P_{wm}\beta \tag{12}$$

(Foley, 1990), where β is the phase ratio,

$$\beta = V_{pm}([SURF] - CMC)/(1 - V_{pm}([SURF] - CMC)) \tag{13}$$

with [SURF] the concentration of surfactant in solution, V_{pm} the partial molar volume of the surfactant micelle (for SDS, $V_{pm} = 0.26$ dm^3 mol^{-1} (Saitoh et $al.$, 1989)). The cmc is the critical micelle concentration for the surfactant. The cmc may vary as a function of the ionic strength of the buffer medium, and although the cmc of SDS is around 8 mM in deionised water, it may be lower in buffer solutions typically used in CE (Tanford, 1980).

To achieve the desired resolution of analytes in MECC there are several parameters which may be varied. Unfortunately, these parameters are often interdependent and if this is the case, predicting the effect of a given change can be quite difficult. The simplest case concerns the analysis of uncharged analytes, when the surfactant concentration, the buffer pH and ionic strength, temperature and buffer modifiers are important variables. A separation by MECC—in fact, the first ever published—is shown in Fig. 7 (Terabe et $al.$, 1984). Capacity factor as well as migration time is shown. Peak 1 is due to methanol, a solute that is unretained in the micelles, and thus provides an electroosmotic flow marker. Peak 6 is due to Sudan III, a mixture of dye compounds which has been used by many workers as a marker of the micelle velocity. In between fall the analytes, with k' values ranging between about 0.5 and 4.

Excellent resolution is seen for the analysis shown in Fig. 7. However, separations are not as good for more hydrophobic analytes that are well included in the micelles. These elute in the latter part of the electropherogram, where small differences in k' will not lead to good resolution. For highly retained solutes, a number of approaches may be used to increase resolution. Firstly, a decrease in electroosmotic flow to more closely match the micelle velocity will spread out the electropherogram, but this will only achieve better resolution at the expense of a considerable increase in analysis time. A preferable alternative is to reduce the partition coefficients of the analytes. This can be done by the use of various additives which increase the affinity of hydrophobic analytes for the buffer phase. Perhaps most frequently reported is the option of adding organic modifiers such as methanol or acetonitrile to the buffer. The addition of urea and cyclodextrin additives can also be effective. In each case, the addition of modifiers affects not only the partition coefficients but also the electroosmotic and micelle mobilities.

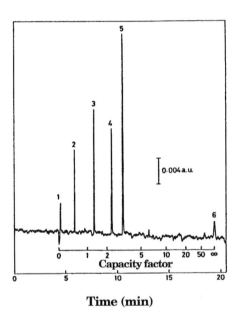

Time (min)

Fig. 7.   An early MECC separation of (2) phenol, (3) *p*-creosol, (4) 2,6-xylenol and (5) *p*-ethylphenol. All solutes are eluted between methanol (1), which is not included in the micelles, (capacity factor = 0) and Sudan III (6), which is completely included (capacity factor = ∞). (Reprinted with permission from Terabe *et al.*, 1984).

With charged analytes, MECC often gives a very different selectivity from a separation achieved at a similar pH by CZE. Not only do partitioning effects into the interior of the micelle play a part, but ion-paring may occur. The charge on the analyte may have a great effect on its ability to associate with the micelle—for example, anionic solutes may be repelled from the highly negatively charged surface of an SDS micelle while being preferentially associated with a cationic surfactant. The behaviour of ionic analytes in MECC has been described by Khaledi *et al.* (1991) and Strasters and Khaledi (1991).

3. Applications in Pharmaceutical Analysis.

In this section applications concerning the analysis of bulk drugs and drugs in dosage forms will be discussed. Usually sensitivity is not a problem, so the concern is mainly with achieving good reproducibility and linear or dynamic range. CE methods, in particular MECC and CZE, have been applied to the analysis of a wide variety of compounds of pharmaceutical interest. Small molecule examples include analgesics (Fujiwara and Honda, 1987; Nishi et al., 1990a), tricyclic antidepressants (Salomon et al., 1991) and β-lactam antibiotics (Hoyt and Sepaniak, 1989; Nishi et al., 1989a). An inter-company cross validation exercise for the chiral analysis of clenbuterol has shown that CE methods can be successfully transferred between laboratories (Altria et al., 1993).

3.1. General Pharmaceutical Applications

An example which contrasts the use of CZE and MECC is the analysis by CE of water-soluble vitamins in standard samples and pharmaceutical preparations. Several papers have been published on the analysis of these compounds by CE (Fujiwara et al., 1988; Nishi et al., 1989b). The effect of pH ranging from slightly acidic to alkaline conditions on the mobility of vitamins in a CZE separation is shown in Fig. 8a, mobility values being calculated from the data in Nishi et al., 1989b. In the pH range from 6 to 9 the analytes ranged from anionic through neutral to cationic species. In the CZE separation shown in Fig. 8a, neutrals migrate together with the electroosmotic flow; optimum resolution for the rest of the analytes appears to be around pH 8 to 9. Separation of all the components is impossible, and so for this some sort of complexation or inclusion method is needed.

Fujiwara et al. (1988) investigated the use of SDS micelles for the complete separation of charged and uncharged water-soluble vitamins, while Nishi et al. (1989b) used SDS, as well as sodium pentane sulfonate as a non-micelle-forming anionic surfactant. The effect of the addition of SDS on the effective mobilities of the solutes with a pH 9 phosphate-borate buffer is shown in Fig. 8b, mobilities again being calculated from data in Nishi et al., 1989b. Methanol was used as a neutral marker of electroosmotic flow. It can be seen that the addition of SDS generally has a minimal effect on the mobility of most of the anionic solutes. This can be explained because their negative charge is likely to repel them from the highly negatively charged micelle surface. Vitamin B_{12} is well included in the micelle, as evidenced by a large increase in effective mobility; the other two neutral analytes nicotinamide and pyridoxamine show less inclusion. The

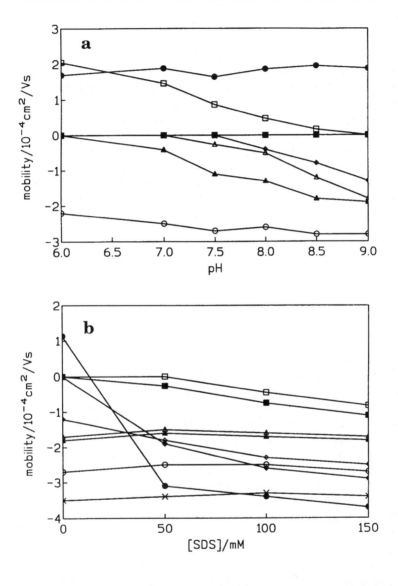

Fig. 8. Effective mobilities of water-soluble vitamins in a) CZE
as a function of pH and b) MECC as a function of SDS
concentration in a pH 9 phosphate-borate buffer. □,
Pyridoxamine; ■, nicotinamide; △, pyridoxal; ▲,
vitamin B6; ◇, vitamin B2 phosphate; ◆, vitamin B12;
O, niacin; ●, vitamin B1; ✕, pyridoxal-5'-phosphate.
Vitamin B2 phosphate and pyridoxamine-5'-phosphate
both have mobilities very similar to that of niacin, and
thus those data have been eliminated for clarity.

greatest change in mobility is seen with the cationic vitamin B_1, which elutes before the electroosmotic flow marker in CZE but after the marker in MECC. In this case, inclusion with the SDS micelle is aided by the electrostatic attraction between the oppositely charged species. The use of the non-micelle-forming surfactant sodium pentanesulfonate resulted in no change in the elution order of vitamin B_1, suggesting that inclusion of this cationic analyte into the micelle occurs, rather than ion-pairing. This example illustrates that the addition of SDS in a separation already partially resolved by CZE can be used to modify resolution, with the effect on a particular component strongly dependent on its charge. If a cationic surfactant were to be used instead of SDS, there might be only a slight modification of the vitamin B_{12} mobility, and greater effects on the anionic solutes.

Fujiwara *et al.* (1988) investigated the quantitation by MECC of a variety of water-soluble vitamins in a commercial vitamin injection. Using an internal standard, they achieved RSDs of 2.1% or lower (n=10) for the B-group vitamins thiamine, nicotinamide, pyridoxine, and riboflavine phosphate, and for ascorbic acid. Absolute peak area results showed higher RSDs, but this was probably due primarily to variability in the injected quantity in the manual apparatus which was used. Under favorable conditions, with automated commercial instrumentation, RSDs of less than 1% (intra-day) to less than 2% (inter-day) should be expected for many analyses (Wätzig and Dette, 1993). The determined analyte concentrations varied from 98.8 to 104% of the expected concentrations. In the above example, the B-group vitamins were present in the pharmaceutical preparation at concentrations varying from 0.5 to 10 mg ml^{-1}, and these were quantitated with direct injection of the untreated sample. Ascorbic acid was present at a concentration of 25 mg ml^{-1} and was assayed separately using a 25-fold diluted sample, presumably due to problems of linear range of the instrument. Limited sensitivity and linear range are potentially important limitations for CE methods if the determination of impurities present at 0.1% of the concentration of the main components in a sample is necessary. Currently available commercial detectors show peak-area responses which are linear with sample concentration over two to three orders of magnitude (Moring *et al.*, 1990). However, peak-height linearity usually covers a considerably smaller range than peak-area linearity, due to electromigration dispersion in the sample zone at high sample concentrations. The need to work with samples at concentrations high enough to cause significant electromigration dispersion may occur because of sensitivity limitations with UV absorbance detection. Typically, limits of detection using UV absorbance fall in the high ng ml^{-1} to tens of µg ml^{-1} range. Limits of quantitation are therefore several times higher, and if impurities at the 0.1% level are to be measured, this implies that the main peaks of interest will be present at high µg ml^{-1} to mg ml^{-1} concentrations. This can impose severe strains on the separation system, especially if the impurities to be measured migrate in close proximity to the main peak(s). As discussed earlier, higher buffer strengths may help to alleviate this problem while trying to maintain a reasonably low separation current or providing good heat dissipation.

3.2. Analysis of stereoisomers

The separation of the optical isomers of bioactive compounds is a topic of considerable interest in HPLC. Frequently, lower separation efficiencies are seen in HPLC chiral separations than in achiral HPLC, and so it is not surprising that there has been a considerable degree of interest in developing CE chiral separation methods. The first chiral separations by CE were demonstrated by Gozel et al. (1987) using a copper (II)-aspartame complex to resolve the enantiomers of various amino acids. Since that time, most of the interest in the resolution of enantiomers by CE has focused on the use of cyclodextrin buffer additives (Snopek et al., 1991). Chiral surfactants have also received some attention as chiral selectors in CE. These have either been bile salts such as sodium taurodeoxycholate, used for the separation of the enantiomers of the calcium-channel blocker diltiazem (Nishi et al., 1990b) or derivatives of amino acids such as sodium N-dodecanoyl-L-valinate, used to separate warfarin enantiomers (Otsuka et al., 1991).

The behaviour of cyclodextrins in CE seems to be broadly similar to their action when used as immobilised chiral selectors in HPLC stationary phases. Cyclodextrins may be used alone as buffer additives, or in conjunction with surfactants. One problem is that solubilities of β-cyclodextrin and to a lesser extent γ-cyclodextrin are very low. This can be overcome with the addition of urea or the use of more soluble derivatised cyclodextrins. The use of charged derivatised cyclodextrins allows separation of neutral chiral molecules and increased resolution for analytes of the opposite charge by increasing the relative mobility difference between the free and complexed forms. An example of the use of cyclodextrins in pharmaceutical analysis is given by Fanali and Bocek (1990), who have shown the analysis of epinephrine enantiomers using di-O-methyl β-cyclodextrin as a chiral selector, in a 0.1 M, pH 2.5 phosphate buffer. In this case, a coated capillary was used, resulting in minimal electroosmotic flow. Thus, migration of the epinephrine was due to its positive charge, and chiral separation took place based on the difference in inclusion of the epinephrine enantiomers in the uncharged cyclodextrin. It was noted that differences in inclusion between two enantiomers into the cyclodextrin can lead to differences in their optical properties. Therefore, a correction may have to be made for this in any attempt at quantitation.

The high resolution of CE can also achieve separations of diastereoisomers which may otherwise be difficult to separate by HPLC. For such compounds, a chiral selector may not be necessary, provided there are significant differences in the physical properties of the two diastereoisomers which allow separation. An example where diastereoisomers can be separated by CE without a chiral selector is the analysis of the diastereoisomers of l-buthionine-(R,S)-sulfoximine (Lloyd, 1992). The drug is administered as a mixture of two diastereoisomers, and these can be separated using MECC with the surfactant SDS (detailed discussion in section 4). The only reported HPLC separation for these diastereoisomers without prior derivatisation uses ligand-exchange chromatography with copper-D-proline complexes as chiral selectors. Achiral CE systems will not separate all diastereoisomers; with compounds where the two chiral centres are well separated, a chiral selector is needed in the CE buffer. An example is the analysis of the diastereoisomers of leucovorin and its active metabolite 5-methyltetrahydrofolate using CE with cyclodextrin additives (Shibukawa *et al.*, 1993). The diastereoisomers of these compounds are not resolved using achiral HPLC, but the separation is possible using a bovine serum albumin HPLC column. These diastereoisomers were not separated by MECC using either anionic or cationic surfactants; however, good resolution can be achieved using CE with cyclodextrin additives. α-, β- and γ-cyclodextrin give resolution of 5-methyltetrahydrofolate (suggesting that the separation is not based on inclusion), but leucovorin is well separated only when using γ-cyclodextrin. Urea is used at high concentrations to solubilise the cyclodextrins. With 250 mM γ-cyclodextrin, leucovorin diastereoisomers are separated with a resolution (R) value of 1.27, and 5-methyltetrahydrofolate diastereoisomers with R = 1.12.

3.3. Protein Analysis

Protein separations are often performed by CZE, although capillary isoelectric focusing is also possible. Surfactant additives are not often used unless denaturation of the proteins is desired. Many proteins have proved difficult to analyse by CE, due to a tendency to adsorb to the fused silica capillary surface. This leads to problems of band-broadening and consequent loss of resolution. The problem is particularly severe when analysing more basic proteins at intermediate pH. Under such conditions, there is an electrostatic attraction to the negatively charged capillary walls and localised regions of high positive charge-density on the protein surface may cause adsorption. Two general approaches to avoiding protein adsorption have been taken. One is to permanently modify the capillary surface by some sort of coating which hides the surface silanol groups. The second is to optimise the separation buffer concentration, pH and additives to reduce protein-wall interactions. Some examples of these approaches are given in Table 1.

Table 1. Methods for reducing protein-wall interactions

Deactivation Method	Example Analytes	Reference
siloxane-bonded polyacrylamide coating	human haemoglobin, human transferrin	Hjertén (1985)
vinyl-bonded polyacrylamide coating	bovine serum albumin, insulin, ovalbumin	Cobb et al. (1990)
octadecylsilane-derivatised surface, nonionic surfactant additives	lysozyme, cytochrome c, ovalbumin, beta-lactoglobulin A, B	Towns and Regnier (1991)
high ionic-strength electrolytes	lysozyme, trypsinogen, myoglobin	Green and Jorgenson (1989)

The potential of CE methods in protein analysis is illustrated by the work of Nielsen et al. (1989) dealing with the separation of biosynthetic human insulin (BHI) and human growth hormone (hGH) and derivatives of these materials. Both these proteins are relatively easy to analyse, with little difficulty due to wall adhesion. Their separations were made by CZE, with a buffer system typically consisting of 10 mM tricine buffer, 5.8 mM morpholine and 20 mM NaCl, adjusted to pH 8 with 1 M NaOH. This pH was above the pI of the proteins studied, and thus wall-adhesion was minimised. hGH was well separated from its desamido and didesamido derivatives in ten minutes, with separation efficiencies greater than 100,000 plates. In a later publication, these authors showed that hGH was best analysed at mildly alkaline pH, and that at moderately acidic pH poor peak shape and low efficiency were observed (Nielsen and Rickard, 1990a). This could easily be explained since the pI of hGH is around 5.2, and thus at low pH there were considerable electrostatic interactions between hGH and the capillary wall. Triplicate repeat analyses of a hGH/desamido-hGH mixture at five injection volumes ranging from 2.44 to 21.98 nl showed RSDs for the five datasets of around 5% for hGH and desamido-hGH peak areas, and for the percentage desamido content. BHI and desamido-BHI were well separated in ten minutes with a similar buffer system to that used for hGH above, and other derivatives were also successfully resolved. Comparison with HPLC in the determination of the desamido content of acid-degraded samples showed results that were generally in agreement between the two methods, although the CE results showed poorer precision. The RSD for BHI and desamido-BHI peak areas was reported to be around 5% (Nielsen et al., 1989). Recently, Lookabaugh et al. (1991) have reported analyses of commercial dosage forms of insulin, and they found problems with poor analytical accuracy and precision using a similar separation system. Comparison of HPLC and CE analyses of insulin injection preparations resulted in significantly different amounts being determined for the same samples by the two techniques, although the analytical precision was reasonable using an internal standard (RSD $\leq 2.5\%$).

As well as having potential for the analysis of intact protein pharmaceuticals, CE methods have proved useful in the analysis of peptides from protein digests (Grossman et al., 1989a). Unlike some intact proteins, small peptides do not usually suffer from adhesion to the capillary wall, and the main analytical challenge is to achieve adequate resolution of a large number of peaks. Both MECC and CZE methods may be useful, and there has been some work reported on the prediction and optimisation of peptide elution in CZE (Grossman et al., 1989b; van de Goor et al., 1991). Nielsen and Rickard (1990b) have described the use of CZE to analyse tryptic peptides from hGH digests. The composition of the expected hGH fragments was known, so the isoelectric points of the fragments could be calculated. The pIs ranged from 3.5 to 10.4, giving little indication of an optimum pH for the separation. Thus, the approach to optimisation was first to analyse the digests over a wide range of pH values (from 2.4 to 10.4) while maintaining roughly constant conductivities by the addition of NaCl to the higher pH buffers. The optimum was found to be 8.1, which is considerably higher than the mean (6.3) and median (5.9) calculated pI values for the peptides. Once the optimum pH was found, other selectivity adjustments were made by alteration of the concentration of the buffer ion, concentration of NaCl added to the buffer, and addition of morpholine. By altering these parameters, the authors found that after pH optimi-sation the buffer concentration and buffer capacity were most impor-tant in obtaining the best resolution. Even under their optimised conditions, not all of the tryptic peptides could be resolved, but analysis at pH 2.4 achieved separation of the remaining compounds.

Cobb and Novotny (1992) have reported the scaling-down of protein digestion reactions to use low-picomolar quantities of protein in low-micromolar volumes, which more appropriately match the microanalytical capabilities of CE. Their procedure for reduction and alkylation is a one-step process, with the simultaneous addition of tributylphosphine as the reducing agent and 2-methylaziridine for the alkylation step. Enzymatic digestion was then performed with immo-bilised trypsin and chymotrypsin, in 40 cm × 1 mm pyrex reactor columns. Tryptic and chymotryptic digestions of proteins with intact or reduced and alkylated disulfide bonds were shown, as well as cyanogen bromide digests. Model proteins included trypsinogen and human serum albumin.

4. Analyses from Biological Fluids

The analysis of xenobiotics in biological fluids is a challenging area of application for capillary electrophoresis. There are an increasing number of publications in this area, and it seems likely that CE methods will prove useful for such analyses. Initially it is likely that it will be applied only in cases where traditional methods fail to provide satisfactory results. If CE instrumentation becomes more common-place in laboratories doing bioanalysis, CE may become a method of choice for reasons of simplicity, reliability and ease of automation.

The first report of a CE method for the determination of a drug in plasma appeared in 1986 in an article by Fujiwara and Honda (1986) on analysis of cinnamic acid and analogs. The metabolite of the anti-ulcerative agent γ-oryzanol, 4-hydroxy-3-methoxycinnamic acid, was determined in canine plasma. The analysis followed the type of procedure familiar in chromatography of first extracting the analyte from the biological matrix (in this case, by liquid-liquid extraction), after which CE was used for quantitation of the solute of interest. A similar approach was taken by Roach et al. (1988), who used solid-phase or liquid-liquid extraction methods to recover methotrexate and its metabolite 7-hydroxymethotrexate from plasma, and then determined the quantities of these compounds using CE with laser-induced fluorescence detection. The impressive feature of this assay was that the high-sensitivity detection method allowed quantitation of methotrexate down to 10^{-10} M levels in serum. Interestingly, it has also been shown that it is possible in some cases to perform CE analyses of plasma without sample treatment. Nagakawa *et al.* (1988) demonstrated the analysis of the antibiotic cefpiramide with direct injection of plasma onto the capillary. In such cases MECC is used, and binding of SDS to plasma proteins leads to reduction of the otherwise troublesome adhesion of proteins to the capillary walls.

One of the first steps when approaching the analysis of a drug or metabolite in a biological matrix by CE is to ask what are the expected analyte concentrations, and how might the analyte be detected. This is particularly important because of the limited concentration sensitivity of UV absorbance detectors for CE. Using HPLC, relatively long wavelengths are often used for UV absorbance detection, because of problems of mobile-phase absorbance at lower wavelengths. However, many compounds have much weaker absorbance coefficients at longer UV wavelengths than at 190 or 200 nm, wavelengths which can often be used in CE. Other analytes that normally are considered not to be detectable by UV absorbance using HPLC may often be analysed using CE with low detection wavelengths. The disadvantage of using short detection wavelengths is that the number of potential interferences is greatly increased, and electropherograms become much more complex. If UV absorbance detection is not likely to achieve adequate sensitivity, there are two options—use another detection technique or perform some form of analyte concentration. Sample concentration may take two forms— on or off column—or both forms may be used together. Detectors other than UV absorption are not generally available on most commercial CE instrumentation, so the former option involves some home construction. A variety of laser-induced fluorescence and electrochemical detectors have been described in the literature and may be suitable for certain applications (see Chapter 2). After answering the question of detection, the choice of separation system must be made. If all analytes are ionic, the separation may be successfully approached using CZE, although ion-pairing agents or even MECC may be needed for adequate resolution. Any non-ionic solute will probably require MECC, or perhaps the use of other complexing agents such as cyclodextrins. The sample matrix must be considered when choosing the separation system. For direct injection of plasma, the presence of surfactants is necessary to reduce interactions between proteins and fused silica capillaries.

The "conventional" approach of performing a sample extraction, followed by CE analysis is illustrated by an assay developed by Lloyd *et al.* (1991) for the analysis of the antileukaemic agent cytosine-β-D-arabinoside (ara-C) in human plasma. An analogue of cytidine, ara-C differs only in the configuration at the 2' position in the sugar moiety. Ara-C has good UV absorbance, with $\varepsilon \approx 10,000$ at 212 nm, and $\varepsilon \approx$

13,000 at 281 nm at pH 2. Cytidine has pK values of 4.2 (cationic) for the amine group and 12.5 (anionic) for the sugar, thus CZE with either low or high pH buffers may be possible for the analysis of ara-C. Because of the very high value of the sugar pK, operation at low pH to protonate the amine group is the favoured approach. Using a 40 mM, pH 2.5 citrate buffer, ara-C is positively charged and is detected after 4.5 minutes with an 8 kV separation potential across a 26 cm–long, 50 μm–diameter capillary (20 cm to detector). A 2.8 nl injection of a 1 μg ml^{-1} solution of ara-C gives a peak that is approximately at the limit of detection, using a Beckman PACE 2000 CE system with UV absorbance detection at 280 nm. For high-dose ara-C therapy, clinically relevant drug concentrations range from hundreds of ng ml^{-1} to tens of μg ml^{-1}. Most of the necessary increase in sensitivity can be obtained by the use of on-capillary peak-stacking. Peak-stacking or compression occurs when the sample is introduced onto the capillary in a matrix that has a low conductivity relative to the separation buffer (Burgi and Chien, 1991). An enhancement in the electric field in the sample zone occurs, leading to more rapid movement of ions in this region, and a consequent sharpening of the analyte zone. The effectiveness of this process is illustrated in Fig. 9, which shows the measured ara-C peak height as

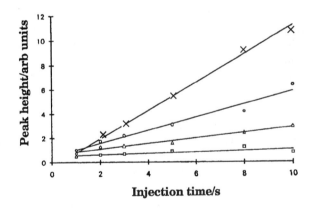

Fig. 9.   Peak-stacking of ara-C analysed by CZE with a field strength of 577 V cm^{-1} using a 20 mM citrate buffer (pH 2.5) with 0 (×), 10 (O), 20 (△) and 40 mM NaCl (□) added to the sample. When the buffer conductivity is higher than the sample conductivity, peak stacking occurs; with aqueous samples the stacking effect is linear up to injection times of 10 s (injected volume 28 nl).

a function of injection volume (ranging from 2.8 to 28 nl), with the sample dissolved in various concentrations of NaCl or in the separation buffer.

It can be seen that the greatest increase in peak height with increasing injection volume occurs with the sample dissolved in water, with no added salt. To take advantage of on-capillary peak-stacking, it is necessary to remove excess electrolytes from the injected sample. In this analysis, solid-phase extraction provided a convenient sample cleanup method, removing both excess electrolytes and plasma proteins. Using C-18 cartridges, 0.2 ml of plasma was treated; ara-C was eluted from the cartridge with acetonitrile, and then reconstituted in 0.1 ml of water for injection onto the capillary. Ara-C in the desalted extracts exhibited good peak stacking. With 28 nl injections, a detection limit for ara-C in plasma of $\approx$120 ng ml^{-1} was achieved (S/N = 3).

A typical electropherogram of ara-C extracted from plasma is shown in Fig. 10. In this case, the original ara-C concentration was 500 ng ml^{-1}. Occasionally, very slow migration of ara-C and broadened peaks with reduced peak areas were seen. If this was the case, the analysis was repeated with another aliquot of the same sample after filtration or centrifugation; this usually resulted in a successful analysis. An internal standard was not used, because it was difficult to find a suitable compound with extraction and electrophoretic properties similar to those of ara-C. Many cytidine analogues were tried, but either were not extracted or co-eluted with ara-C.

Another example of CE analysis following extensive cleanup of biological samples is the analysis of the enantiomers of the antihypertensive agent cicletanine in human plasma and urine reported by Pruñonosa et al. (1992). γ-Cyclodextrin is used as a chiral selector. Once again, both off- and on-capillary concentration are used to achieve the required sensitivity in the ng ml^{-1} range. In this example, a tenfold concentration occurred during the liquid-liquid extraction step, with the sample being reconstituted in 1:9 acetonitrile:water. The extracted (and desalted) sample (46 nl) was then injected onto a 57 cm long (50 cm to the detector), 75 μm i.d. capillary, in a Beckman PACE 2000 CE system. The separation buffer was 100 mM sodium borate (pH 8.6), with 110 mM SDS and 25 mM γ-CD as a chiral selector. It is apparent from the electropherograms (Fig. 11) that peak-stacking is occurring; the peak width at half-height at the detector was of the order of 3 mm, while the injected zone width was 10 mm. This contributes to the good limit of detection of 10 ng ml^{-1}. Inter-day RSD was found to be 11% or less for each enantiomer of cicletanine for concentrations of 25 ng ml^{-1} or above, using an internal standard.

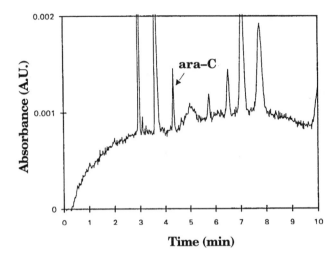

Fig. 10. Electropherogram showing the separation of ara-C spiked into pooled human plasma at a concentration of 500 ng ml^{-1}. Buffer, 20 mM citrate, pH 2.5; field strength, 308 V cm^{-1}; capillary 26 cm fused silica, 50 μm i.d. (20 cm to detector); detection, UV absorbance at 280 nm.

The above examples illustrate that CE may be used in place of HPLC after extraction of the analyte from plasma. However, of more interest is the possibility of making direct analyses of biological samples, with minimal sample treatment. The high resolving power of CE and the open-tubular nature of CE columns with CE separations occurring entirely in the liquid-phase invite direct analysis of biological samples. There are a variety of potential advantages. The removal of an extraction step also removes a major potential source of error in variable and incomplete extraction, and there are concomitant benefits in the saving of time and labor. Sample size may also be important—if several ml of a matrix needs to be taken and extracted, the microanalytical abilities of CE are not being exploited. If biological fluids with little or no pretreatment are analysed, much smaller samples may be used. This ability may prove beneficial in sample-limited situations, such as pediatric patients, or with experimental animals. A number of examples of direct-injection CE assays exist, often with very respectable analytical reproducibility due at least in part to the reduction of the number of sample-processing steps. However, the main difficulty to be faced with direct injection CE assays is that the absence of sample concentration steps must be compensated by the use of high-sensitivity detectors, if available, or else higher detection limits must be accepted.

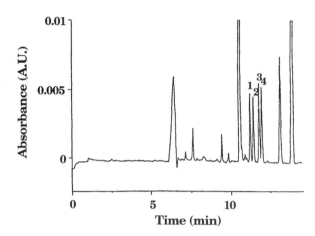

Fig. 11. Separation of enantiomers of cicletanine in extracted human plasma. 1,2, R/S internal standard; 3, S-cicletanine; 4, R-cicletanine. Buffer, 100 mM sodium borate, pH 8.6, 110 mM SDS, 25 mM γ-cyclodextrin; field strength, 263 V cm^{-1}; capillary 57 cm fused silica, 75 μm i.d. (50 cm to detector), detection, UV absorbance at 214 nm (reproduced with permission from Pruñonosa *et al.*, 1992).

The analysis of protein-containing matrices such as plasma is generally unsuccessful by CZE if no precautions are taken to avoid adhesion of proteins to the capillary wall. Protein adhesion usually results in large parts of the electropherogram being obscured by extremely broad protein peaks, and considerable variation in electro-osmotic flow as the capillary surface is modified. One possible solution would be to use coated capillaries designed to reduce the adsorption of proteins. The elegant method used by Nakagawa *et al.* (1988) to avoid this problem in the analysis of drugs in plasma was to use MECC as an analytical method. In this case, SDS binds strongly to proteins (Tanford, 1980), denaturing them and reducing their tendency to adsorb to the silica capillary. Sample preparation then consists simply of filtration or centrifugation to remove gross particulate matter, preferably just before injection. Occasionally with plasma samples the capillary may become blocked, and a normal wash cycle by application of pressure or vacuum does not succeed in unblocking the capillary. A high pressure wash from an HPLC pump in the constant pressure mode is usually very effective in clearing plugged capillaries. Otherwise, it is sometimes possible to clear a blockage by electrophoresing a concentrated acid or base through the capillary, and then using a pressure-driven wash after the cleaning solution.

The use of MECC with direct injection of plasma is illustrated by the analysis of the diastereoisomers of L-buthionine-(R,S)-sulfoximine (BSO, structure shown in Fig. 12) in human plasma (Lloyd, 1992). BSO is an inhibitor of glutathione synthesis, and is undergoing clinical trials as a radiosensitizing agent for cancer therapy. A previously reported HPLC assay for the diastereoisomers is relatively complex, with postcolumn fluorescence derivatisation, and was not validated for the analysis of BSO in biological matrices. As mentioned earlier, aqueous samples of L-(R,S)-BSO are readily analysed using MECC with SDS. In fact, the same analysis can be used for the analysis of BSO in human plasma. Fig. 12 shows the determination of the diastereoisomers of BSO at concentrations of ≈ 25 μg ml^{-1} in plasma.

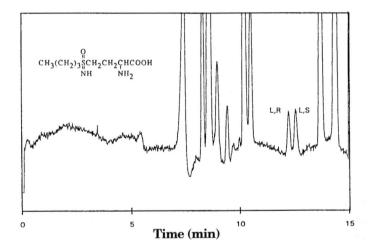

Fig. 12. Analysis of the diastereoisomers of L-buthionine-(R,S)-sulfoximine with direct injection of human plasma. Peaks are 22.7 μg ml^{-1} L,R-BSO and 27.3 μg ml^{-1} L,S-BSO. Buffer, 20 mM phosphate, pH 6.8, with 170 mM SDS; field strength, 236 V cm^{-1}; capillary, 72 cm fused silica, 50 μm i.d. (50 cm to detector), detection, UV absorbance at 190 nm; instrument, ABI 270A.

Despite the presence of SDS in the running buffer, some modification of the capillary surface does occur when directly injecting plasma.

Twenty consecutive injections of plasma spiked with BSO were made on a capillary that was previously unused for direct injections of plasma, although it had previously been used for some days for analyses of aqueous standard samples of BSO. Increasing modification of the capillary surface (presumably by plasma proteins) lead to a reduction of the electroosmotic flow, as evidenced by the increase in migration times of the analytes. For L,R-BSO the migration time went from 9.92 minutes (1st injection) to 11.71 minutes (20th injection), and for L,S-BSO the migration time drifted from 10.12 to 12 minutes. However, the effective mobility of the BSO diastereomers showed no drift, with average values of -1.87×10^{-4} cm^2 V^{-1} s^{-1} (L,R-BSO) and -1.94×10^{-4} cm^2 V^{-1} s^{-1} (L,S-BSO), and mobility RSDs of 0.6% in each case. Although there is some modification of the capillary surface, this effect is considerably reduced compared to a separation without SDS. The assay was validated for the determination of BSO in plasma over a concentration range of 5 to 1000 µg ml^{-1}. Intra-day (n = 10) and inter-day (n = 15) RSDs were found to be always <4% at a concentration of ≈500 µg ml^{-1} for each stereoisomer, and < 8% at a concentration of ≈25 µg ml^{-1}.

Direct analysis of urine without sample preparation is simpler than analysis of plasma in that extensive modification of the capillary by endogenous material is generally not a problem. Therefore, both CZE and MECC analyses of unextracted urine may be performed. However, the huge number of components present in urine makes the resulting electropherograms very complex; consequently, there may be problems of peak identification and resolution. This is obvious from Fig. 13, which shows an MECC electropherogram with UV absorbance detection at 280 nm of human urine from a healthy subject. Detection at 200 nm gives considerably more complex electropherograms, and peak identification becomes very problematic. Certain endogenous compounds such as uric acid are easily recognised in urine electropherograms because of their very high concentrations, and indeed may be useful as marker peaks during method development.

Despite the considerable complexity of electropherograms of urine, it is quite possible to develop robust CE methods that are useful for answering biomedical problems. The following examples are CE methods used for determining the metabolic phenotype of an individual for two different metabolic pathways. The oxidative metabolism of the antitussive agent dextromethorphan (DX) to dextrorphan (DR) is genetically controlled, with approximately 5–10% of the population being so-called "poor metabolisers," with urinary DX/DR ratios around 0.5. The rest are classified as "extensive metabolisers," with urinary DX/DR ratios ranging from 0 to around 0.3. The electropherogram shown in Fig. 14 is of urine spiked with

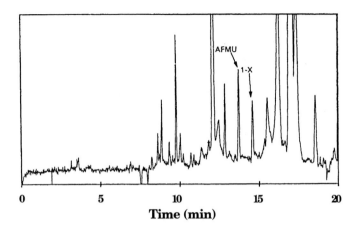

Fig. 13. MECC electropherogram of human urine with UV
absorbance detection at 280 nm.   Peaks are 5-
acetylamino-6-formylamino-3-methyluracil (AFMU)
and 1-methylxanthine. Buffer, pH 8.43 phosphate-
borate, 70 mM SDS; field strength, 236 V cm^{-1};
capillary, 72 cm, 50 μm i.d. (50 cm to detector);
instrument, ABI 270A.

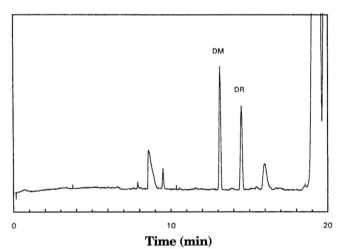

Fig. 14. CZE electropherogram of dextromethorphan (DM) and
dextrorphan (DR) spiked into human urine; conc. 2.25
μg ml^{-1} each. Buffer, 175 mM sodium borate, pH 9.3;
field strength, 167 V cm^{-1}; capillary, 72 cm fused silica,
50 μm i.d. (50 cm to detector); detection, UV absorbance
at 200 nm; instrument, ABI 270A.

DX and its metabolite DR at concentrations of 2.25 µg ml^{-1} each. To determine an individual's phenotype, 15 mg of DX is given orally and urine is collected for analysis after four hours. DX and DR are present in the urine mainly in the form of glucuronides, and so enzymatic deconjugation must be done before analysis. With a pH 9.3, 175 mM borate buffer, DX and DR elute early in the electropherogram and are well resolved from any interferences (Li *et al.*, 1993). Detection is by UV absorbance at 200 nm to achieve adequate sensitivity, with relatively large (3 s) injections and some peak-stacking in the high-concentration buffer. In this analysis the short-wavelength detection causes no problems, as the analytes are well resolved from other compounds. Quantitation is excellent, with inter-day RSDs in the determination of 2.5 µg ml^{-1} DX and DR of 6.2 and 5.6%, respectively. This CE analysis is a very useful method for the determination of phenotypes in the oxidative pathway responsible for metabolism of DX. Its main advantage over HPLC methods is speed. Both CE and HPLC methods require deconjugation of DX and DR, but with CE no sample extraction is necessary. HPLC methods reported in the literature require the use of liquid-liquid or solid-phase extraction, adding considerably to the overall time and effort necessary for the phenotype determination.

A second example is the use of MECC for the determination of an individual's N-acetylation metabolic phenotype (Lloyd *et al.*, 1992). This involves measuring the peak-area ratio of two caffeine metabolites that are excreted in the urine, 5-acetylamino-6-formylamino-3-methyluracil (AFMU) and 1-methylxanthine (1-X). Measured using UV absorbance detection at 280 nm, subjects with AFMU:1-X ratios greater than 0.5 are classified as "fast" acetylators, while those with lower ratios are "slow" acetylators. In this case, the determination is achieved using a pH 8.43 phosphate-borate buffer system, with 70 mM SDS. Urine is injected directly, without pretreatment. The analysis takes less than 15 minutes, using a 72 cm capillary (50 cm to the detector) at a separation potential of 17 kV. A typical electropherogram for a fast acetylator after ingestion of caffeine is shown in Fig. 13. Because of the complexity of the electropherograms, peak identification can be a problem. In such cases, on-column multiwavelength detection can be most useful (Thormann et al., 1991, 1992). Both AFMU and 1-X are anions at the pH of the separation, and their negative charge causes repulsion from the anionic SDS micelles, resulting in poor micellar inclusion. However, the presence of SDS is useful because of its effect on other urine components, which otherwise interfere with 1-X. In both of the above analyses, analytical sensitivity was not a problem for most subjects. A high proportion of extensive metabolisers of DX had no detectable DX in their urine samples, but this is also often the case with data obtained by other analytical procedures. Likewise, AFMU was sometimes difficult to analyse. If this was the case it was usually because of a rather dilute urine sample, and longer injections were used in order to achieve detectable levels.

5. Conclusions

A wide variety of pharmaceutical and biomedical applications of CE have been reported. Increasing attention is being given to the quantitative abilities of CE methods, and validated CE methods are appearing in the literature. Considerable research effort is being expended on the development of CE methods to give quantitative, high-quality protein separations. The continuing increase in the publication of research articles on CE indicates that this is still a young and growing technique. A great deal of work remains to be done in the refinement of instrumentation and separation methodologies, as well as in the development of applications. One area of primary interest for the wider use of capillary methods is the improvement of detection sensitivity, and the commercialisation of more sensitive detection systems. This is particularly a concern for the direct analysis of biological fluids, where at present CE can often give the required resolution for an analysis but not always the required sensitivity.

Acknowledgments

I would like to acknowledge the contributions of my coworkers, A. Cypess, A. Dahlan, K. Fried, N. Markoglou, V. Sandor, A. Shibukawa and I. Wainer, and to thank T. Flarakos for assistance in preparing the manuscript.

6. References

Albin, M., P. D. Grossman and S. E. Moring (1993). *Anal. Chem., 65,* 489A–497A.

Altria, K. D., R. C. Harden, M. Hart, J. Hevizi, P. A. Hailey, J. V. Makwana and M. J. Portsmouth (1993). *J. Chromatogr., 641,* 147–153.

Beckers, J. L., F. M. Everaerts and M. T. Ackerman (1991). *J. Chromatogr., 537,* 407-442.

Bruin, G. J. M., P. P. H. Tock, J. C. Kraak and H. Poppe (1990). *J. Chromatogr., 517,* 557-572.

Burgi D. S. and R-.L. Chien (1991). *Anal. Chem., 63,* 2042-2047.

Cobb, K. A., V. Dolnik and M. Novotny (1990). *Anal. Chem., 62,* 2478-2483.

Cobb K. A. and M. V. Novotny (1992). *Anal. Chem., 64,* 879-886.

Cohen, A. S., D. R. Najarian, A. Paulus, A. Guttman, J. Smith and B. L. Karger (1988). *Proc. Nat. Acad. Sci. USA, 85,* 9660-9665.

Edmonds, C. G., J. A. Loo, C. J. Barinaga, H. R. Udseth and R. D. Smith (1989). *J. Chromatogr., 474,* 21-37.

Everaerts, F. M., J. L. Beckers and T. P. Verheggen (1976). *Isotachophoresis: Theory, Instrumentation and Applications,* Elsevier, New York.

Fanali S. and P. Bocek (1990). *Electrophoresis, 11,* 737-760.

Foley, J. P. (1990). *Anal. Chem., 62,* 1302-1307.

Foret, F. and P. Bocek (1990). In *Advances in Electrophoresis 3,* (A. Chambrach, M. J. Dunn and B. J. Radola, eds.), VCH, Weinheim, pp. 273-347.

Fujiwara S. and S. Honda (1986). *Anal. Chem., 58,* 1811-1814.

Fujiwara, S. and S. Honda (1987). *Anal. Chem. 59,* 2273-2276.

Fujiwara, S., S. Iwase and S. Honda (1988). *J. Chromatogr., 447,* 133-140.

Gozel, P., E. Gassmann, H. Michelsen and R. N. Zare (1987). *Anal. Chem., 59,* 44-49.

Green J. S. and J. W. Jorgenson (1989). *J. Chromatogr., 478,* 63-70.

Grossman, P. D., J. C. Colburn, H. H. Lauer, R. G. Nielsen. R. M. Riggin, G. S. Sittampalam and E. C. Rickard (1989a). *Anal. Chem., 61,* 1186-1194.

Grossman, P. D., J. C. Colburn and H. H. Lauer (1989b). *Anal. Biochem., 179,* 28-33.

Hjerten, S. (1985). *J. Chromatogr., 347,* 191-198.

Hjerten, S. (1967). *Chromatogr. Rev., 9,* 122-219.

Hjerten, S., K. Elenbring, F. Kilar, J.-L. Liao, A. C. Chen, C. J. Siebert and M.-D. Zhu (1987). *J. Chromatogr., 403,* 47-61.

Hoyt, A. M., Jr. and M. J. Sepaniak (1989). *Anal. Lett., 22,* 861-873.

Huang, X. H., M. J. Gordon and R. N. Zare (1988). *Anal. Chem., 60,* 375-377.

Huang, X. H., W. F. Coleman and R. N. Zare (1989). *J. Chromatogr., 480,* 95-110.

Jorgenson, J. W. and K. D. Lukacs (1981a). *Anal. Chem., 53,* 1298-1302.

Jorgenson, J. W. and K. D. Lukacs (1981b). *Clin. Chem., 27,* 1551-1553.

Khaledi, M. G., S. C. Smith and J. K. Strasters (1991). *Anal. Chem., 63,* 1820-1830.

Klepárník, K. and P. Bocek (1991). *J. Chromatogr., 565,* 3-42.

Knox, J. H. and I. H. Grant (1991). *Chromatographia, 32,* 317-328.

Lambert, W. J. and D. L. Middleton (1990). *Anal. Chem., 62*, 1585-1587.

Li, S., K. Fried, I. W. Wainer and D. K. Lloyd (1993). *Chromatographia 35*, 216-222.

Lloyd, D. K., K. Fried and I. W. Wainer (1992). *J. Chromatogr., 578*, 283-291.

Lloyd, D.K. (1992). *Anal. Proc., 29*, 169-170.

Lloyd, D. K., A. M. Cypess and I. W. Wainer (1991). *J. Chromatogr., 568*, 117-124.

Lookabaugh, M., M. Biswas and I. S. Krull (1991). *J. Chromatogr., 549*, 357-366.

Mazzeo, J. R. and I. S. Krull (1991). *Anal. Chem., 63*, 2852-2857.

Mikkers, F. E. P., F. M. Everaerts and T. P. E. M. Verheggen (1979). *J. Chromatogr., 169*, 11-20.

Moring, S. E., J. C. Colburn, P. D. Grossman and H. H. Lauer (1990). *LC-GC 8*, 34-46.

Muck, W. M. and J. D. Henion (1989). *J. Chromatogr., 495*, 41-59.

Nagakawa, T., Y. Oda, A. Shibukawa and H. Tanaka (1988). *Chem. Pharm. Bull., 36*, 1622-1625.

Nielsen R. G. and E. C. Rickard (1990a). *ACS Symp. Ser.*, No. 434, 50-59.

Nielsen R. G. and E. C. Rickard (1990b). *J. Chromatogr., 516*, 99-114.

Nielsen, R. G., G. S. Sittampalam and E. C. Rickard (1989). *Anal. Biochem., 177*, 20-26.

Nishi, H., N. Tsumagari, T. Kakimoto and S. Terabe (1989a). *J. Chromatogr., 477*, 259-270.

Nishi, H., N. Tsumagari, T. Kakimoto and S. Terabe (1989b). *J. Chromatogr., 465*, 331-343.

Nishi, H., T. Fukuyama, M. Matsuo and S. Terabe (1990a). *J. Chromatogr., 498*, 313-323.

Nishi, H., T. Fukuyama, M. Matsuo and S. Terabe (1990b). *J. Chromatogr., 515*, 233-243.

Otsuka, K., J. Kawahara, K. Tatekawa and S. Terabe (1991). *J. Chromatogr., 559*, 209-214.

Pruñonosa, J., R. Obach, A. Diez-Cascón and L. Gouesclou (1992). *J. Chromatogr., 574*, 127-133.

Roach, M. C., P. Gozel and R. N. Zare (1988). *J. Chromatogr., 426*, 129-140.

Saitoh, T., H. Hoshino and T. Yotsuyangi (1989). *J. Chromatogr., 469*, 175-181.

Salomon, K., D. S. Burgi and J. C. Helmer (1991). *J. Chromatogr., 549*, 375-385.

Schwer C. and E. Kenndler (1991). *Anal. Chem., 63*, 1801-1807.

Shaw, D. J. (1969). *Electrophoresis*, Academic Press, London.

Shibukawa, A., D. K. Lloyd and I. W. Wainer (1993). *Chromatographia* *35*, 419-429.

Snopek, J., H. Soini, M. Novotny, E. Smolkova-Keulemansova and I. Jelinek (1991). *J. Chromatogr., 559*, 215-222.

Strasters, J. K. and M. G. Khaledi (1991). *Anal. Chem., 63*, 2503-2508.

Sustácek, V., F. Foret and P. Bocek (1991). *J. Chromatogr., 545*, 239-248.

Tanford, C. (1980). *The Hydrophobic Effect: Formation of Micelles and Biological Membranes*, 2nd Ed., Wiley, New York.

Terabe, S,. K. Otsuka, K. Ichikawa, A. Tsuchiga and T. Ando (1984). *Anal. Chem., 56*, 111-113.

Thormann, W., A. Minger, S. Molteni, J. Caslavska and P. Gebauer (1992). *J. Chromatogr., 593*, 273-288.

Thormann, W. P. Meier, C. Marcolli and F. Binder (1991). *J. Chromatogr., 545*, 445-460.

Thormann, W. (1990). *J. Chromatogr., 516*, 211-217.

Towns J. K. and F. E. Regnier (1991). *Anal. Chem., 63*, 1126-1132.

Turner, K. (1991). *LC-GC Intl., 4*, 32-38.

van de Goor, T. A. A. M., P. S. L. Janssen, J. W. van Nispen, M. J. M. van Zeeland and F. M. Everaerts (1991). *J. Chromatogr., 545*, 379-389.

Wallingford, R. A. and A. G. Ewing (1989). *Adv. Chromatogr., 29*, 1-76.

Wallingford, R. A. and A. G. Ewing (1987). *Anal. Chem., 59*, 1762-1766.

Wätzig, H. and C. Dette (1993). *J. Chromatogr., 636*, 31-38.

Weinberger, R. and M. Albin (1991). *J. Liq. Chromatogr., 14*, 953-972.

Wieme, R. J. (1975). In *Chromatography*, (E. Heftmann, ed.), pp. 228-281, Van Nostrand Reinhold, New York.

Wren, S. (1991). *J. Microcol. Sep., 3*, 147-154.

CHAPTER 2

Novel Approaches to the Liquid Chromatographic

Analysis of Primary Amines, Amino Acids and Peptides

CHRISTOPHER M. RILEY, JOHN F. STOBAUGH
and SUSAN M. LUNTE

*Center for BioAnalytical Research, and Department of
Pharmaceutical Chemistry, University of Kansas,
Lawrence, KS 66047, U.S.A.*

1. Introduction

The primary amino functional group is present in a wide variety of compounds of pharmaceutical and biomedical interest. Therefore, the analysis of primary amines in biological samples and other relevant media has received a great deal of attention. In particular, a number of important derivatization reagents have been described, such as o-phthalaldehyde/thiol (OPA/thiol), fluorescamine and ninhydrin, to enhance the detectability of primary amines. In the 1980s scientists at the University of Kansas first described a new reagent, naphtha-lene-2,3-dicarboxaldehyde/cyanide, for the fluorogenic derivatization of primary amines (Carlson *et al.,* 1986; de Montigny 1986; de Montigny *et al.,* 1987, 1990; Matuszewski *et al.,* 1987; Sternson *et al.,* 1988; Lunte and Wong 1989, 1990; Soper and Kuwana 1989). Originally this reagent was intended for the precolumn, fluorogenic derivatization of amino acids and peptides. However, subsequent work has extended the applicability of this reagent system (Kennedy *et al.,* 1989; Lunte *et al.,* 1989; Oates and Jorgenson 1989a,b, 1990; Oates *et al.,* 1990; Dave 1991; Dave *et al.,* 1992a,b; Nussbaum *et al.,* 1992). Of particular significance in this regard were the observations that the products of the reaction of NDA/CN and primary amines are electrochemically active (Kennedy *et al.,* 1989; Lunte *et al.,* 1989; Oates and Jorgenson, 1989, 1990; Oates *et al.,* 1990; Nussbaum *et al.,*

1992) and that replacement of cyanide by thiol allows rapid postcolumn derivatization of primary amines (Dave *et al.*, 1992b).

2. Naphthalene-2,3-dicarboxaldehyde

2.1. *Chemical properties and Reaction Kinetics*

A number of factors must be taken into account for the successful application of a fluorogenic reaction to precolumn or postcolumn derivatization in liquid chromatography. Those factors include the rate and chemical yield of the reaction, the fluorescence intensity of the product, and the stability of the reagent and of the product. Ignoring for the moment the question of reagent stability, the important reaction in the derivatization of an analyte may be conveniently expressed by two sequential first order processes that describe the formation of the derivative (k_f) and its subsequent degradation (k_d) (Eq. 1):

$$\text{Analyte} \xrightarrow[\text{Reagent}]{k_f} \text{Derivative} \xrightarrow{k_d} \text{Degradation Product} \qquad (1)$$

The kinetic expression that describes the time course of the derivative is then given by Eq. 2:

$$C = \frac{C_0 k_f}{k_d - k_f}\left(e^{-k_f t} - e^{-k_d^{-t}}\right) \qquad (2)$$

where C is the concentration at time t and and C_0 is the concentration of product if it were all produced instantly. Four reaction profiles (curves a to d) have been simulated in Fig. 1 to demonstrate the various kinetic possibilities.

Ideally, the reaction should be complete within 1 to 2 min and the product should be stable for several hours (curve a). In this case the reaction is suitable for either pre- or postcolumn systems. Derivatization reactions that proceed rapidly but produce unstable products (*e.g.*, curve b) are more suitable for postcolumn systems because it is only necessary that the fluorescent product be produced in time for it to be detected in the eluant stream. Type b reactions can be used in the precolumn mode, but the addition of the reagents and the injection of the sample must be carefully timed to provide adequate precision. In contrast, reactions that are slow but produce stable products (curve c) are suitable only for precolumn methods because substantial band broadening will result during the long delay that is needed between the end of the column and the detector to ensure complete reaction. Clearly, slow reactions that produce unstable products (curve d) are unsuitable for pre- or postcolumn methods. The curves have been simulated using Eq. 1 with the following values for the formation and degradation rate constants: Curve a: $k_f = 2.31 \times 10^{-2}$ s^{-1} ($t_{0.5}=30$ s), $k_d = 1.6 \times 10^{-6}$ s^{-1} (120 h); curve b: $k_f = 2.31 \times 10^{-2}$ s^{-1} (30 s), $k_d = 1.16 \times 10^{-3}$ s^{-1} (10 min); curve c: $k_f = 3.85 \times 10^{-4}$ s^{-1} (30 min), $k_d = 1.6 \times 10^{-6}$ s^{-1} (120 h); curve d: $k_f = 3.85 \times 10^{-4}$ s^{-1} (30 min), k_d 1.16×10^{-3} s^{-1} (10 min).

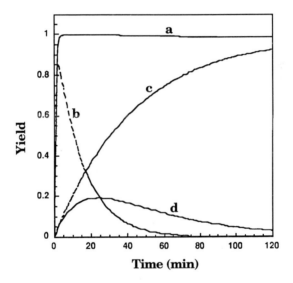

Fig. 1. Theoretical relationships between the yield of a derivatization reaction and time for four different types of reaction. Curve a: rapid derivatization to produce a stable product; curve b: rapid derivatization to produce an unstable product; curve c: slow derivatization to produce a stable product; and curve d: slow derivatization to produce an unstable product.

The reagent system naphthalene-2,3-dicarboxaldehyde/cyanide (NDA/CN) (Fig. 2) was rationally designed in an attempt to overcome some of the disadvantages of the o-phthalaldehyde/thiol (OPA/thiol) system originally described by Roth for the fluorogenic derivatization of primary amines (Roth, 1971). The main disadvantage of OPA/thiol is the poor chemical stability of the derivatives, which arises from auto-oxidation (Stobaugh *et al.*, 1983, 1985; Sternson *et al.*, 1985; Alvarez-Coque *et al.*, 1989; Mifune *et al.*, 1989). The poor stability of OPA/thiol adducts of primary amines makes that reagent more suitable for postcolumn derivatization than for precolumn systems. In addition, OPA/thiol fails to produce fluorescent products with terminal amino groups of peptides, and can only be used for the fluorogenic derivatization of lysyl side chains (Sternson *et al.*, 1985).

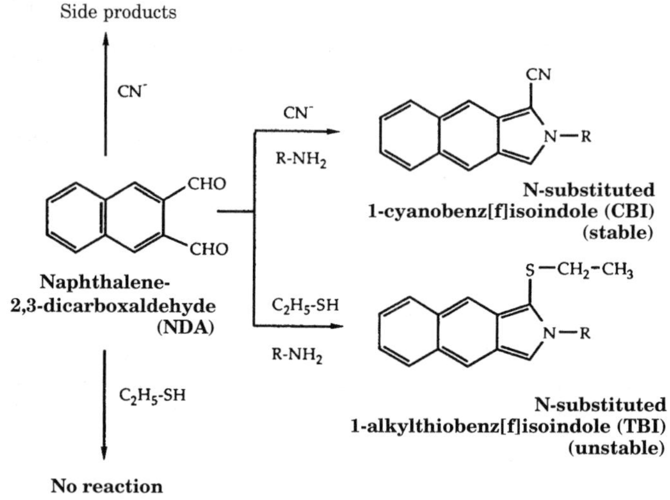

Fig. 2. Important reactions in the derivatization of primary amines (with NDA and CN⁻ or a thiol) to produce the corresponding CBI or TBI derivatives.

Figure 3 shows that stable products are produced within 15 min following the reaction of NDA/CN with ^{5}met-enkephalin and the synthetic opioid peptide D-^{2}ala-^{5}met-enkephalin, exemplifying ideal properties of the NDA/CN system for the precolumn derivatization of amines in general and peptides in particular (Mifune *et al.*, 1989). The rate and the yield of the reaction of primary amines with NDA/CN depend on the pH of the reaction and the nature of the solvent system. Studies by de Montigny and coworkers (de Montigny, 1986; de Montigny *et al.*, 1987, 1990) have shown that the rate and the yield of the reaction of primary amines with NDA/CN are maximum when the pH of the solution is equal to the pKa of the amine. This pH dependency may permit some degree of kinetic selectivity for the derivatization of the N-terminal amino group of peptides, which has a pKa value around 6.8 to 7.2 compared with a typical primary amine pKa value of 9.0 to 10.5. However, the N-terminal amino group of most peptides is less reactive than an unhindered amino group of a lysine side chain. Therefore, even if the reaction is conducted at a pH close to the pK_a of the terminal amino group (*i.e.*, 7), derivatization of lysine side chains is difficult to avoid (Nicholson *et al.*, 1990).

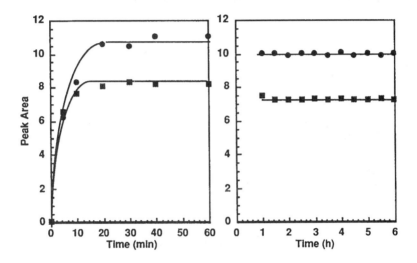

Fig. 3.   Formation kinetics (left) and product stability (right) for the reaction of [5]met-enkephalin (■) and D-[2]ala-[5]met-enkephalin (●) with NDA/CN. See ref. for experimental details. (Adapted and reproduced with permission from Mifune *et al.*, 1989.)

The problem of multiple products for peptides containing two or more derivatization sites occurs in precolumn derivatization strategies because those products are likely to be separated by the chromatographic system (Lunte *et al.*, 1989; Nicholson *et al.*, 1990). This leads to an unsatisfactory analytical method because the analyte is then distributed among several peaks. By appropriate adjustment of the reaction conditions, it may be possible to produce a single product from an analyte with several derivatization sites; however, multiple derivatization of analytes with several available primary amino groups with NDA/CN generally gives rise to internal quenching of fluorescence and concomitant loss of sensitivity (Matuzewski *et al.*, 1987; Lunte *et al.*, 1989; Nicholson *et al.*, 1990). The problem of loss of fluorescence with multiple derivatization can be avoided by using electrochemical detection (Sec. 4.3), in which case sensitivity is roughly proportional to the number of CBI groups on the analyte (Kennedy *et al.*, 1989; Lunte *et al.*, 1989; Oates *et al.*, 1989a,b, 1990; Oates *et al.*, 1990; Nussbaum *et al.*, 1992). An alternative approach is to trap a single derivative by quenching the reaction after a fixed period of time. Quenching of the reaction is conveniently achieved by addition of an excess of a very reactive polar amine such as taurine (Kristjansson 1987; Mifune *et al.*, 1989; Riley *et al.*, 1989; Sprancmanis, 1989; Nicholson *et al.*, 1990; Sprancmanis *et al.*, 1990; Miundi *et al.*, 1991). Quenching of the reaction with taurine also serves to prevent the production of fluorescent side products arising from the reaction of NDA with CN^-.

The solubility of NDA in water is only about 1 mM; therefore, derivatizations with NDA are generally conducted in a mixture of acetonitrile-water. In addition to increasing the solubility of NDA, the presence of acetonitrile in the reaction mixture may also help to reduce the adsorption of peptides to glass surfaces (Nicholson *et al.*, 1990). The rate, but not the yield, of the reaction of amines with NDA/CN is reduced slightly by addition of an organic solvent such as acetonitrile. A slight reduction in the rate of the reaction may be advantageous in precolumn strategies because it may allow better control of the reaction; in particular, precise timing of the addition of quenching reagents may be less critical (Nicholson *et al.*, 1990).

2.2. *Precolumn and postcolumn strategies*

The reagent NDA/CN is only suitable for precolumn strategies because its reaction with peptides requires 10 to 15 min at a temperature of 4°C, which is too slow (Fig. 3) for postcolumn methods (de Montigny *et al.*, 1987, 1990; Mifune *et al.*, 1989; Nicholson *et al.*, 1990). Reactions with amino acids take even longer (20–30 min at room temperature). The reactions proceed more rapidly at higher temperatures, but the side products of the reaction of NDA with CN^- are more significant (de Montigny *et al.*, 1987, 1990; Mifune *et al.*, 1989; Nicholson *et al.*, 1990). Reaction kinetics are not the only factors that must be considered in the design of a pre- or postcolumn system. Reagent stability and production of side products are also important. NDA is known to react with CN^- to produce fluorescent products (de Montigny *et al.*, 1987, 1990; Mifune *et al.*, 1989; Nicholson *et al.*, 1990). In precolumn methods these products can be minimized by quenching of the reagent system, and those small peaks that are produced can be separated from the analytes of interest by appropriate selection of the chromatographic conditions. In contrast, in postcolumn systems fluorescent degradation products of the reagent will contribute to the background signal. A small constant background signal in itself is not a problem because it can be offset electronically. However, variations in postcolumn mixing of the column eluant with the derivatization reagent superimposed on a high background signal will result in increased noise.

Recently, Dave et al. (1992b) explored the idea of replacing CN^- with a thiol when derivatizing of primary amines with NDA. As expected from the previous experiences with OPA/thiol, the derivatives were much less stable (de Montigny et al., 1987). However, the reaction of opioid peptides, chosen as model amine-containing analytes, with NDA/β-mercaptoethanol proceeded much more rapidly than with NDA/CN. In addition to rapid reaction with amines, the reagent system was much more stable and produced much less background fluorescence. Thus, it was concluded that NDA/CN should be used as a precolumn reagent for primary amines and NDA/thiol should be reserved for postcolumn systems (Dave et al., 1992b).

Figure 4A shows that a single peak corresponding to a CBI-derivative is produced if excess ^{6}lys-^{5}met-enkephalin is reacted with an equimolar mixture of NDA and CN^-. In a real analytical sample the concentration of the analyte is unknown and a substantial excess of the derivatization reagent is necessary to achieve complete reaction of the analyte and to ensure that the reagent is not exhausted by reaction with other analytes in the medium. Figure 4B shows the side products of the reaction of NDA with CN^- that must be separated from the analytes of interest when excess reagent is used. Figure 4B also exemplifies the problem of double-derivatives that can occur when the analyte, in this case ^{6}lys-^{5}met-enkephalin, contains more than one reactive functional group (Nicholson et al., 1990).

Although excess reagent should be used to ensure complete reaction, the amount of derivatization reagent should be kept as small as possible to minimize both the production of side products and dilution of the sample, which will compromise the sensitivity of the method. Typically, the concentration of NDA for precolumn derivatization methods is in the range 0.1–2 mM. Generally, the molar ratio of NDA to CN^- is the range 0.5:1.0–2:1 for peptides. For amino acids, a ratio of 1:1 is most commonly employed. de Montigny has described an optimized derivatization system for the analysis of leu-enkephalin in spiked plasma samples (de Montigny et al., 1990). The peptide is preconcentrated by solid phase extraction to a dry residue that is then reconstituted in 200 μl of phosphate buffer (pH 6.8) and derivatized by addition of NDA (25 μl, 1 mM) in acetonitrile and sodium cyanide in water (25 μl, 0.5 mM). This reaction medium should prove suitable for the analysis of amines in general; however, the pH should be adjusted with an appropriate buffer to be approximately

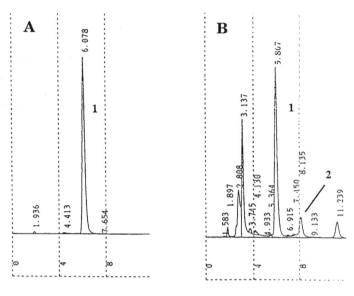

Time (min)

Fig. 4. Chromatograms obtained after (A) reacting NDA/CN with a 1000-fold excess of lys-met-enkephalin and (B) reacting lys-met-enkephalin with a 1000-fold excess of NDA/CN. (1) monoderivative; (2) diderivative. See reference for experimental details. (Reproduced with permission from Nicholson *et al.*, 1990.)

equal to the pKa of the amine. The optimized system of de Montigny *et al.* enabled detection of leu-enkephalin at levels of 310 nM with an error of <4% for 25 pmol injected. The detection was ultimately limited by the sensitivity of the detector and by interferences from side products of the reagents. Improvements in sensitivity beyond what they described require more sensitive detection systems such as laser-induced fluorescence (LIF) and multidimensional separations (column switching) to improve the separations.

Dave *et al.* have described a miniaturized postcolumn reactor for fluorogenic derivatization of opioid peptides (Dave *et al.*, 1992b). In that system, the peptides were derivatized by the postcolumn addition of a mixture of NDA (11 mM) and β-mercaptoethanol (22 mM) in a solvent system comprising acetonitrile-phosphate buffer (20 mM, pH 6.8, 1:1, v/v). The reagent was added to the mobile phase (aceto-nitrile-phosphate buffer (20 mM, pH 6.8, 16:84, v/v) in equal proportions and the peaks were detected by LIF (see Sec. 4.2). Two types of mixers of equal internal volume (23 μl) were compared—a crocheted Teflon tubing (0.012 in i.d.) and a modified serpentine design of PEEK tubing (0.005 in i.d.) (Fig. 5). The modified serpentine system contributed substantially less extra-column band broadening than the crocheted design. However, this was believed to be due to the reduced i.d. of the PEEK tubing rather than the actual configuration of the postcolumn reactor.

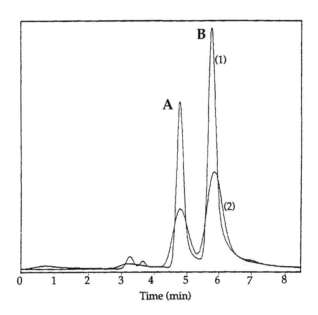

Fig. 5.   Comparison of the band broadening caused by (1) modified serpentine and (2) a crocheted postcolumn reactor using laser-induced fluorescence detection. Analytes (A) [5]met-enkephalin sulfoxide; (B) [5]met-enkephalin sulfone. See Dave *et al.*, 1992b for other experimental details. (Reproduced with permission of Pergamon Press)

3. Detection Systems

Naphthalene-2,3-dicarboxaldehyde (NDA) forms the basis of a very flexible reagent system for the derivatization of primary amines because it can be used in the precolumn mode with cyanide as the nucleophile or in the postcolumn mode with a thiol such as β-mercaptoethanol (Fig. 1). The versatility of the reagent system is further enhanced by the fact that it is suitable for the three most popular modes of detection in liquid chromatography, UV/visible absorbance, fluorescence and electrochemical.

3.1. *UV/Visible Absorbance*

Although NDA/CN was originally designed as a fluorogenic reagent, it has been used occasionally for the liquid chromatographic analysis of primary amines with UV/visible detection. Visible wavelength detection has typically been employed in those cases where fluorescence quenching is a problem. An example is the determination of desmosine, which contains four primary amine groups. The CBI derivatives absorb at 420 and 440 nm, making it possible to detect the analytes selectively at these wavelengths. Absorbance detection has also been employed for the determination of amino acids following separation by capillary electrophoresis. Fig. 6 shows the detection of aspartate, glutamate and other amino acids by CE-UV in a rat brain homogenate.

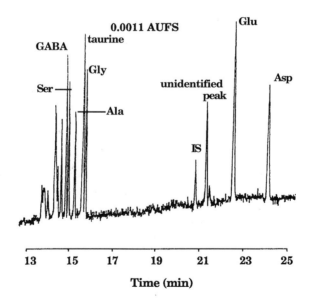

Fig. 6. Electropherogram of CBI-derivatized rat brain homogenate. Separation conditions: 115 cm × 50 μm i.d. column in 20 mM sodium borate, pH 9.0 operated at 30 kV; 420 nm detection. (Reproduced with permission from Weber *et al.*, in press.)

3.2. Fluorescence

3.2.1. Spectral properties

The fluorescence properties of several CBI derivatives of simple amines and amino acids have been characterized extensively (de Montigny 1986; de Montigny *et al.*, 1987; Matuszewski *et al.*, 1987). The analytical characterization with purified standards showed that the excitation and emission spectra of simple CBI derivatives are essentially the same with excitation maxima at 420 nm (molar absorptivity, a = 5,600) and 440 nm (a = 5,300) and an emission maximum at approximately 490 nm (Fig. 7) The quantum yields of fluorescence, Φf, are generally in the range 0.6 to 0.9 (Eq. 3) (de Montigny 1986; de Montigny *et al.*, 1987; Matuszewski *et al.*, 1987).

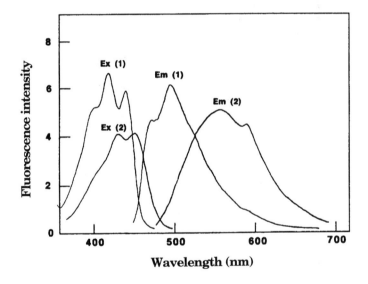

Wavelength (nm)

Fig. 7.   Fluorescence excitation (Ex) and emission (Em) spectra produced by the reaction of [5]leu-enkephalin with NDA and (1) KCN or (2) β-mercaptoethanol. Ambient temperature (22±1°C). Solvent: acetonitrile-phosphate buffer (pH 6.8, 50 mM). Excitation and emission wavelengths for the cyanide and β-mercaptoethanol adducts were 440 and 490 nm and 460 and 560 nm, respectively. (Adapted and reproduced with permission from Dave *et al.*, 1992b)

The overall fluorescence intensity is described by the following equation:

$$F = 2.303 \, \Phi_f I_0 abc \tag{3}$$

where F is the fluorescence intensity, Φ_f is the quantum yield of fluorescence, I_0 is the intensity of the absorbed radiation, a is the molar absorptivity, b is the pathlength in cm and c is the molar concentration. Another important feature of CBI derivatives from the chromatographic standpoint is that their fluorescence properties appear to be independent of the pH of the solvent and of the presence of organic solvents. Therefore, detection limits by liquid chromatography are independent of the composition of the mobile phase.

The fluorescence properties of more complex CBI derivatives (*e.g.*, CBI peptides) appear to be the same as those of smaller CBI-amines. However, this has not been confirmed unambiguously because attempts to isolate pure CBI derivatives of peptides with more than three amino acid residues have proved unsuccessful. Thus it is not possible to separate the derivatization yield from the fluorescent yield for a large CBI derivative generated *in situ*. The detection limits for larger CBI-peptides are of the same order of magnitude as the detection limits of smaller CBI amines, leading one to conclude that the spectroscopic properties of the CBI ring system are generally independent of its extended structure. There are some reports that the sensitivity of peptides derivatized at the ε-amino group of a lysyl side chain is about five times the sensitivity of a similar peptide derivatized at the terminal α-amino group (Nicholson *et al.*, 1990). However, it is not clear whether this apparent increased sensitivity following reaction of NDA/CN is due increased fluorescence (Eq. 3) or increased yield of the reaction, because the apparent quantum yield of fluorescence, Φ_{app}, is equal to:

$$\Phi_{app} = \Phi_f . \Phi_r \tag{4}$$

where Φ_r is the relative yield of the derivatization reaction.

Detection limits for CBI-derivatives of around 10 to 100 fmol can be achieved readily by excitation with a 150 W Xe lamp (see Sec. 5 below). One of the basic spectroscopic properties of CBI-derivatives that ultimately limits their detectability by fluorescence is their relatively low molar absorptivities (5,000 to 6,000 for λ_{ex} = 420–440 nm) (Carlson *et al.*, 1986; de Montigny 1986; de Montigny *et al.*, 1987; Matuszewski *et al.*, 1987). The fluorescence intensity of CBI derivatives can be enhanced by about an order of magnitude by excitation in the UV region of the electromagnetic spectrum at approximately 250 nm where the molar absorptivity of CBI derivatives is approximately

50,000 (Carlson *et al.*, 1986; de Montigny 1986; de Montigny *et al.*, 1987; Matuszewski *et al.*, 1987). Unfortunately, excitation at lower wavelengths substantially reduces the selectivity of the detection system. Other than matrix peaks, some of the chromatographic factors that limit the detectability of CBI-derivatives include interference from side products of the reagent and the reduced fluorescence of multiple derivatives. These can be partially eliminated by quenching of the reagent (see Sec. 2.2), column switching (see Sec. 4.1), or in the latter case by the use of electrochemical detection (see Sec. 3.3).

3.2.2. Laser-induced fluorescence detection

a. Spectral considerations. One of the ways in which the sensitivity of fluorescence detection can be enhanced is to increase the intensity of the light source (Eq. 3). This can be achieved by using a laser rather than a conventional xenon or deuterium lamp. In addition to being high powered, laser emissions are monochromatic, collimated, coherent and polarized (Rossi, 1990). Therein lie both the advantages and the disadvantages of laser-induced fluorescence (LIF) detection in liquid chromatography and capillary electrophoresis. The collimated nature of laser emission is a distinct advantage for detection in miniaturized systems such as capillary electrophoresis or capillary liquid chromatography because the light source may be focused on at a point within a very small flow cell. The monochromatic nature of laser emissions is a distinct disadvantage because the wavelength of the laser line must coincide with the excitation spectrum of the analyte. At present there are only two lasers in common use for the LIF detection in liquid chromatography: the He-Cd laser, which has excitation lines available at 325 nm and 442 nm and the argon-ion laser which has excitation available at 351, 362, 457.9, 488 and 514.5 nm. The lack of laser lines below 300 nm is an important problem because it means that the native fluorescence of endogenous peptides containing tryptophan or phenylalanine cannot be detected by LIF. At present, wavelengths below 300 nm are available by the use of frequency doubling; however, very high power lasers are needed to access lower wavelengths because about 90% of the power is lost with each doubling. The availability of commercial diode lasers will ultimately help to solve the problem of the lack of lasers with emission lines in the UV region of the electromagnetic spectrum.

Roach and Harmony (1987) were among the first to appreciate the need to tailor the excitation maxima of fluorescent tags to the laser lines that are available with commercial lasers. They demonstrated detection limits for CBI-amino acid standards in the range 110–430 amol using conventional liquid chromatography and LIF detection with the 457.9 argon-ion laser line (Roach and Harmony, 1988). Even lower detection limits were demonstrated using this detection system and microbore liquid chromatography. Although the absorption maxima of the CBI derivatives are at 420 and 440 nm, there is sufficient absorption of energy at 457.9 for the argon-ion laser to be useful for LIF detection of CBI derivatives. The He-Cd line at 442 nm is coincidental with one of the absorption maxima of the CBI ring system and provides a much better match of excitation spectra and laser excitation lines. Thus, the majority of reports of LIF detection of CBI-derivatives have used the He-Cd laser as an excitation source even though the commercially available He-Cd lasers are much less powerful than the commercially available argon-ion lasers. Another attractive feature of the He-Cd laser is its relatively low cost compared with the argon-ion laser. The main disadvantage of the He-Cd laser is the high noise that arises from fluctuations in the source; however, the problem of source noise has recently been substantially overcome by the use of electronic signal ratioing techniques. Detection limits of 1–10 amol can be obtained for pure standards of simple CBI-derivatives using either conventional liquid chromatography, microbore liquid chromatography, capillary liquid chromatography or capillary electrophoresis. However, detection limits are several orders of magnitude greater in real samples due to interferences from matrix peaks and the slower reaction kinetics with small amounts of analyte.

The concept of matching the excitation spectra of the fluorescent tags to the 442 nm He-Cd laser line has been explored extensively by Novotny and coworkers, who have described a range of reagents based on the NDA/CN system (Beale *et al.*, 1988, 1989, 1990; Novotny, 1988; Hseih *et al.*, 1989; Liu *et al.*, 1990, 1991a,b) (Fig. 8). These reagents have been used mainly for LIF detection of amino acids in protein hydrolysates following separation by capillary electrophoresis and capillary liquid chromatography. The kinetic and spectroscopic properties are broadly the same as those of NDA/CN and have the same disadvantages arising from side products due to the reaction of the reagent with cyanide ion.

The combination of NDA and a thiol for the derivatization of primary amines was originally discarded because the products of the reaction were too unstable to be useful (Dave *et al.*, 1992b). Indeed, it was not clear from the original studies whether the lack of fluorescence of the TBIs (Fig. 2) produced *in situ* was due to low reactivity (Φ_r) or low quantum yield of fluorescence (Φ_f) (Eq. 4) (de Montigny 1986; de Montigny *et al.*, 1987). More recently, Dave *et al.* have described a miniaturized system for the postcolumn derivatization of peptides that uses the combination of NDA and a thiol (typically

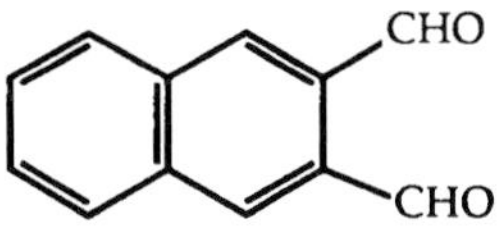

Naphthalene-
2,3-dicarboxaldehyde
(NDA)

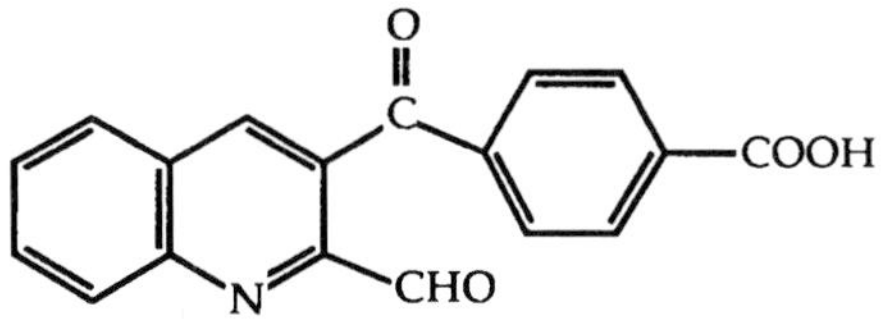

3-(4-Carboxybenzoyl-2-quinoline-
carboxaldehyde
(CBQCA)

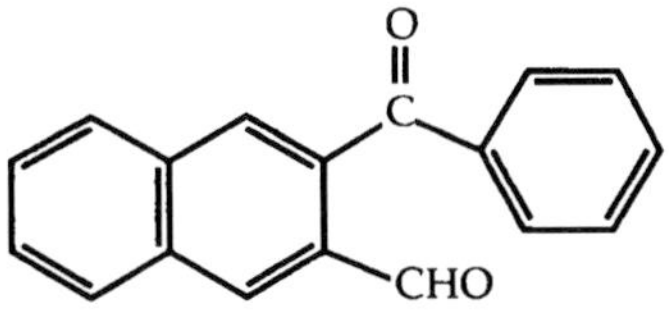

3-Benzoyl-2-naphthaldehyde
(BNA)

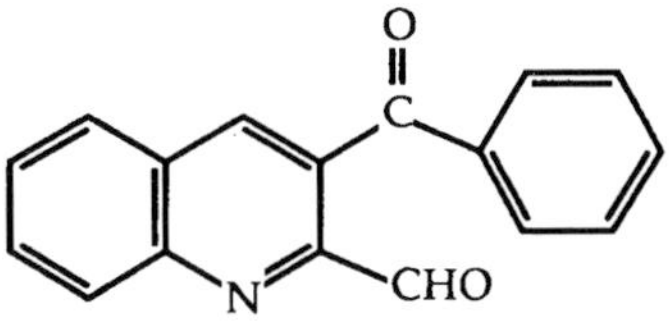

3-Benzoyl-2-quinolinecarboxaldehyde
(BQCA)

Fig. 8.   Structures of NDA and NDA analogs used for the derivatization primary amines in the presence of cyanide ion.

β-mercaptoethanol) and LIF detection (Dave *et al.*, 1992b). They found that highly fluorescent TBI-derivatives form rapidly when opioid peptides are reacted with NDA/thiol but that the products of the reaction were much less stable than the corresponding CBI-derivatives (type c, Fig. 1). The background fluorescence from the reagent was much reduced in comparison to NDA/CN.

In addition to the kinetic advantages of NDA/thiol compared with NDA/CN for postcolumn derivatization, the former also provides a number of other attractive analytical features. In particular, Fig. 7 shows that the emission spectrum of TBI-[5]met-enkephalin is shifted to longer wavelengths by about 20 nm compared with CBI-[5]met-enkephalin such that one of its excitation maxima at 460 nm coincides precisely with the 459.7 nm line of the argon-ion laser. Fig. 7 also shows that the emission maximum of TBI derivatives is about 560 nm compared with 490 nm for the corresponding CBI-derivatives. This shift in emission maximum to higher wavelengths is particularly important for LIF detection because it moves the emission away from the Raman scattering of acetonitrile-water mixtures, which are commonly used as mobile phase for the separations of the analytes.

Figure 9 shows an isocratic separation of four opioid peptides by conventional liquid chromatography with UV detection at 210 nm and by microbore liquid chromatography with postcolumn NDA/thiol derivatization and LIF detection. The detection limit for [5]leu-enkeph-

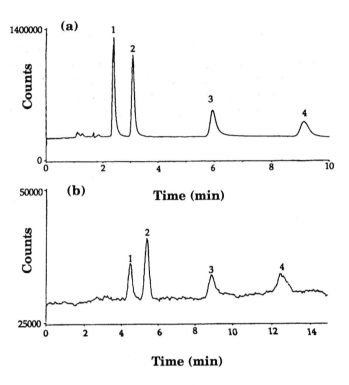

Fig. 9.   Liquid chromatography separation of four opioid
peptides with (a) UV (210 nm) detection, 50 pmol
injected and (b) LIF detection using an Ar-ion
laser (459.7 nm), 50 fmol injected. 1, met-
enkephalin sulfoxide; 2, met-enkephalin sulfone;
3, met-enkephalin; 4, leu-enkephalin. (Reproduced with permission from Dave *et al.*, 1992b).

alin following post-column derivatization and argon-ion (459.7 nm)
LIF detection was approximately 36 fmol. Although the background
fluorescence of NDA/thiol reagent was much less than that of
NDA/CN, detection in the postcolumn NDA/thiol system was
ultimately limited by trace background fluorescence arising from the
postcolumn system. Fig. 10 shows the increase in background signal
and noise that arose in the LIF detector when the postcolumn
reagents were added to the eluant. It is not clear whether the background fluorescence in the postcolumn NDA/thiol reagent system
arose from fluorogenic degradation of the reagent or reaction of the
reagent with trace amines in the solvents. Unfortunately, fluctua-

tions in the background fluorescence due to variations in mixing contribute significantly to the noise and seriously compromise detection limits of the postcolumn NDA/thiol system. Nevertheless, postcolumn NDA/ thiol does represent an attractive alternative to precolumn NDA/CN because it is easier to automate and there is no precolumn dilution of samples or sample handling. These are particularly attractive features for the analysis of very small samples.

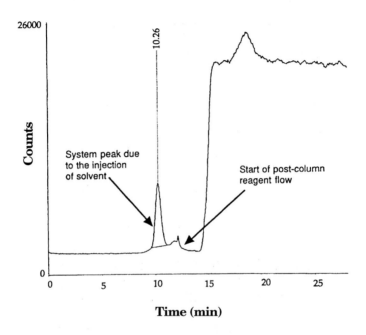

Fig. 10. Effect of adding postcolumn NDA/thiol to the column eluant on the noise and background signal of an Ar-ion laser (459.7 nm). (Reproduced with permission from Dave *et al.*, 1992b.)

b. Flow cell designs. Two basic types of flow cell have been described for LIF detection systems in liquid chromatography and capillary electrophoresis (Fig. 11). Both systems use a quartz capillary tube as the flow cell. The original LIF flow cell described by Yeung (1981) utilized fiber optics for the collection of the fluorescence. One end of a fused-silica optical fiber is inserted into a 1 mm (i.d.) quartz capillary flow via a tee and the column eluant is introduced into the flow cell through the second arm of the tee (Fig. 11a). The opposite end of the optical fiber is terminated at the focal point of a collimating lens that is connected to a photomultiplier tube via a monochromator or an interference filter. In this configuration, the laser beam is focused on the quartz cell by a series of mirrors; external vibrations are minimized by mounting the whole system on an optical table.

In an alternative configuration (Fig. 11b) the laser output is brought into the flow cell by a quartz-silica optical fiber and the fluorescence is captured by placing a photomultiplier tube in close proximity to the flow cell (Soper and Kuwana, 1989; Soper *et al.*, 1989). An interference filter is generally placed in front of the photomultiplier tube. The efficiency of light capture can be improved by about 100% by placing a concave mirror on the side of the flow cell opposite the photomultiplier tube; however, the S/N ratio is not affected (Bostick *et al.*, 1992).

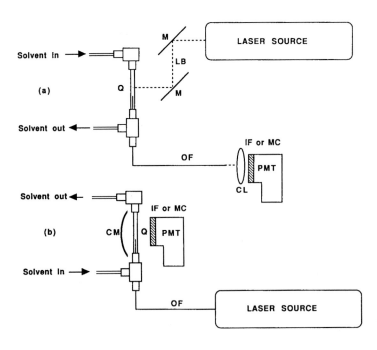

Fig. 11. Laser-induced fluorescence detector flow cells showing (a) the Yeung cell and (b) an alternative configuration. IF, interference filter; MC, monochromator; CM, concave mirror; OF, optical fiber; M, mirror; CL, collimating lens.

3.3 Electrochemical Detection of CBI Derivatives

3.3.1 Electrochemical properties

The electrochemical properties of the CBI derivatives of several peptides and amino acids have been characterized by both cyclic and hydrodynamic voltammetry (Lunte and Wong, 1989; 1990; Nussbaum *et al.*, 1992). Figure 12 shows cyclic voltammograms of the mono- and di-derivatized CBI-lysine. Two features are apparent from these voltammograms. The first is that the oxidation potential for these derivatives is relatively low (approximately +750 mV for the mono-derivative and +600 mV vs. Ag/AgCl for the di-derivative) and, therefore, the selectivity of electrochemical detection for these derivatives is good. NDA itself is not electroactive and thus does not interfere in the detection of the derivatives. The second thing to note is that the

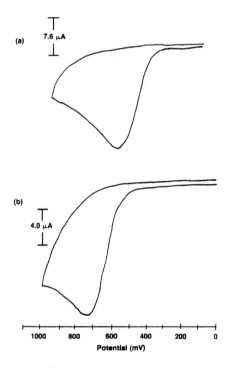

Fig. 12. Cyclic voltammograms of (a) (CBI)$_2$-lysine and (b) CBI-lysine at a glassy carbon electrode. Scan rate: 100 mV/s; reference: Ag/AgCl. (Reproduced with permission from Lunte and Wong, 1989.)

oxidation potentials for the mono- and di-derivatized compounds are different. This means that it is possible to distinguish these two compounds based on their voltammetry.

The voltammetry of several different CBI derivatives has been investigated and substantial differences in redox potential have been found, depending on the molecular composition of the compound being derivatized. For example, CBI-aspartate and CBI-glutamate were found to be the most difficult to oxidize, while CB-lysine and CBI arginine were the most easily oxidized amino acid derivatives. Table 1 shows the wide variation in half-wave potentials for some amino acids and peptides derivatized with NDA/CN.

Table 1. Half-wave potentials of selected CBI-amino acids and CBI-peptides relative to CBI-Gly

Compound	$E_{1/2}$ vs. CBI-Gly (mV)
Peptides	
Gly-His-Lys	−180
Gly-Gly-Arg	−100
Leu-Ala	+90
Gly-Gly-Phe-Leu	−70
Arg-Gly-Gly	+30
Try-Gly	+20
Gly-Gly-Phe	+15
Amino Acids	
Lys	−150
Arg	−105
Met	−65
Thr	−40
Val	−30
Gln	−5
Gly	0
Asp	+15
Glu	+65

This difference in voltammetry can be used to verify peak identity. This is especially useful in CE, where migration times are less reproducible than in LC. To determine the peak purity, hydrodynamic voltammetry is employed. A hydrodynamic voltammogram (HDV) is a plot of current vs. potential obtained by changing the potential step-wise and measuring the peak height of the compound of interest. The shape is generally sigmoidal, reaching a plateau where the current is no longer dependent on potential. This is in contrast to cyclic voltammetry (unstirred system) where the response decreases after the peak potential. For verification of peak identity, it is not necessary to obtain the entire HDV but only to look at that part of the curve where the slope of the current-potential curve is steepest. This can be accomplished by ratioing the current response at a potential where the current is changing most dramatically with potential to that obtained at the current-limited plateau.

An example of the use of voltammetry for peak identity verifica-tion is given in Figure 13. This figure shows a CE-EC separation of several amino acid neurotransmitters found in a brain microdialy-

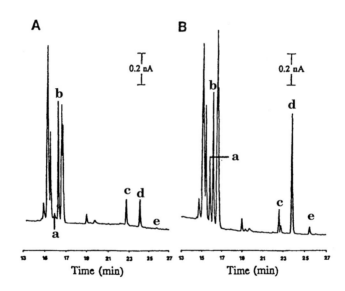

Fig. 13. Derivatized brain dialysate samples obtained by
        perfusion with (A) normal Ringer's solution and (B)
        high K+ Ringer's solution. Peaks: a, GABA; b, Ala;
        c, internal standard; d, Glu; e, Asp. (Reproduced
        with permission from O'Shea *et al.*, 1992.)

sate sample. Peaks a, d and e were tentatively identified as GABA, Ala and Glu, respectively (O'Shea *et al.*, 1992). By running the same electropherogram at three different potentials and comparing it to the standard, it was possible to verify the identity of these peaks.

Table 2 shows the resulting data. Clearly, peak a is impure, since the ratio does not agree well with that of the standard. On the other hand, the identities of Asp and Glu are further substantiated by the voltammetric data. If this same separation had been accomplished by laser-induced fluorescence detection, there would have been no easy way to verify peak identity.

Table 2. Voltammetric Characterization

Component	Retention time (min)	Current response +0.45 V/0.80 V	+0.65 V/0.80 V
Peak a	15.9	0.21	0.72
GABA	15.7	0.12	0.65
Peak b	16.2	0.12	0.64
Alanine	16.0	0.10	0.62
Peak d	24.4	0.04	0.55
Glutamate	24.1	0.03	0.55
Peak e	25.7	0.03	0.54
Aspartate	25.9	0.03	0.56

The oxidation potentials of the CBI derivatives have been found to be independent of pH (Nussbaum *et al.*, 1992). This means that greater selectivity can be obtained at lower pH values, where the oxidation potential for compounds undergoing proton dependent processes would be relatively high. This was illustrated with a group of phenolic compounds, where the selectivity for the CBI derivatives was greatly enhanced at pH 2 (Nussbaum *et al.*, 1992).

3.3.2 Applications

One advantage of amperometric detection is its compatibility with microcolumn-based separation methods. In contrast to optical detection methods, the cell volume can be made very small without a loss in sensitivity. Electrochemical detection has therefore been employed as the detection method of choice for the determination of CBI-labeled amino acids and peptides by open tubular liquid chromatography and capillary electrophoresis.

Because of the small volumes needed for these techniques, it is possible to analyze minute amounts of material. Jorgenson and coworkers have taken advantage of this fact by employing NDA/CN in conjunction with OTLC for the amino acid analysis of submicrogram quantities of protein. Using amperometric detection, the amino acid composition of 0.1 ng of chymotrypsin was determined with an error of less than 10% (Oates and Jorgenson, 1990b; Oates *et al.*, 1990). The final derivatization volume was only 25 nl. Derivatization with NDA/CN prior to analysis by OTLC-EC has also been employed for the determination of amino acids in single cells (Kennedy *et al.*, 1989; Oates *et al.*, 1990). Fig. 14 shows a chromatogram of NDA-labeled amino acids from a single E4 neuron from *Helix aspera*. Aspartate and glutamate have also been determined in brain microdialysates by capillary electrophoresis with electrochemical detection following derivatization with NDA/CN (O'Shea *et al.*, 1992).

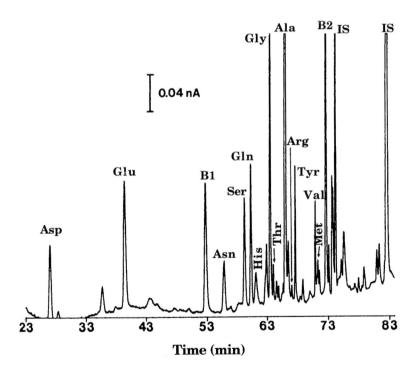

Fig. 14.   Chromatogram of NDA-tagged amino acids from a single E4 neuron from *Helix aspersa*. B1 and B2 are present in blank runs; IS, internal standards; unlabeled peaks, unknowns. (Reproduced with permission from Oates *et al.*, 1990.)

Another advantage of electrochemical detection is the ability to detect multiderivatized compounds. These compounds generally exhibit fluorescence quenching and are cannot be determined at low levels by LC-fluorescence. Some of the compounds which fall in this category include lysine, desmosine and the polyamines. Desmosine, which contains four primary amine groups, was determined in an elastin hydrolysate by liquid chromatography with electrochemical detection (Fig. 15) (Lunte *et al.*, 1989). Detection limits were approximately 100 fmol injected (5×10^{-9} M). In general, these multi-derivatized compounds are more easily oxidized, provide a larger response and elute later than their mono-derivatized counterparts. These properties make LC-EC the method of choice for the detection of compounds containing more than one primary amine group.

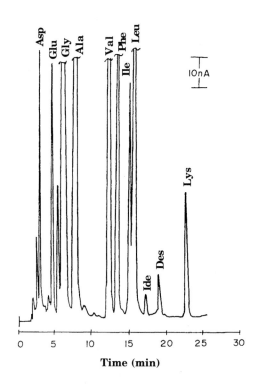

Fig. 15.   Detection of desmosine and isodesmosine in the elastin hydrolysate using gradient elution LC-EC.   Mobile phase A, 85% 0.005 M sodium citrate, 10% methanol, 5% THF; mobile phase B, 90% methanol, 10% citrate. Both the methanol and citrate buffers contain 0.05 M sodium perchlorate.   Elastin concentration, 40 μg ml^{-1}. (Reproduced with permission from Lunte *et al.*, 1989.)

3.4. Chemiluminescence

The chemical activation of fluorescence of the CBI-ring system represents an attractive alternative to light-activated generation of emission. In principle, chemiluminescence should be more sensitive than conventional fluorescence, and detection limits may be further enhanced by the lack of background emissions (noise). Recognizing these potential advantages, a number of workers have explored the use of chemiluminescence for detection of CBI-derivatives in liquid chromatography (Alvarez et al., 1986; Givens and Schowen, 1989; Givens et al., 1989, 1990). The chemiluminescence system that has been most widely investigated for the activation of the CBI ring system is the oxalate ester-hydrogen peroxide system (Fig. 16) originally described by Kobayashi and Imai (1980). Various oxalate esters have been investigated, but bis-trichlorophenyloxalate (TCPO) remains the most commonly used reagent (Fig. 16).

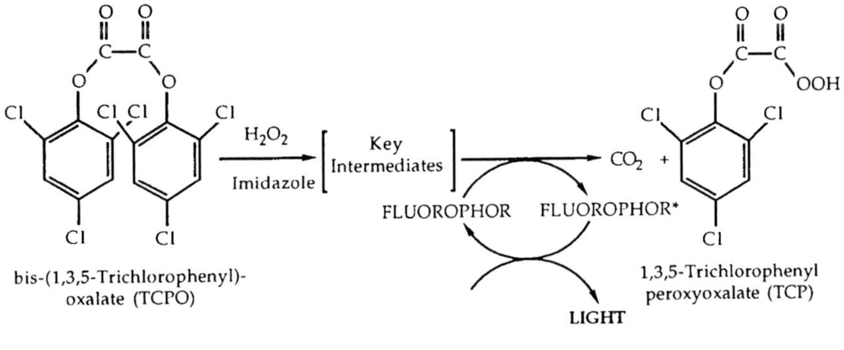

Fig. 16.   The peroxyoxalate/hydrogen peroxide chemiluminescence reaction.

3.4.1. Kinetic considerations

The reactions by which an oxalate such as TCPO and hydrogen peroxide generate fluorescence in the presence of a fluorophore is complex. The original theory proposed by McCapra and Rauhut (Rauhut et al., 1967; McCapra et al., 1981) involved the formation of a four-membered "dioxetanedione" which was postulated to be the key reactive intermediate responsible for the activation of the fluorophore. This theory (Rauhut et al., 1967; McCapra et al., 1981) is now believed to be incomplete because Givens and co-workers (Alvarez et al., 1986; Givens et al., 1989, 1990) have proposed a more complex model that incorporates three reactive intermediates of which two give rise to chemiluminescence and a third serves as a bridge between the other two. This model gives rise to a complicated light intensity-time profile in which two maxima are observed corresponding to the sequential activation of the fluorophore by an early and a late intermediate. Fortunately, under the conditions of the mixed organic-aqueous solvents system encountered in typical reversed-phase liquid chromatography systems, the two maxima collapse into a reasonably simple biexponential relationship between the emission of light and time.

A simple model for the TCPO-hydrogen peroxide chemiluminescence process is shown in Fig. 17. The light intensity-time profile (I_{CL} vs. t) shown in Fig. 17 can be represented by a pooled-intermediate

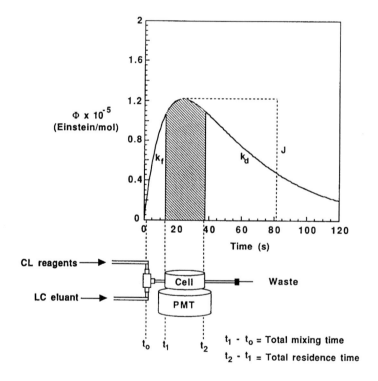

Fig. 17. Relationship between the chemiluminescence intensity-time profile and the timing of the post-column mixing of the reagents. (Reproduced with permission from Givens *et al.*, 1989.)

model using a kinetic expression (Eq. 5) which contains a first order rate constant for the formation (k_f) and the decline (k_d) of the chemiluminescence:

$$I_{CL} = \frac{Mk_f}{k_d - k_f} (e^{-k_f t} - e^{-k_d t}) \qquad (5)$$

where M is the (theoretical) initial maximum intensity. The intensity of the maximum (J), the time to reach maximum intensity (t_{max}) and the quantum efficiency (F_{CL}) are given by Eqs. 6 – 8.

$$J = M(\frac{k_d}{k_f})^{(k_d/[k_f - k_d])} \qquad (6)$$

$$t_{max} = (\ln \frac{k_d / k_f}{[k_f - k_d]}) \qquad (7)$$

$$\Phi_{CL} = \int_0^\infty I_{CL} dt = \frac{M}{k_d} \qquad (8)$$

3.4.2. Practical applications of chemiluminescence to liquid chromatography

The practical application of chemiluminescence to detection in liquid chromatography requires optimization of the reaction profile so that the maximum amount of light is generated in the flow cell (Fig. 17). Therein lies the main disadvantage of chemiluminescence: the kinetics of light production and the quantum efficiency of chemiluminescence are related to variety of extrinsic and intrinsic factors including pH, the nature of the solvent system, the concentration of hydrogen peroxide, the nature and the concentration of the oxalate ester, and the nature and the concentration of the imidazole catalyst. In addition, it is now clear that the oxalate ester-hydrogen peroxide reagents system itself produces significant background chemiluminescence, which, because of incomplete mixing in the postcolumn reactor, contributes to the background noise.

In general, the maximum intensity of chemiluminescence (J, Eq. 6) increases with increasing oxalate ester concentration. Unfortunately, the solubility of oxalates such as TCPO in typical mobile phases using reversed-phase liquid chromatography is very low. Therefore, the oxalate ester must be added in an organic solvent such as ethyl acetate. Even with the use of an organic solvent for postcolumn addition of oxalate esters, precipitation of oxalates is a serious problem, particularly when using mobile phases with a high water content. Although its function in the reaction is unclear, the presence of a basic catalyst is essential; imidazole or imidazole derivatives are vastly superior to other aromatic or aliphatic amines such as Tris, diethyl amine, triethylamine, pyridine or aniline. The effects of other factors influencing TCPO-hydrogen peroxide generated chemiluminescence have been reviewed extensively by Givens *et al.* (1990).

Despite the significant practical problems of chemiluminescence detection in liquid chromatography, some very impressive detection limits have been achieved which approach those that can be achieved by laser-induced fluorescence detection. Figure 18 shows the 25-fold

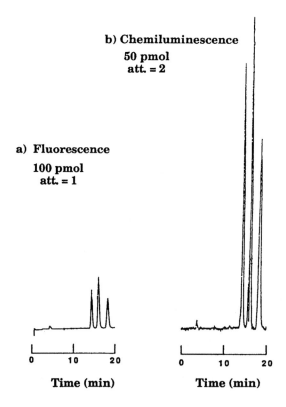

Fig. 18.   Comparison of (a) fluorescence and (b) chemiluminescence detection of three opioid peptides. In order of
elution, the analytes are the CBI-[5]met-enkephalin,
CBI-[5]met-[2][D]-ala-enkephalin and CBI-[5]leu-enkephalin.
(Reproduced with permission from Givens *et al.*, 1990.)

increase in sensitivity achieved in the detection of opioid peptides (as their CBI-derivatives) by chemiluminescence compared with conventional fluorescence detection (Givens *et al.*, 1990). Similarly, Lunte and Wong (1989) have shown an order of magnitude improvement in the detectability of a neuropeptide, phenylalanine-neurotensin in blood by chemiluminescence compared with conventional fluorescence detection (Fig. 19).

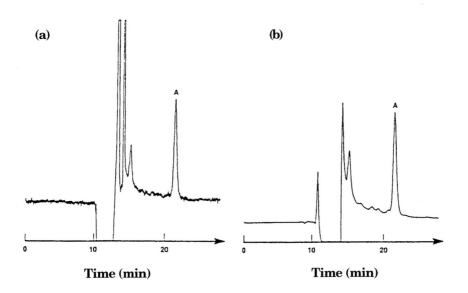

Fig. 19. Detection of phenylalanine-neurotensin in human blood by (a) LIF fluorescence and (b) chemiluminescence. See Lunte *et al.*, 1989 for experimental details. (Reproduced with permission from Lunte *et al.*, 1989.)

4. Separation Approaches

4.1. Liquid Chromatography

Retention prediction, structure-retention relationships and optimization of separations of amino acids and peptides have been studied too extensively to be within the scope of this chapter, which will concentrate mainly on the important chromatographic properties of amine analytes following derivatization with NDA. Isocratic elution is the most convenient method for the separation of CBI-derivatives by liquid chromatography because it requires the least sophisticated equipment and generally results in the shortest analysis times. Gradient elution is only required for the analysis of complex samples containing analytes with a wide range of polarity, such as protein hydrolysates, which give rise to the general elution problem when analyzed by isocratic elution. Gradient elution may also useful for the optimization of isocratic separations of CBI-peptides using the approach proposed by Snyder and co-workers (Snyder *et al.*, 1979; Snyder *et al.*, 1983; Quarry *et al.*, 1984, 1986). Multidimensional systems are often required for the trace analysis of peptides using highly sensitive detection systems such as laser-induced fluorescence or chemiluminescence because the peak capacities of single column systems may be inadequate for the resolution of the analyte of interest from components of the matrix.

4.1.1. Isocratic elution

Simple isocratic reversed-phase systems are usually adequate for the analysis of single CBI-derivatives or of mixtures of CBI-derivatives that are relatively similar in polarity, provided that the desired detection limits are in excess of about 1 to 10 nM. Detection limits of less than 1 nM usually require sophisticated detectors and multi-dimensional separations. In addition to the enhancement of detection, precolumn derivatization of primary amines with NDA/CN may improve the chromatographic properties of the analytes because it will effectively eliminate a basic functional group that might otherwise contribute to a mixed mechanism of retention through interactions with residual silanol groups on the surface of the hydrocarbonaceous support. Such interactions with silanol groups are known to result in peak tailing and variability in retention times.

In addition to the removal of amino groups that might otherwise interact with silanol groups, the addition of a bulky CBI ring system substantially increases the hydrophobicity of the analytes. The contribution of the CBI ring system to the retention of opioid peptides has been assessed by Dave *et al.* (1992a) using a functional group approach. The contribution of a particular functional group, τ_j, to retention may be calculated from the equation

$$\log k'_j = \log k'_i + \tau_j \tag{9}$$

where k'_j and k'_i are the capacity ratios of the substituted and unsubstituted molecules, respectively. Using this approach (Eq. 9) Dave *et al.* (1992a) have shown that the functional group contribution for the CBI ring system, τ_{CBI}, is approximately 2. Therefore, the addition of the CBI ring system increases the retention of the analyte by a factor of approximately 100 and stronger eluant systems are required to elute CBI derivatives from reversed-phase columns compared with the corresponding free amine.

Figure 20 represents an alternative method of demonstrating the effect of adding a CBI-ring system on the retention of peptides. In this figure, the log k' values of series of opioid peptides and their corresponding CBI derivatives are plotted against the sum of the hydrophobic fragmental constants (Σf) of the constituent amino acids in the peptides. The log k' values for the unreacted peptides and their CBI derivatives lie along two separate lines that are essentially parallel. In this case the value of τ_{CBI} is equal to the vertical distance between the two lines. The fact that the two lines are parallel indicates that the contribution of the CBI derivative is independent of structure. The good linearity between log k' for the peptides and their Σf further suggests that solute hydrophobicity may be used to predict the retention of the opioid peptides. This is generally true; however, the relationship is significantly perturbed by peptides containing arginyl residues and other functional groups that may interact with residual silanol groups. Thus, those opioid peptides with residues that interact with residual silanol groups were omitted from the analysis.

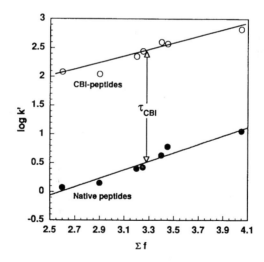

Fig. 20.   Relationship between the isocratic log k' values (at $\Psi_{MeCN} = 0.25$) of seven opioid peptides and their CBI derivatives and the sum of their hydrophobic fragmental constants (Σf). (Reproduced with permission from Dave *et al.*, 1992a.)

The relationship between retention and the volume fraction of the organic modifier in the mobile phase, Ψ, is given by Schoenmakers and Mulholland (1988):

$$\log k' = -S.\Psi + \log k'_w \qquad (10)$$

where S is the slope coefficient and k'_w is the capacity ratio of the analyte in a completely aqueous mobile phase. Substituting Eq. 9 into Eq. 10 gives:

$$\tau_{CBI} = -(S_{R-CBI} - S_{R-H})\Psi + (\log k'_{w,CBI} - \log k'_{w,R-H}) \qquad (11)$$

Assuming that the effect of adding the functional group has a much greater effect on the value of k'_w than it does on the value of S, then it follows that:

$$\tau_{CBI} = \log k'_{w,R-CBI} - \log k'_{w,R-H} \qquad (12)$$

Similarly, Eq. 13 may be used to calculate the difference in mobile phase organic modifier concentration, $\Delta\Psi_{CBI}$, required to elute the CBI derivative with the same value of k' as the corresponding unreacted amine, because

$$\Delta\Psi_{CBI} = \frac{(\log k'_{w,R-CBI} - \log k'_{w,R-H})}{S} \qquad (13)$$

and

$$\Delta\Psi_{CBI} = \frac{\tau_{CBI}}{S} \qquad (14)$$

For small molecules (m.w. $\leq$ 1,000), the value of S for acetonitrile is approximately 4. Therefore, by assuming a value of 2 for τ_{CBI} we arrive at a value of 0.5 for $\Delta\Psi$, which represents the amount by which the volume fraction of acetonitrile may have to be increased to achieve the same k' value for the CBI-derivative that was achieved for the corresponding unreacted amine. In practice, the value of $\Delta\Psi$ is in the range 0.35 to 0.40, which is less than predicted from theory because log k'_w as well as S tend to increase with increasing hydrophobicity. Nevertheless, this $\Delta\Psi$ approach is useful in method development and for locating peaks for the derivatives of interest.

The increase in hydrophobicity of amines following reaction with NDA/CN is particularly useful for the analysis of very polar compounds because it will substantially increase their retention on reversed-phase columns. Conversely, further increasing the retention of compounds that are already very hydrophobic may make it difficult to elute such compounds from reversed-phase columns. Thus, very hydrophobic CBI derivatives may be more conveniently separated on less hydrophobic columns such as phenyl, octyl or cyano bonded phases rather than phases with a high carbon loading such as octadecyl bonded phases. An alternative approach for the analysis of very hydrophobic peptides would be the use of postcolumn derivatization with NDA/thiol (Sec. 2.2), where the chromatography is not a function of the physicochemical properties of the derivatization reagent.

The development of an isocratic system for CBI-derivatives is generally unremarkable and the optimization strategy typically involves the investigation of the effects of organic modifier type and concentration. Precolumn derivatization of amines, amino acids or peptides generally renders the analyte neutral or acidic. Adjusting the pH of the mobile phase will have little effect on the retention of neutral derivatives, but it may change the retention of potential interferences in the matrix. Thus, investigation of mobile phase pH may help to resolve the analyte of interest from components of the matrix. If the derivatives are acidic, the retention of the analyte will also be influenced by the pH of the mobile phase.

The liquid chromatographic analysis of compounds with two or more basic functional groups presents a particularly difficult challenge. It may be possible by appropriate choice of reaction conditions to kinetically control the extent of derivatization. Thus for compounds with more than one basic group, the resultant CBI-derivative may still be basic, in which case its retention may be function of the mobile phase pH. The retention of ionic derivatives can also be controlled by ion-pairing techniques. For example, the anticancer drug 15-deoxyspergualin (DSG) (Fig. 21) contains three basic functional groups, a primary amine, a secondary amine and a guanodinium group (Sprancmanis 1989;. Sprancmanis *et al.*, 1990; Miundi *et al.*, 1991). Only the primary amino group will react with NDA/CN thus the resultant CBI derivative has a net charge of +2. Figure 22 shows how the retention of CBI-DSG can be selectively enhanced and thus resolved from the components of plasma by the addition of an ion-pairing agent, sodium octylsulfate, to the mobile phase.

While apparently not an issue in the chromatography of DSG, the interactions of basic groups, and guanodinium groups in particular, with residual silanols can cause substantial modification of chromatographic behavior, resulting in excessive retention, peak tailing or both. This can occur in both isocratic and gradient elution systems. For example, Dave *et al.* were unable to elute the CBI derivatives of

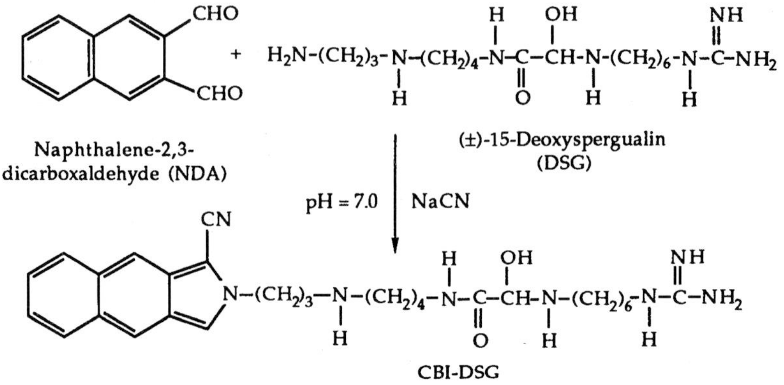

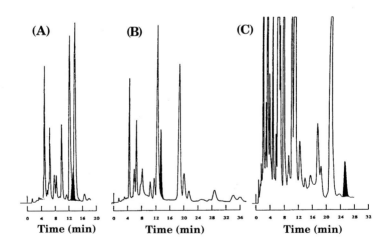

Fig. 21. Precolumn derivatization of the anticancer drug (±)-15-deoxyspergualin with NDA/CN.

Fig. 22. Effect of adding (A) 2 mM, (B) 5 mM and (C) 8 mM sodium octylsulfate to the mobile phase on the separation of CBI-DSG (Fig. 16) from plasma ultrafiltrate. (Reproduced with permission from Sprancmanis *et al.*, 1990.)

β-endorphin, α-neo-endorphin and dynorphin A, which contain several lysyl and arginyl residues, under any mobile conditions from a cyano-bonded column that contained a high surface concentration of residual silanol groups (Dave 1991; Dave *et al.*, 1992a). The effect of residual silanols on the isocratic elution of CBI-peptides has been studied extensively by Kristjansson (1987), using the important neuropeptide substance P as a model compound. That study con-cluded that retention of peptides by residual silanols could be con-trolled and effectively manipulated by appropriate choice of competing ions and the use of THF as a mobile-phase modifier.

4.1.2. Gradient elution

Reversed-phase gradient elution is predominantly used for the resolution of complex mixture of components that have a wide range of polarity.   For example, Fig. 23 shows the gradient elution separation of 18 CBI-amino acids on a Hypersil ODS column (de Montigny *et al.*, 1987). The gradient conditions were a linear ramp from 100% solvent A to 60% solvent B over 60 min, where solvent A was THF–phosphate buffer (pH 6.8) (10:90, v/v) and solvent B was acetonitrile-methanol-phosphate buffer (pH 6.8) (55:10:35 v/v/v). That particular composition of the mobile phase was crucial for the resolu-

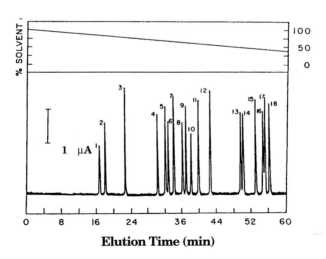

Fig. 23.   Gradient separation of 18 amino acids as their CBI-derivatives See text and de Montigny *et al.*, 1987 for experimental details. (Reproduced with permis-sion from de Montigny *et al.*, 1987.)

tion of certain CBI-amino acids. For example, a simple linear gradient of 15 to 40% acetonitrile over 60 min resulted in partial resolution of the pair cya/asp, the triplet his/gln/ser and the triplet phe/ile/leu and the complete overlap of the pair gly/thr.

In addition to studying the effects of structure on the retention of CBI-peptides (see Sec. 4.1.2), Dave *et al.* (1992a) also explored the utility of Snyder's theory (Snyder *et al.*, 1979, 1983; Quarry *et al.*, 1984, 1986) of gradient elution to predict the optimum isocratic elution conditions for opioid peptides and their corresponding CBI derivatives. They concluded that the ability of the theory to predict isocratic conditions for reversed-phase chromatography of opioid peptides depended upon the nature of the retention mechanism, which in turn was a function of the primary structure of the peptide. Thus, the agreement between the observed and predicted isocratic retention was excellent for those peptides that did not interact with residual silanols (Fig. 24a). In contrast, peptides which contained α-amino, ε-amino, guanidino or sulfoxide functional groups were retained due to a combination of silanophilic and solvophobic interactions.

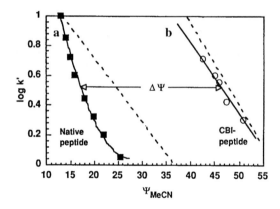

Fig. 24. Relationship between log k' for [6]lys-5-met-enkephalin (■) and its CBI derivative (○) and the volume fraction of acetonitrile (Ψ_{MeCN}) in the mobile phase. The symbols and solid lines represent experimental data and the dotted lines represent the relationships between log k' and Ψ predicted from gradient theory. (Reproduced with permission from Dave *et al.*, 1992a).

Interactions with residual silanol groups caused significant deviations from linearity in the relationship between log k' and Ψ, resulting in substantial differences between the observed and predicted values of k' (Fig. 24a). Interestingly, introduction of a bulky CBI ring system appeared to reduce the contribution of silanophilic interactions to the overall retention, even if that CBI ring system was not attached to the functional group responsible for the silanol interaction (Fig. 24b). Consequently, the relationship between log k' and Ψ was linear for all CBI-opioid peptides studied, and the agreement between observed isocratic retention times and those predicted from gradient elution was excellent. Fig. 24 shows that an acetonitrile concentration of 10–25% is required to elute the native peptide [6]lys-[5]met-enkephalin with a k' value of 1 to 10 compared to an acetonitrile concentration of approximately 40–55% required to elute the corresponding CBI-derivative with the same k' value. This translates to a $\Delta\Psi$ value of 0.35 (Eq. 13).

Interactions between amines and residual silanol groups are well documented. However, the apparent interaction between residual silanols and the sulfoxide group in the oxidized form of [5]met-enkephalin (*i.e.*, [5]met[O]-enkephalin) was quite surprising. As with the other peptides studied, introduction of the bulky CBI ring system changed the relationship between the log k' of [5]met[O]-enkephalin and Ψ from parabolic to linear (Fig. 25). It interesting to note that above $\Psi_{MeCN} = 0.35$ the native peptide met[O]-enkephalin was more retained than its CBI-derivative as a result of the strong silanophilic interactions between the native peptide and the stationary phase (Fig. 25).

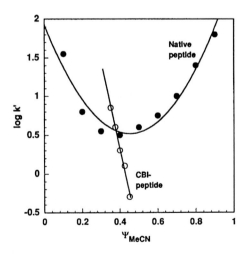

Fig. 25. Relationship between log k' for [5]-met[O]-enkephalin (●) and its CBI-derivative (○) and the volume fraction of acetonitrile (Ψ_{MeCN}) in the mobile phase. (Reproduced with permission from Dave et al., 1992a.)

4.1.3. *Multidimensional systems*

The probability of a minor component in the sample being detected at the same elution point as the analyte increases as the required limit of detection for a given analyte decreases. Multidimensional liquid chromatography (column switching) is a valuable approach to the ultra-trace analysis of compounds in complex samples.

a. Selectivity considerations. There are two major advantages to a multidimensional approach to the resolution of complex mixtures compared with the use of a single column. The first arises from the ability to change the selectivity of the system in mid-run by abruptly changing the physicochemical properties of the stationary phase and the mobile phase. This principle is demonstrated in Fig. 26, which shows the resolution of three opioid peptides by multidimensional liquid chromatography (Mifune *et al.,* 1989). The three peptides co-elute from the first column, but are separated from the excess reagent and other side products in the sample. The peaks of interest are then switched to a second column containing a different stationary phase, where they are separated from each other. This particular type of column switching is often referred to as "heart-cutting."

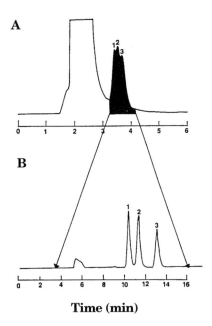

Fig. 26. Multidimensional separation of three opioid peptides. (1) CBI-[5]met-enkephalin, (2) CBI-[5]met-[2][D]-ala-enkephalin and (3) CBI-[5]leu-enkephalin. (Reproduced with permission from Mifune *et al.,* 1989.)

b. *Peak capacity.* The second advantage of multidimensional liquid chromatography arises from the increase in peak capacity that can be achieved compared with a single column or two columns linked in series. The peak capacity of a column in liquid chromatography, N_c, is given by:

$$N_c = 1 + \frac{\sqrt{N}}{4 R_s} \ln (k' + 1) \qquad (15)$$

where k' is the capacity ratio of a given peak, N is the number of theoretical plates and R_s is the resolution factor for each pair of adjacent peaks in the chromatogram. A value of 1.5 for R_s represents baseline resolution with less than 1% overlap of adjacent peaks. Substituting $t_r/t_o - 1$ for k' and L/u for t_o (where L and u are the column length and the linear velocity) gives an expression (Eq. 16) that relates peak capacity to column efficiency (N) and analysis time (t_r).

$$N_c = 1 + \frac{\sqrt{N}}{4 R_s} \ln \left(\frac{t_r\, u}{L}\right) \qquad (16)$$

If we assume that the peak capacities of two columns are additive then the total peak capacity in a multidimensional system, $N_{c,md}$, is given by:

$$N_{c,md} = 2 + \frac{1}{4 R_s}\left\{\sqrt{N_1} \ln \left(\frac{t_r\, u_1}{L_1}\right) + \sqrt{N_2} \ln \left(\frac{(t_r - t_s)\, u_2}{L_2}\right)\right\} \qquad (17)$$

where t_s is the time that the peak is switched from column 1 to column 2.

Figure 27 compares the peak capacities achieved with (a) a single column with 10,000 theoretical plates, (b) two columns each with 10,000 theoretical plates that are coupled in series with (c) a multidimensional (heart-cutting) system consisting of two columns each with 10,000 theoretical plates. The curves were simulated (Eqs. 16 and 17) using values of 0.25 cm s^{-1} for u_1 and u_2, a value of 1.5 for R_s and value of 600 s for t_s. Two important conclusions may be drawn from the curves in Fig. 27. First, the value of the peak capacity of the two columns coupled in series is only about 10% greater than the single column system at t_r = 20 min. Secondly, and more importantly, Fig. 27 shows that the peak capacity achieved after 20 min by the multidimensional system is 45% greater than the peak capacity of the two columns coupled in series. Figure 28 shows a practical example of this approach in which multidimensional liquid chromatography has been applied to the analysis of [5]leu-enkephalin and [5]met-enkephalin in the striatal region of the rat brain (Mifune *et al.*, 1989).

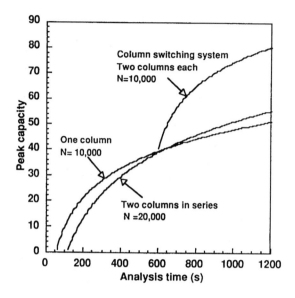

Fig. 27.   Relationship between peak capacities and capacity ratios.

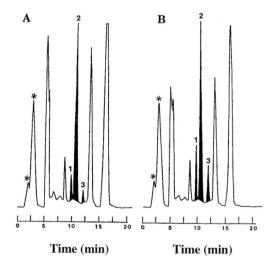

Fig. 28.   Multidimensional separation of **1,** [5]met-enkeph-
alin and **3,** [5]leu-enkephalin as their CBI derivatives in rat striatal tissue. **2,** internal standard
[5]met-[2][D]-ala-enkephalin. **A,** brain sample; **B,**
brain sample spiked with [5]met-enkephalin and
[5]leu-enkephalin. (Reproduced with permission
from Mifune *et al.,* 1989.)

c. Optimization strategies. The main disadvantage of multidimensional liquid chromatography is that the separation of the analyte of interest must be optimized on both columns. In addition, it is very important that the mobile phase that is transferred from column 1 is compatible with the chromatographic conditions being used for the elution of column 2; otherwise, substantial peak distortion will occur. On the other hand, if the mobile phase conditions are adjusted so that the eluant being transferred from column 1 is weaker than the eluant used to elute column 2, then zone compression will occur and peak dispersion arising from column 1 will be eliminated (Mifune *et al.*, 1989). Therefore, the apparent column efficiency of the multidimensional system will be greater than if the two columns had been coupled in series.

The prediction of retention and the optimization of the separation of CBI-peptides by multidimensional liquid chromatography have been studied by Nicholson *et al.* (1990). Their study was stimulated by the observation that the observed total retention time in a multidimensional separation, t_{tot}, was not simply equal to the sum of the retention times on the individual columns (t_{11} and t_{22}) plus the time spent in the switching value and associated tubing (t_{ext}). Thus, substantial errors were obtained if Eq. 18 was used to predict t_{tot}:

$$t_{tot} = t_{11} + t_{22} + t_{ext} \qquad (18)$$

They showed that these errors (Eq. 18) arose from the chromatographic processes that occurred during the transfer of solutes from the first column to the second and that components were subjected to the effects of mobile phase 1 for significant periods of time during their elution on column 2. They were also able to show that the actual overall retention time in a multidimensional system is given by:

$$t_{tot} = t_{0,1}(1 + k'_{11}) + t_s\left(\frac{k'_{12} - k'_{22}}{k'_{12}}\right) + t_{0,2}(1 + k'_{22}) + t_{ext} \qquad (19)$$

where $t_{0,1}$ and $t_{0,2}$ are the unretained peak times on columns 1 and 2, respectively, and k_{12} is the capacity ratio of the peak on column 2 when eluted with the mobile phase from column 1.

Having established a reliable method of predicting the retention times of CBI-peptides in a multidimensional system from data generated from the results of single column systems, Nicholson *et al.* (1990) defined a multidimensional chromatographic optimization function (COF) (Eq. 20). The COF approach allowed the optimal multidimensional conditions to be predicted from experiments run on single columns (Fig. 24):

$$COF = \sum_{i=1}^{n-1} A \ln \left[\left(\frac{\sqrt{N}}{4 \, R_s} \right) \left(\frac{k^*_{i+1}}{1 + k^*_{i+1}} \right) \right] + B \, (k^*_{max} - k^*_n) \qquad (20)$$

where A and B are arbitrary weighting factors and n is the total number of peaks in the chromatogram. The apparent multidimensional capacity ratio (k^*) was defined as:

$$k^* = \frac{t_1 + t_2}{t_{01} + t_{02}} - 1 \qquad (21)$$

where t_1 and t_2 are the retention times on columns 1 and 2, respectively. Figure 29 shows two equivalent, optimized separations of four CBI-peptides in multidimensional, both employing CPS Hypersil as column 1 and Spherisorb Phenyl as column 2.

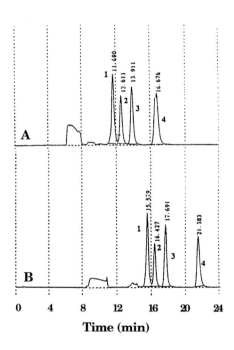

Fig. 29. Optimized multidimensional separation of four opioid peptides as their CBI-derivatives. The analytes are (1) CBI-[5]met-enkephalin, (2) CBI-[5]met-[2][D]-ala-enkephalin, (3) CBI-[5]leu-enkephalin and (4) CBI-[6]lys-[5]met-enkephalin. (Reproduced with permission from Nicholson et al., 1990.)

4.2. Capillary Electrophoresis

Recently, there has been a considerable amount of effort expended on the separation of CBI amino acids by capillary electrophoresis. Initial studies using CE employed LIF detection and were concerned with the amino acid analysis of a single cell (Kennedy *et al.*, 1989). Using a run buffer of pH 9.5 borate (0.01 M) containing 0.04 M KCl, it was possible to separate several of the neurochemically important amino acids, in particular, aspartate and glutamate, from other amines present in a neuron of *Helix aspera*. Similar separation conditions were employed for the CE-EC analysis of brain microdialysis samples derivatized with NDA/CN (O'Shea *et al.*, 1992).

An improved separation of 14 amino acids was accomplished by micellar electrokinetic chromatography (MECC) with laser-induced fluorescence detection (Ueda, *et al.* 1991; 1992). A helium-cadmium laser was employed as the excitation source. Limits of detection were approximately one amol on-column for all the amino acids. Addition of β-cyclodextrin and SDS to the run buffer enhanced the selectivity of the system for the CBI-amino acids. The separation of several amino acid enantiomers has also been accomplished through the addition of γ-cyclodextrin. Figure 30 shows the separation of several CBI-amino acid enantiomers by cyclodextrin-modified MECC.

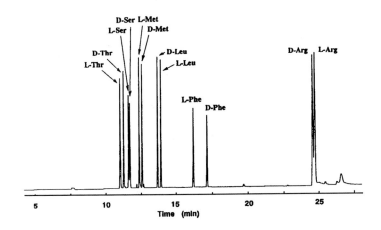

Fig. 30. Electropherogram of a mixture of six CBI-DL-amino acids, all 200 nM. Electrolyte composition: 10 mM γ–CD, 50 mM SDS, 100 mM borate buffer, pH 9.0. Applied voltage, 15 kV, current 35 μA. (Reproduced with permission from Ueda *et al.*, 1992.)

5. Applications

5.1. Amino Acids

The first studies using NDA/CN involved the separation of amino acids, which was reported in a early paper describing the reagent by de Montigny and coworkers (1987). Since that time, numerous other separations have been developed, including those by capillary electrophoresis (see Section 4.2). In general, a C-18 column is employed, with either UV, electrochemical, or fluorescence detection. Laser-induced fluorescence provides the lowest limits of detection (Roach and Harmony, 1987; Soper et al., 1989). However, UV or electrochemical detection permit the determination of multiderivatized amino acids such as lysine and desmosine (Lunte et al., 1987).

Since the optimum pH for the derivatization of amines with NDA/CN is normally the pKa of the amine to be derivatized, most amino acid analyses are performed in 10 mM sodium borate, approximately pH 9.5. If trace analysis is to be done, the buffer should not be adjusted to pH 9.5 with sodium hydroxide, since that has been shown to add impurities which show up as spurious peaks. It is best to conduct the derivatization in 10 mM borate at a pH of approximately 9.3. As with all trace methods, ultra-clean glassware and analytical grade or better reagents should be employed. The final concentration of the NDA and cyanide in the reaction mixture must be in excess of the total amines; however, the appearance of fluorescent side products becomes a problem if a large excess of NDA is employed. Typical concentrations for amino acid analysis are 1 mM NDA and 1 mM cyanide. If cyanide concentrations are too high, the rate of the reaction is decreased. The reaction typically takes between 20 and 30 minutes to complete. It is important to add the NDA last to minimize the formation of side products. CBI-derivatives of most amino acids have been shown to be stable at 4°C in the dark; the exception is methionine, which slowly oxidizes to the sulfoxide (Karn et al., 1992).

The resulting derivatives are very hydrophobic and are separated most effectively by reversed-phase HPLC. Some work has been done on the separation optimization, including the effect of methanol, pH and ionic strength on the separation of the amino acids commonly occurring in protein hydrolysates (de Montigny, 1986; Karn et al., 1992).

The CBI moiety is included into cyclodextrins and chiral separation of the CBI derivatives of numerous amino acids have been reported. This has been accomplished by HPLC using a cyclodextrin bonded phase (Duchateau, 1992) and by MECC with cyclodextrin additives (Ueda et al., 1991; 1992).

5.2. Other Compounds

Most of the applications presented in this chapter center on the determination of peptides by derivatization with NDA/CN. However, this reagent has been used successfully for several other types of primary amines. Some representative examples are discussed below.

Several important pharmaceuticals possess a primary amine functionality and are therefore amenable to detection after derivatization with NDA/CN. Phenylpropylamine (Hayakawa et al., 1989) 15-deoxyspergualin (Sprancmanis, 1990) and amphetamine (Konig et al., 1990) have all been successfully derivatized with NDA/CN. Phenylpropanolamine, benzylamine and phenylethylamine were determined at the sub-fmol level using chemiluminescence detection.

The analysis of the anticancer agent 15-deoxyspergualin is particularly challenging because it does not have any unique chromophore that will allow detection at biologically relevant levels. Current methods of analysis employ GC-MS or postcolumn derivatization. Using NDA/CN, it was possible to develop a method for the determination of this compound at a level of 5 ng ml^{-1}, which is a substantial improvement over currently available methods (see Fig. 22).

Catecholamines have been detected using precolumn derivatization with NDA/CN followed by fluorescence or chemiluminescence (CL) detection (Kawasaki et al., 1989) with CL detection limits in the fmol range. Dopamine, 3,4-dihydroxybenzylamine and norepinephrine were determined in human urine using this technique.

A method for the detection of tyramine, a physiological marker for depression, was developed by Konig et al. (1990). Tyramine is present in cheddar cheese at a level of 180 µg g^{-1}. It was found that the tyramine level in the urine of a healthy human almost doubled after eating 100 g of cheddar cheese. Tyramine was determined in human urine and rat brain dialysates with a minimal number of samples and preparation steps.

The artificial sweetener aspartame and its degradation products have been monitored by reaction with NDA/CN and liquid chromatography with fluorescence detection (Hayakawa et al., 199). The CBI derivatives of L-α-aspartyl-L-phenylalanine, 2,4-disubstituted-diketopiperazine, aspartic acid and phenylalanine can all be determined by reversed-phase chromatography with fluorescence detection. The derivatization step is particularly important for the determination of aspartic acid and phenylalanine, which are difficult to determine with adequate sensitivity by UV detection.

6. Summary

It has been demonstrated that naphthalene-2,3-dicarboxdialdehyde reacts with primary amines in the presence of a nucleophile (thiol or cyanide) to produce highly fluorescent isoindole derivatives. One of the most significant advantages of NDA over *ortho*-phthalaldehyde is that derivatives of peptides maintain their fluorescent properties. The reaction with CN⁻ is fairly slow, but also produces the most stable derivatives. Therefore, cyanide is most applicable to precolumn derivatization. The CBI derivatives exhibit an excitation maximum which exactly matches the 442 nm line of the He-Cd laser system. Low fmol detection limits have been obtained for a variety of CBI derivatives of primary amine analytes using precolumn derivatization followed by laser-induced fluorescence or chemiluminescence detection.

Reaction with mercaptoethanol is faster, but yields products that are not very stable. Therefore, this method has been limited to post-column derivatization. This approach is not as sensitive as reaction with cyanide, but there are no problems with sideproducts, which are inherent in the CN-based derivatization method. Only fluorescence detection has been employed for the detection of thiol derivatives. The excitation maximum coincides with the 457.9 line of the argon ion laser, making it possible to detect low levels of primary amines using this technique.

In conclusion, NDA has many advantages for the analysis of primary amines and peptides, in particular. The choice of nucleophile used in the derivatization is dependent in part on the objective of the analytical method. However, in both cases, a highly sensitive and selective procedure can be developed based on this methodology.

7. References

Alvarez, F., N. Parekh, B. Matuszewski, R. Givens, T. Higuchi and R. Schowen (1986). *J. Amer. Chem. Soc., 108,* 6435–6437.

Alvarez-Coque, M. C. G., M. J. M. Hernandez, R. M. V. Camanas and C. M. Fernandez (1989). *Anal. Biochem., 178,* 1–7.

Beale, S., J. Savage, S. M. Weitstock and M. Novotny (1988). *Anal. Chem., 60,* 1765–1768.

Beale, S. C., Y.-Z. Hseih, J. C. Savage, D. Wiesler and M. Novotny (1989). *Talanta 36,* 321–325.

Beale, S. C., Y.-Z. Hseih, D. Wiesley and M. Novotny (1990). *J. Chromatogr. 499,* 579–587.

Bostick, J. M., J. W. Strojek, T. Metcalf and T. Kuwana (1992). *Appl. Spectrosc., 46,* 1532–1539.

Carlson, R., K. Srinivasachar, R. Givens and B. Matuszewski (1986). *J. Org. Chem.*, *51*, 3978–3983.

Dave, K. J. (1991). Liquid chromatographic methods for the analysis of opioid peptides. Ph.D. Thesis, University of Kansas.

Dave, K. J., J. F. Stobaugh and C. M. Riley (1992a). *J. Pharm. Biomed. Anal.*, *10*, 49-60.

Dave, K. J., J. F. Stobaugh, T. M. Rossi and C. M. Riley (1992b). *J. Pharm. Biomed. Anal.*, *10*, 965–977.

de Montigny, P. (1986). Naphthalene-2,3-dicarboxaldehyde (NDA) and cyanide ion: A new precolumn derivatization reagent system for the analysis of amino acids and peptides by high performance liquid chromatography. Ph.D. Thesis, University of Kansas.

de Montigny, P., C. M. Riley, L. A. Sternson and J. F. Stobaugh (1990). *J. Pharm. Biomed. Anal.*, *8*, 419–430.

de Montigny, P., J. Stobaugh, R. Givens, R. Carlson, K. Srinivasachar, L. Sternson and T. Higuchi (1987). *Anal. Chem.*, *59*, 1096–1101.

Duchateau, A. L. L., G. M. P. Heemels, L. W. Maesen and N. K. de Vries (1992) *J. Chromatogr.*, *603*, 151–156.

Givens, R. S., D. A. Jencen, C. M. Riley, J. F. Stobaugh, H. Chokshi and N. Hanaoka (1990). *J. Pharm. Biomed. Anal.*, *8*, 477–491.

Givens, R. S. and R. L. Schowen (1989). In: *Chemiluminescence and the Photochemical Reaction in Chromatography* (J. Birks, ed.), pp. 125–147, VCH, New York.

Givens, R. S., R. L. Schowen, J. Stobaugh, T. Kuwana, F. Alvarez, N. Parekh, B. Matuszewski, T. Kawasaki, O. Wong, M. Orlovi´c, H. Chokshi and K. Nakashima (1989). In: *Luminescence Applications in Biological, Chemical, Environmental, and Hydrological Sciences* (M. C. Goldberg, ed.), ACS Symposium Series 383, American Chemical Society, Washington, DC.

Hayakawa, K., K. Hasegawa, N. Imaizumi, O. S. Wong and M. Miyazaki (1989). *J. Chromatogr.*, *464*, 343–352.

Hseih, Y.-Z., S. C. Beale, D. Wiesler and M. Novotny (1989). *J. Microcolumn Sep.*, *1*, 96–100.

Karn, J. A., T. Mohabbat, R. D. Greenhagen, C. J. Decedue and S. M. Lunte (1992) *Current Separations*, *11*, 57–60.

Kawasaki, T, T. Higuchi, K. Imai and O. S. Wong (1989) *Anal. Biochem.*, *180*, 279–285.

Kennedy, R. T., M. Oates, B. R. Cooper, B. Nickerson and J. Jorgenson (1989). *Science*, *246*, 57–63.

Kobayashi, S. and K. Imai (1980). *Anal. Chem.*, *52*, 424–427.

Kongi, H., H. Wolf, K. Venema and J. Korf (1900) *J. Chromatogr.*, *533*, 171–178.

Kristjansson, F. (1987). Fluorogenic derivatization of lysine containing peptides with naphthalene-2,3-dicarboxaldehyde and cyanide ion for trace analysis by high performance liquid chromatography. Ph.D. Thesis, University of Kansas.

Liu, J., K. Cobb and M. Novotny (1990). *J. Chromatogr. 519*, 189–197.

Liu, J., Y.-Z. Hseih, D. Wiesler and M. Novotny (1991a). *Anal. Chem., 63*, 408–412.

Liu, J., O. Shirota and M. Novotny (1991b). *Anal. Chem., 63*, 413–417.

Lunte, S. M., T. Mohabbat, O. S. Wong and T. Kuwana (1989). *Anal. Biochem., 178*, 202–207.

Lunte, S. M. and O. S. Wong (1989). *LC/GC 7*, 11.

Lunte, S. M. and O. S. Wong (1990). *Current Separations 10*, 19–26.

Matuszewski, B., R. Givens, K. Srinivasachar, R. Carlson and T. Higuchi (1987). *Anal. Chem., 59*, 1102–1105.

McCapra, F., K. Perring, R. J. Hard and R. A. Hann (1981). *Tetrahedron Lett.*, 5087–5091.

Mifune, M., D. K. Krehbiel, J. F. Stobaugh and C. M. Riley (1989). *J. Chromatogr. 496*, 55–70.

Miundi, H. B., S.-J. Lee, L. Baltzer, A. Jakubowski, H. I. Scher, L. A. Sprancmanis, C. M. Riley, D. V. Velde and C. W. Young (1991). *Cancer Res. 51*, 3096–3101.

Nicholson, L. M., H. B. Patel, F. Kristjansson, S. C. C. Jr, K. Dave, J. F. Stobaugh and C. M. Riley (1990). *J. Pharm. Biomed. Anal. 8*, 805–816.

Novotny, M. (1988). *Anal. Chem. 60*, 500A–510A.

Nussbaum, M. A., J. E. Przedwiecki, D. U. Staerk, S. M. Lunte and C. M. Riley (1992). *Anal. Chem., 64*, 1259–1263.

Oates, M. D. and J. W. Jorgenson (1989). *Anal. Chem., 61*, 432–435.

Oates, M. and J. Jorgenson (1989). *Anal. Chem., 61*, 1977–1980.

Oates, M. D., B. R Cooper and J. W. Jorgenson (1990). *Anal. Chem., 62*, 1573–1577.

Oates, M. D. and J. W. Jorgenson (1990). *Anal. Chem., 62*, 1577–1580.

O'Shea, T. J., P. L. Weber, B. P. Bammel, M. T. Smyth, C. E. Lunte and S. M. Lunte (1992) *J. Chromatogr., 608*, 189–195.

Quarry, M. A., R. L. Grob and L. R. Snyder (1984). *J. Chromatogr. 285*, 1–18.

Quarry, M. A., R. L. Grob and L. R. Snyder (1986). *Anal. Chem., 58*, 907–917.

Rauhut, M. M., L. J. Bollyky, B. G. Roberts, M. Loy, R. H. Whitman, A. V. Iannotta, A. M. Sensel and R. A. Clarke (1967). *J. Am. Chem. Soc., 89*, 6515–6522.

Riley, C., J. Stobaugh, C. Kindberg, M. Slavik and T. Jefferies (1989). In: *Proceedings of the 16th International Symposium on Controlled Release* (R. Pearlman and J. A. Miller, eds.), pp. 26–27, Controlled Release Society, Chicago.

Roach, M. and M. D. Harmony (1987). *Anal. Chem., 59,* 411–415.

Roach, M. and M. D. Harmony (1988) *J. Chromatogr., 455,* 332–335.

Rossi, T. M. (1990). *J. Pharm. Biomed. Anal., 8,* 469–476.

Roth, M. (1971). *Anal. Chem., 43,* 880–882.

Schoenmakers, P. J. and M. Mulholland (1988). *Chromatographia 25,* 737–748.

Snyder, L. R., J. W. Dolan and J. R. Gant (1979). *J. Chromatogr. 165,* 3–30.

Snyder, L. R., M. A. Stadalius and M. A. Quarry (1983). *Anal. Chem. 55,* 1412A–1430A.

Soper, S., S. M. Lunte and T. Kuwana (1989). *Anal. Sci. 5,* 29–36.

Soper, S. A. and T. Kuwana (1989). *Appl. Spectrosc. 43,* 883–886.

Sprancmanis, L. A. (1989). Determination of (-)-15-deoxyspergualin in plasma ultrafiltrate by high performance liquid chromatography using precolumn derivatization with naphthalene-2,3-dicarbox–aldehyde/cyanide. Ph.D. Thesis, University of Kansas.

Sprancmanis, L. A., C. M. Riley and J. F. Stobaugh (1990). *J. Pharm. Biomed. Anal. 8,* 165–175.

Sternson, L. A., A. J. Repta and J. F. Stobaugh (1985). *Anal. Biochem. 144,* 233–245.

Sternson, L. A., J. F. Stobaugh, J. Reid and P. de Montigny (1988). *J. Pharm. Biomed. Anal. 6,* 657–668.

Stobaugh, J. F., A. J. Repta, L. A. Sternson and K. W. Garren (1983). *Anal. Biochem., 135,* 495-501.

Stobaugh, J. F., A. J. Repta and L. A. Sternson (1984). *J. Org. Chem., 49,* 4306-4309.

Ueda, T., F. Kitamura, R. Mitchell, T. Metcalf, T. Kuwana and A. Nakamoto (1991). *Anal. Chem., 63,* 2979-2981.

Ueda, T., R. Mitchell, F. Kitamura, T. Metcalf, T. Kuwana and A. Nakamoto (1992). *J. Chromatogr., 593,* 265-274.

Weber, P. L., T. J. O'Shea and S. M. Lunte *J. Pharm. Biomed. Anal.,* in press.

Yeung, E. S. (1986). In: *Analytical Applications of Lasers,* (E. H. Piepmeir, ed.), pp. 273–290, Wiley-Interscience, New York.

CHAPTER 3

Fast Liquid Chromatography

for the Analysis of Enantiomers

SCOTT R. PERRIN

Regis Technologies, Inc., 8210 Austin Avenue
Morton Grove, IL 60053, U.S.A.

1. Introduction

Over the past ten years there has been an ever-increasing demand by regulatory agencies throughout the world for the regulation of drug chirality (Caldwell, 1989; De Camp, 1989; Testa and Trager, 1990; Cayen, 1991; Hutt, 1991; Shindo and Caldwell, 1991). Consequently, there has been a corresponding increase in demand for the development of chiral selectors—chiral stationary phases (CSP) that can discriminate between enantiomers. These are being used in part to evaluate chiral drugs for their different pharmacological activities. Therefore, in the analytical laboratories of many industrial pharmaceutical companies, the need for rapid screening of a wide variety of potentially pharmacologically active chiral drugs becomes pertinent.

Biological systems within the human body govern themselves predominantly by stereospecific mechanisms, so it is not surprising that many drug mechanisms are also stereospecific. As a result, a racemic drug must be seen as a mixture of two different stereoselective chemical species, each with its own and often completely different pharmacological action.

Stereoselectivity in both the action of a given chiral drug and the conversion of the drug to its metabolite requires either knowing or discovering the dynamics and kinetics of action of each chiral drug. The study of these actions is intended to elucidate the pre-requisites to minimize adverse chiral drug interactions, potential teratogenic effects, toxic reactions, and side effects by using a pure enantiomer

91

rather than the racemic drug. For these purposes, stereospecific analytical methods are indispensable.

Many of today's drugs are chiral. In the past, partly because of economic factors such as the expense of stereoselective synthesis, many pharmacologically active chiral drugs were marketed as racemates. Also, many of these chiral drugs contain more than one stereogenic center, and thus the presence of more than two stereoisomers compounded the difficulty of the stereoselective synthesis.

In drug design, new emphasis has been placed on the stereoselective synthesis, production, and isolation of enantiomerically pure drugs (Crosby, 1991). Fortunately, the technology for resolving enantiomers has been improved considerably by the introduction of new chiral selectors and by a better understanding of chiral recognition mechanisms.

The direct resolution of enantiomers by a chromatographic process upon a CSP solves a variety of stereoselective issues in both analysis and synthesis. In analysis, an efficient CSP makes possible the rapid and accurate assessment of enantiomeric purity of a given chiral compound. In stereoselective synthesis, an appropriate CSP permits the monitoring of a synthetic reaction. In addition, accurate determination of enantiomeric excess, especially in trace analysis, generally requires baseline resolution of both enantiomers. Consequently, the CSP column should be highly efficient (such efficiency requires that the column be well-packed with small silica particles ($d_p < 5$ μm) and narrow size distributions) and general in its ability to resolve enantiomers, with or without prior achiral derivatization.

As yet, very few CSP have been developed that are both efficient and general in applicability. Most present CSP are not available on silica supports of high efficiency, that is, of small silica particle size. Further, although several present CSP can resolve a wide range of chiral compounds without the need for achiral derivatization, the chiral recognition mechanisms of these CSP still remain complex in nature and not well understood. Without such an understanding, these CSP cannot be brought to full use.

The large number of commercially available chiral stationary phases presents the analyst with a broad range of possibilities. Deciding which CSP is best for a given application is difficult. To assist, Wainer has grouped CSP according to chiral recognition mechanisms (Wainer, 1988; Wainer and Alembik, 1988). Each mechanism has interacting parts: how the solute-CSP complexes are formed and how the stereoselective differences are expressed during and after the formation of these complexes. With a basic understanding of these mechanisms, one can make a rational choice of a chiral selector for a given application.

From studies in nuclear magnetic resonance (NMR) at the University of Illinois during the mid-1970s, Pirkle and coworkers observed unequal NMR shifts of diastereomeric complexes formed between a racemic π-acid sulfoxide and a chiral solvating agent (Pirkle and Sikkenga, 1975). These observations led to the development of the first generation of Pirkle-Concept CSP. Pirkle-Concept CSP, derived from either natural or synthetic amino acids, were among the first commercially available.

2. History

In 1987, the study of the applications of fast liquid chromatography was introduced in the papers of Danielson and Kirkland on the separation of macromolecules and the practical uses of small particles (d_p < 5 μm) (Snyder and Kirkland, 1979; Danielson and Kirkland, 1987). Their studies indicated that when small particles are packed into columns of reduced lengths, improved mass transfer enables them to separate macromolecules rapidly (Danielson and Kirkland, 1987).

In 1986, Unger *et al.* developed non-porous 1 μm silica microspheres and stated that fast separation of proteins was possible. However, the low sample loading with non-porous bonded-phase silicas was an issue (Unger *et al.*, 1986). Again in 1986, Pearson had concluded that a column less than 1 cm long could provide adequate resolution for molecules than are structurally different from each other (Pearson, 1986). In 1988, Szczerba confirmed Pearson's results and reported on the use of 3 μm, 300 Å, bonded phase, spherical silicas packed into 1 cm columns (Szczerba *et al.*, 1988). With a fast gradient but only standard commercial instrumentation, a standard protein mixture was baseline-resolved in less than 90 s (Szczerba *et al.*, 1988). In 1991, Perrin reported that, when necessary, enantioselectivity and retention can be reduced by using higher concentrations of polar organic modifiers, elevated temperatures, or shorter columns (Perrin, 1991).

3. Fast Liquid Chromatography

In this day of high technology and rapid growth, time is of the essence. A fast answer can mean the difference between actually attaining a goal and merely coming close to it. In the analytical laboratory, the need to screen several CSP both effectively and rapidly becomes crucial.

The first and uppermost demand on fast liquid chromatography (FLC) is that it provide the desired separation. For instance, with simple separations resolution may be many times larger than required and can be traded for greater speed. Such an exchange is especially valuable for repetitive chiral separations.

In comparison to either stereoselective synthesis or resolution by direct recrystallization with a resolving agent, chiral chromatography may be regarded as a rapid separation technique. Nevertheless, only a small fraction of its potential speed is being used. Hence, there is a need for development of methods to reduce analysis times.

In FLC, one must be prepared to deal with a wide variety of special circumstances. Optimizing for speed (rather than resolution and enantioselectivity) may require using suboptimal conditions. A researcher may prefer to use a column designed for purposes other than speed, but may still wish to shorten analysis times as much as possible. In this particular case, the column length, the particle size, and the nature of the chiral selector are fixed, but temperature, solvent composition, and flow rates can still be optimized. In other situations, limits on temperature, excess back pressures from high solvent polarity and flow rate, column length, and mobile phase selection may restrict the free choice of the optimum conditions.

In the work described here, the emphasis is on high speed chiral separations. However, the research was not restricted to the use of columns of reduced length, particle size, chiral selector, temperature, solvent composition, and flow rate. It will be shown that considerable increases in separation speed can be gained at the expense of large reductions in resolution. An alternative for increasing separation speed is the use of increased column temperatures. An increase in column temperature will generally reduce resolution but increase separation speed. If resolution and enantioselectivity can often be sacrificed for speed, the compromise is often very profitable.

3.1. Advantages

The advantages of FLC become apparent when one considers the many benefits. A reduction in the amount of solvents used means a reduction in total solvent cost. The decrease in analysis times using these shorter columns means that more samples can be analyzed each day. The ability to screen a wide variety of CSP rapidly means a reduction in the cost of method development (Perrin, 1991).

With conventional 25 cm columns, much of the overall elution time of each enantiomer can be spent in waiting for the dead volume to elute. Clearly an advantage in terms of speed could be gained with short columns that also retain the high efficiency of the Pirkle-Concept columns (Perry *et al.*, 1983).

In this chapter, the rapid analysis of a variety of chiral compounds is reported and the feasibility of using FLC techniques with columns of reduced lengths (< 25 cm) that have been packed with a variety of CSP covalently bonded to 3 μm silica particles is discussed. These results are compared with those obtained with conventional 25 cm columns packed with the same CSP covalently bonded to 3 μm silica particles. Some techniques described herein might be intentionally used to gain speed at the expense of both resolution and enantio-selectivity, speed being of the greatest essence.

3.2. Practicality

The practicality of FLC was demonstrated by Doyle (1991) with the use of columns of reduced lengths. His studies indicate that the trend in recent years in analytical laboratories has been the use of columns of reduced lengths for the economic benefits of shorter analysis times and reduced mobile phase consumption.

Research in Doyle's laboratory has shown that 10 cm × 4.6 mm i.d. columns packed with 3 μm silica particles covalently bonded *in situ* with (S)-1-(1-naphthyl)ethylisocyanate to aminopropyl silica ((S)-NEU) can resolve a variety of chiral compounds in a short time compared to the conventional 5 μm, 25 cm columns. For four chiral compounds studied, analysis required only 8 min and resolution factors were sufficient for trace analysis (Table 1) (Doyle, 1991).

Table 1. Chiral HPLC on short (10 cm) columns

Solute	α	R_s	N[a]	Time[b]
Amphetamine, DNB	1.21	3.1	4500	8.0
Metamphetamine, DNPU	1.25	6.7	5280	3.9
Tryptophan, methyl ester	1.89	8.2	3850	6.9
Ibuprofen, DNAn	2.49	11.1	4650	3.3

[a]for 3 μm silica particle size support
[b]for elution of all components at 2 ml min⁻¹
(Reproduced with permission from Doyle, 1991, American Chemical Society)

It was found that with short columns packed with 3 μm silica particles, CSP produce chromatographic separations comparable to those obtainable with conventional 25 cm × 4.6 mm i.d. columns packed with 5 μm particles. However, it was found that for success in FLC, both the enantioselectivities and resolution factors obtained previously with the 25 cm column under optimized conditions had to be high (Perrin, 1991).

3.3. Chiral Selectors

Most of our experimentation has been with Pirkle-Concept or brush-type selectors. The synthesis of brush-type CSP has been fully investigated by Pirkle (Pirkle *et al.*, 1981; Pirkle and Finn, 1983; Pirkle *et al.*, 1984). Of these, a new series of both π-donor and π-acceptor CSP has received considerable attention (Pirkle and Pochapsky, 1986; Pirkle *et al.*, 1986, 1987, 1991; Pirkle and McCune, 1988a,b; Pirkle and Pirkle, 1991). These CSP show high enantio-selectivities for a variety of chiral compounds.

Here, three of these Pirkle-Concept CSP were used. One was the second generation phenylglycine (S,S) (β-GEM 1) derived from *N*-3,5-dinitrobenzoyl-3-amino-3-phenyl-2-(1,1,-dimethylethyl) propanoate co-valently bonded through an ester linkage to 3 μm 11-undecyl silica. The β-GEM is available in both the (R,R) and (S,S) configurations. A π-acceptor, the β-GEM 1 CSP, is named after Pirkle's graduate student John E. McCune (JEM), who first prepared it. Often superior to its analogous precursor phenylglycine phase, this CSP separates the enantiomers of the anilide derivatives of many chiral carboxylic acids, including nonsteroidal anti-inflammatory agents (Pirkle and McCune, 1988a,b; Perrin and Pirkle, 1991) (Fig. 1).

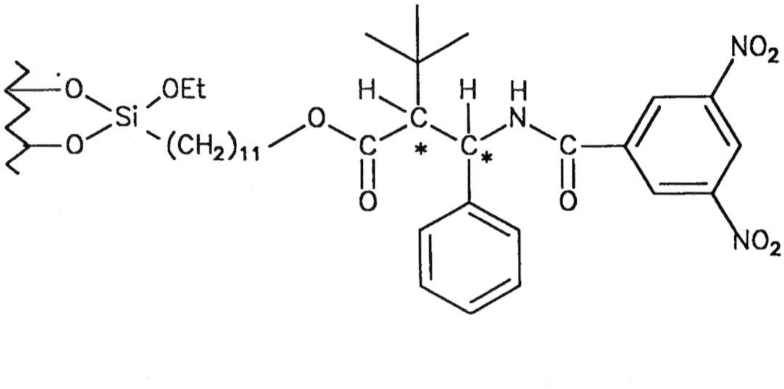

Fig. 1. The chemical structure of the β-GEM 1 CSP.

The second of the CSP used, the most recent π-acceptor (α-Burke 1), is derived from dimethyl N-3,5-dinitrobenzoyl-α-amino 2,2-dimethyl-4-pentylphosphonate covalently bonded to either 3 or 5 µm mercaptopropyl silica. This phase was also named after one of Pirkle's graduate students, J.A. Burke III, who first prepared it. It is available in only the (R) configuration. The α-Burke not only resolves the enantiomers of several underivatized β-blockers but also exhibits unique chromatographic behavior. Using an ethanol-dichloromethane-ammonium acetate mobile phase, reduction of the column temperature reduces the retention of the more retained enantiomer of β-blockers without appreciable band broadening (Perrin and Pirkle, 1991; Pirkle and Burke, 1991; Perrin and Bhat, unpublished results from this laboratory) (Fig. 2).

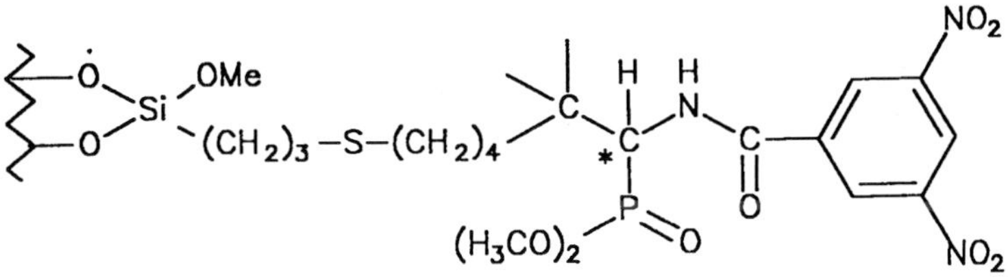

Fig. 2. The chemical structure of the α-Burke 1 CSP.

The third CSP, the π-donor naphthylleucine ((S)-N-1-N Leu), is based on the (S)-N-(1-naphthyl)-derivative of leucine, covalently bonded through an ester linkage to either 3– or 5–µm 11-undecyl silica. This π-donor CSP resolves 3,5-dinitrobenzoyl (DNB) derivatives of amino acids in a reversed-phase mode as the free acid. In the normal phase mode, this CSP resolves the amides and esters of DNB-amino acids with selectivity factors (α) that typically range between 10 and 40 (Perrin and Pirkle, 1991; Pirkle et al., 1991) (Fig. 3).

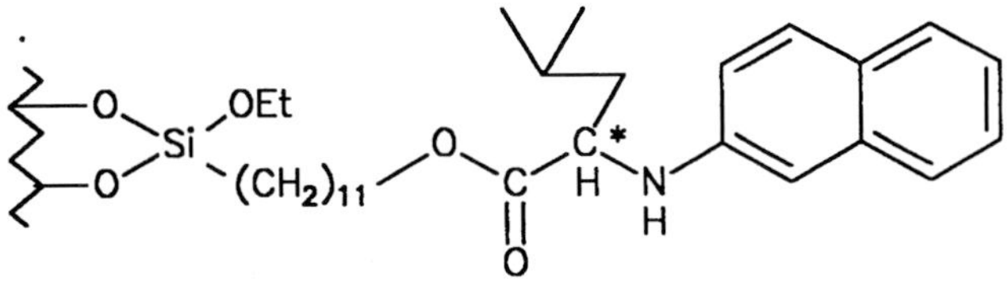

Fig. 3. The chemical structure of the N-1-N Leu CSP.

4. Chromatographic Parameters

Here, only those aspects of chromatographic theory that are pertinent to FLC are discussed, concentrating on surface density, reduced column length, and temperature.

First, the effect on enantioselectivity of varying surface densities was investigated. Second, the effect of varying column length on enantioselectivity, efficiency, retention, and resolution was investigated. Finally, the effect of varying column temperature on retention, enantioselectivity, and efficiency (N) was evaluated .

Various interactions and effects were studied, including the interaction of retention and analysis time; and the effects of relative retention on enantioselectivity for two solutes (α), and of the resolving power on resolution (R_s).

4.1. Surface Density

Enantioselectivity is a function of both the surface density and the enantiomeric purity of the CSP. An increase in surface density sometimes but not always improves enantioselectivity, and it appears that there is an optimum surface density for a given solute.

For a given CSP, Pirkle studied the influences of surface density and of the distance between adjacent strands on enantioselectivity (Pirkle and Hyun, 1985; Finn, 1988; Macaudiere, 1989). For high (0.4 mmol g[-1]), medium (0.2 mmol/g), and low (0.1 mmol/g) surface densities, the resolution of a series of homologous 1-aryl-1-aminoalkyl-3,5-DNB derivatives was evaluated. When alkyl-substituents are short, CSP surface density has little effect on enantioselectivities. However, as the length of the alkyl substituent increases, enantioselectivity decreases. The decrease is most profound with the CSP of highest surface density (Pirkle and Hyun, 1985). In practice, depending on the solute to be resolved and the chiral selector, enantioselectivity may vary as a function of CSP surface density.

The effect on enantioselectivity of varying surface density of a given CSP is shown in Tables 2-6. The CSP surface density was the sole parameter varied. In this study, no end-capping reagents were used. Held constant in the study were all chromatographic conditions such as flow rate, mobile phase composition, injection volume, sample concentration, and temperature. The solute was the 3,5-dinitroanilide (DNAn) derivative of ibuprofen.

The columns used were packed with 3 and 5 μm silica particles. All silica particles were covalently bonded with (S)-N-1-N Leu. Using a proprietary technique, the 3 μm material was slurry packed into the

Table 2. Resolution of the DNAn derivative of ibuprofen[a]

Column Size	Particle Diameter	k'	k'_2	α	R_s
1 × 3.0	3	3.73	5.61	1.50	0.91
3 × 4.6	3	3.04	4.59	1.50	2.20
5 × 4.6	3	4.00	6.02	1.50	4.87
10 × 4.6	3	3.85	5.45	1.41	4.56
10 × 4.6	3	5.41	9.12	1.68	6.95
5 × 4.6	5	4.24	7.34	1.73	4.52
25 × 4.6	5	4.91	8.35	1.70	9.14

[a]Column size: cm × mm i.d.; particle size: microns (μm); mobile phase: using 2-propanol–hexane (5:95, v/v); flow rate: 1.0 ml min^{-1}; 10 μl injection; UV: 254 nm; temperature: 24°C

Table 3. Experimental data of surface density for both the 5 and 3 μm (S)-N-1-N Leu CSP[a]

Column	Lot #	Particle Diameter	α	Carbon	Nitrogen
1 × 3.0	10289	3	1.50	0.190	0.170
3 × 4.6	10289	3	1.50	0.190	0.170
5 × 4.6	10289	3	1.51	0.190	0.170
10 × 4.6	92989	3	1.41	0.170	0.160
10 × 4.6	51491	3	1.65	0.329	0.321
5 × 4.6	06289	5	1.73	0.275	0.243
25 × 4.6	06289	5	1.70	0.275	0.243

[a]Column size: cm × mm i.d.; lot #: month:day:year; particle size: microns (μm); mobile phase: 2-propanol-hexane (5:95, v/v); flow rate: 1.0 ml min^{-1}; 10 μl injection; UV: 254 nm; temperature: 24°C; carbon analysis in mmol g^{-1}; nitrogen analysis in mmol g^{-1}.

SCOTT R. PERRIN

Table 4. Measurement of column efficiency and analysis times for the resolution of the DNAn derivative of ibuprofen[a]

Column Size	Particle Diameter	N	H	N/m	Analysis Time
1 × 3.0	3	170	0.590	16,900	0.67
3 × 4.6	3	1040	0.290	34,800	2.60
5 × 4.6	3	4520	0.100	90,400	4.61
10 × 4.6	3	5530	0.180	55,300	7.97
10 × 4.6	3	6250	0.160	62,500	11.87
5 × 4.6	5	2380	0.210	47,500	5.10
25 × 4.6	5	9900	0.250	39,600	25.54

[a]Column size: cm × mm i.d.; particle size: microns (μm); mobile phase: 2-propanol-hexane (5:95, v/v); flow rate: 1.0 ml min^{-1}; 10 μl injection; UV: 254 nm; temperature: 24°C; analysis time: min; H: (mm × 10).

Table 5. Measurement of column efficiency and analysis times for the resolution of the DNAn derivative of ibuprofen[a]

Column Size	Particle Diameter	N	H	N/m	Analysis Time
1 × 3.0	3	160	0.620	16,100	0.39
3 × 4.6	3	800	0.370	26,700	1.35
5 × 4.6	3	3470	0.140	69,500	2.41
10 × 4.6	3	6170	0.160	61,700	5.81
10 × 4.6	3	6060	0.160	60,600	6.37
5 × 4.6	5	194	0.260	38,800	2.97
25 × 4.6	5	9000	0.280	35,600	12.95

[a]Column size: cm × mm i.d.; particle size: microns (μm); mobile phase: 2-propanol-hexane (10:90, v/v); flow rate: 1.0 ml min^{-1}; 10 μl injection: 254 nm; temperature: 24°C; analysis time: min; H: (mm × 10)

Table 6.   Resolution of the DNAn derivative of ibuprofen[a]

Column Size	Particle Diameter	k'	k'$_2$	α	R$_S$
1 × 3.0	3	0.78	1.14	1.46	0.53
3 × 4.6	3	1.40	2.01	1.43	1.42
5 × 4.6	3	1.78	2.58	1.45	3.29
10 × 4.6	3	1.67	2.31	1.38	3.74
10 × 4.6	3	2.58	4.11	1.59	5.86
5 × 4.6	5	2.23	3.70	1.65	2.38
25 × 4.6	5	2.33	3.75	1.60	6.98

[a]Column size: cm × mm i.d.; particle size: microns (μm); mobile phase: 2-propanol-hexane (10:90, v/v); flow rate: 1.0 ml min^{-1}; 10 μl injection; UV: 254 nm; temperature: 24°C

following configurations: 1 cm × 3.0 mm i.d., 3 cm × 4.6 mm i.d., 5 cm × 4.6 mm i.d., 10 cm × 4.6 mm i.d.; the 5 μm material, 5 cm × 4.6 mm i.d., and 25 cm × 4.6 mm i.d.

Enantioselectivity is not a function of column length, given constant CSP surface density (Fig. 4). However, as shown in Fig. 5, and in agreement with previous observations (Pirkle and Hyun, 1985) and Table 3, enantioselectivity is not only a function of CSP surface density but also goes through a maximum. An increase in carbon coverage does not always improve enantioselectivity. The use of end-capping reagents should equalize enantioselectivity. No experimental data to date have substantiated that theory, although we have observed improved enantioselectivity with the use of end-capping reagents (S.R. Perrin, J.D. Rateike and G. Bhat, unpublished data).

4.2. Column Length

As previously mentioned, columns of reduced length can provide the required performance. Here we will evaluate the effect of varying column length on enantioselectivity, efficiency, capacity factor and resolution.

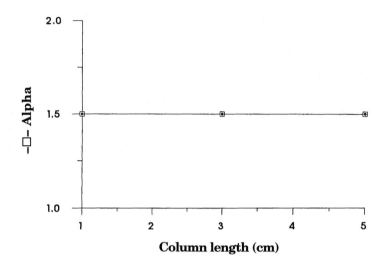

Fig. 4. Enantioselectivity is independent of column length, at constant CSP surface density.

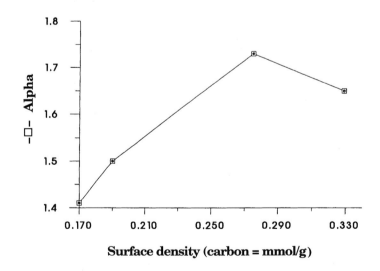

Fig. 5. Enantioselectivity as a function of CSP surface density.

4.2.1. Enantioselectivity

It has been demonstrated that enantioselectivity is independent of column length. With Pirkle-Concept CSP used under normal phase conditions (2-propanol-hexane), changing only the solvent polarity (the concentration of 2-propanol in the mobile phase) has minimal effect upon the degree of enantioselectivity (Pirkle *et al.*, 1985) (Fig. 6). To illustrate this effect, in Fig. 7 we plotted α versus surface density for two different mobile phase compositions (Tables 2 and 6). Decreasing the concentration of 2-propanol in hexane in the mobile phase from 10% to 5% slightly improves the degree of enantioselectivity for the DNAn derivative of ibuprofen.

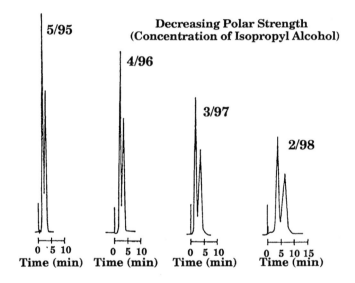

Fig. 6. Decreasing the mobile phase concentration of 2-propanol increased retention and improved resolution of DNB-phenethylamine separated on the 1 cm, 5 μm (S)-N-1-N Leu column. However, minimal change in enantioselectivity (α) was observed.

4.2.2. Efficiency

The plate number, N, describes the number of theoretical plates that are being brought to bear on the separation. The following equation is used to express N:

$$N = 5.54 \ (t_R/w_{1/2})^2 \qquad\qquad (1)$$

Here $w_{1/2}$ is the width of the chromatographic band at half-height and t_R is the retention time for the first eluting enantiomer.

It is important to note that N increases with increasing column length. In order to compare the efficiencies of columns of different lengths, the height of a theoretical plate (H) is used instead of N.

$$H \ (mm) \ = \ L \ (mm)/N \qquad\qquad (2)$$

In theory, the plate height (N/L) is independent of column length. In practice, column length has a clear effect on efficiency, which is particularly pronounced as the column length is reduced below 3 cm (Tables 4 and 5). The effect principally reflects imperfect extracolumn instrumentation, for instance, large detector flow cells (flow cell volumes and pathlengths), dead volumes in the tubing, and injection valves. Other factors which contribute to extra-column effects include large sample concentrations, and injection volumes.

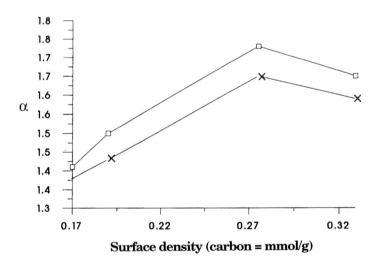

Fig. 7. Decreasing the concentration of 2-propanol in hexane in the mobile phase from 10% (×) to 5% (□) produces only a slight improvement in the enantioselectivity (α) of the DNAn derivative of ibuprofen. Mobile phases: 2-propanol-hexane (5:95, v/v); 2-propanol-hexane (5:95, v/v).

Guiochon (1980) characterized the requirements necessary for the achievement of separations of at least 5,000 plates in less than 5 min. He felt that three criteria must be met. First, the column must be well-packed because the minimum pressure drop is much lower for an efficient column. Second, the column must contain high-quality silica particles with narrow particle size distributions that will achieve the desired pressure drop. Third, the column length must correspond to the minimum pressure drop. Guiochon also noted (1980): "It is remarkable that all the analytical results discussed here can be obtained with approximately 10-cm-long columns packed with particles of about 5 μm in diameter. This means, among other things, that the results will be largely influenced by the exact column length and particle diameter."

Commercial chiral columns are often considered to have low chromatographic efficiency, but this is not true for Pirkle-Concept chiral columns. Depending on mobile phase composition and solute, most of these have plate counts of at least 30,000 plates m^{-1} (Doyle, 1991). For instance, in a separation of the 3,5-dinitroanilide derivative of ibuprofen with a 5 cm column packed with 3 μm silica particles bonded with (S)-N-1-N Leu CSP, a plate count of 90,400 plates m^{-1} was obtained (Fig. 8, Table 4). (We find that we can pack a 5 cm column more efficiently than a 10 cm column with 3 μm silica particles.)

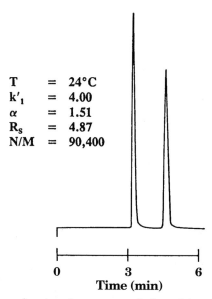

Fig. 8. The short column used for this chromatogram showed acceptable efficiency: 90,400 plates m^{-1}.

4.2.3. Retention

Retention can be manipulated for a given compound by changing the mobile phase solvent strength or the temperature. Capacity factors that decrease over time can result from changes in the volume ratios between the mobile and the stationary phase.

One of the most dramatic effects of reduced column length is reduction in retention, which can be most beneficial in reducing the time per analysis (Tables 4 and 5). In Fig. 9, as the length of the column is decreased, analysis times decreased from 2.4 min to under 24 s.

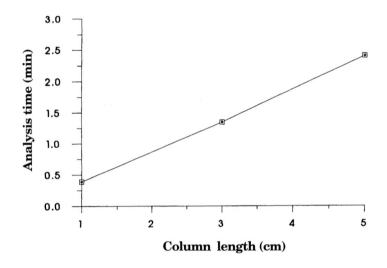

Fig. 9. Decreasing the length of the column from 5 to 1 cm decreases analysis times from 2.4 min to under 0.4 min.

4.2.4. Resolution

The three fundamental parameters that allow one to control resolution are α, N, and k'. To control resolution in FLC, these three parameters must be utilized and it must be determined how they vary under different chromatographic conditions. The three terms of the equation act independently, and may be optimized separately. Although the usual goal in FLC is the sufficient resolution of a given chiral compound, a resolution that is more than adequate is simply time-consuming. With columns of reduced lengths, a major reduction in resolution is generally observed (Tables 2 and 6). However, this need not be critical if resolution factors were previously high on the original 25 cm column (Fig. 10).

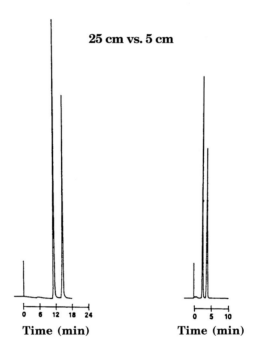

25 cm vs. 5 cm

Time (min)   Time (min)

Fig. 10. More than complete baseline resolution of DNB-phenethylamine is obtained on both the 25 cm and the 5 cm 5 μm (*S*)-*N*-1-*N* Leu columns. Note that with the short column, the separation has been completed in 4 min, well before the first enantiomer has been eluted in 11 min from the longer column.

4.3. *Temperature*

The importance of temperature in FLC separations should not be underestimated. In addition to reducing retention, increased column temperatures usually increase the overall efficiency of any chromatographic system. Although relatively few researchers have reported using temperature to optimize chiral separations (Pirkle and Schreiner, 1981; Pirkle and Tsipouras, 1984; Papadopoulou-

Mourkidou, 1989; Perrin, 1991; Pirkle, in press), several have reported on its effect on achiral separations under both normal and reversed-phase conditions (Diasio and Wilburn, 1979; Gilpin and Sisco, 1980; Henderson and O'Connor, 1984; Hirukawa et al., 1987; Jinno et al., 1988; Tchapla et al., 1988). In this section the effects of both elevated and sub-ambient column temperatures on the chromatographic parameters will be reviewed.

4.3.1. Elevated column temperatures

From the standpoint of the chiral chromatographer utilizing FLC techniques, one of the most important experimental parameters is temperature. The use of elevated column temperature, in general, is not a well-recognized technique in chiral chromatography, usually because most chiral compounds are resolved upon CSP with low enantioselectivities and resolving powers. However, given relatively high enantioselectivities and resolution values, either elevated temperatures or temperature gradients can improve peak shape (Perrin, 1991).

4.3.2. Capacity factor

The dependence of the capacity factor k' on temperature is described by the following relationship:

$$\ln k' = \frac{-\Delta H^{\circ}}{RT} + \frac{\Delta S^{\circ}}{R} + \ln \phi \qquad (3)$$

where ΔH° is the molar enthalpy change for the transfer of the solute from the mobile phase to the stationary phase, ΔS° is the accompanying change in entropy where the concentration of the solute in each phase is in molar units, and ϕ is the phase ratio (Henderson and O'Connor, 1984). Generally, a 10°C increase in temperature decreases retention by a factor 2 or 3 (Schmitt et al., 1971; Majors, 1975). Here in Fig. 11, the analysis times for DNB-aspartic acid-dimethyl ester are shown plotted against temperature (Table 7) (Perrin, 1991). Increasing column temperatures shortens analysis times.

If it is assumed that ΔH°, ΔS°, and ϕ are independent of temperature, the relationship takes the form of the well-known van't Hoff plot (ln k' versus 1/RT), commonly used to estimate the enthalpy and entropy of transfer. If the retention mechanism and the enthalpy change are constant over the range of temperatures examined, then under normal phase conditions the relationship of ln k' vs. 1/T should

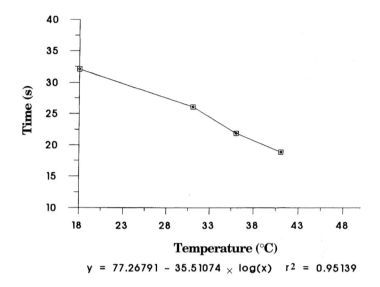

$$y = 77.26791 - 35.51074 \times \log(x) \quad r^2 = 0.95139$$

Fig. 11. With DNB-aspartic acid-dimethyl ester, increasing column temperature reduces analysis time.

Table 7. Separation of DNB amino acid esters at 41°C[a]

Analyte	41°C		36°C		31°C		18°C	
	Rt_2	S.F.	Rt_2	S.F.	Rt_2	S.F.	Rt_2	S.F.
DNB-aspartic acid-dimethyl ester	18.9	1.3	21.9	1.4	26.1	1.5	32.1	1.8
DNB-leucine-octyl ester	21.8	2.1	25.3	2.4	30.7	2.9	33.9	4.1

[a]Column: 1 cm × 3.0 mm i.d.; mobile phase: 2-propanol-hexane (60:40, v/v); flow rate: 1.5 ml min[-1]; UV: 254 nm; Rt_2: retention time for the second eluting enantiomer in s; S.F.: enantioselectivity factor, S.F. = (Rt_1/Rt_2)

be linear. Finding unexpected non-linearity (Equation 3), with use of the (R)-N-(3,5-dinitrobenzoyl)phenylglycine CSP and a conformationally rigid spirolactam, Pirkle attributed it to a temperature-dependent interaction of 2-propanol with either the chiral selector or the solute

(Pirkle, in press). With low concentrations of 2-propanol in the mobile phase, retention decreases, then increases, then decreases again as the column temperature is increased (Pirkle, in press).

In summary, the effect of temperature on the capacity factor is complex in normal phase chiral chromatography. Capacity factors are generally dependent on the relative polarity of both solute and chiral selector and on the degree of adsorption-desorption of both the organic modifier and the solute.

4.3.3. Enantioselectivity

When transient diastereomeric solute-CSP complexes differ in stability, stereoselectivity results. The more strongly retained enantiomer of greatest stability is usually the one that interacts either with more adsorption sites or fewer steric repulsion sites. Both enthalpic and entropic contributions to their binding energies must be considered. In general, the enantiomer adsorbed with the greatest exothermicity also loses the greatest entropy, a situation that reduces enantioselectivity (Perrin and Pirkle, 1991).

Adsorbed complexes may also differ in adsorption isotherms. If each enantiomer is retained by a different adsorption mechanism, a change in temperature may cause a corresponding change in enantioselectivity (Pirkle, 1991; Pirkle and Burke, 1991).

In separating the diastereomeric compound fenvalerate, Papadopoulou-Mourkidou (1989) found that increasing column temperature from 2°C to 35°C caused rapid decreases in retention times and enantioselectivity, but did not affect diastereomeric selectivity.

A similar effect of increased column temperatures on both enantioselectivities and retention times was observed in our use of 1 cm columns. Retention times and enantioselectivities were reduced as column temperature was increased from 18°C to 41°C (Perrin, 1991). Resolution and enantioselectivity can be sacrificed to reduce analysis times as low as 21.8 s for DNB-leucine-octyl ester (Fig. 12, Table 7).

4.3.4. Efficiency

In FLC, increasing column temperature can lead to improvements in peak shapes (Perrin, 1991). As the column temperature increases, several major factors change in a manner that is advantageous for column performance and reduced analysis times. Especially when mobile phase concentrations of 2-propanol are high, an increase in

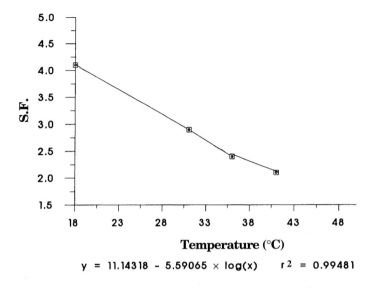

$$y = 11.14318 - 5.59065 \times \log(x) \qquad r^2 = 0.99481$$

Fig. 12. With DNB-leucine-octyl ester, increasing column
temperature decreased separation factors.

column temperature simultaneously reduces the inlet pressure
required to achieve the desired flow rate and increases the solute
diffusion rate, thus improving mass transfer between the mobile and
stationary phases.

4.3.5. Sub-ambient column temperatures

If retention is not already excessive, decreasing column temperature usually increases enantioselectivity and capacity factors. A
change in resolution is often associated with differences in enantioselectivities (Papadopoulou-Mourkidou, 1989). The extent of change
in the capacity factor with change in temperature varies with compound and chiral selector. With β-blocking agents, however, using the
α-Burke 1 CSP and a mobile phase of methylene chloride-ethanol
(19:1, v/v) containing $CH_3CO_2NH_4$ (0.5 g l[-1]), decreasing column temperature also decreased retention. This unusual effect occurred
without any measurable deterioration in peak shape (Table 8) (Pirkle
and Burke, in press).

Table 8. The effect of temperature upon retention and enantioselectivity for some β-blockers and analogs using α-Burke 1

Analyte	21°C α^a	21°C $k'_1{}^b$	0°C α^a	0°C $k'_1{}^b$	-24°C α^a	-24°C $k'_1{}^b$
Metoprolol	1.16	2.57	1.21	1.05	1.48	0.64
Oxprenolol	1.00	2.28	1.00	0.75	1.03	0.50
Pronethalol	1.13	5.14	1.21	2.21	1.31	1.50
Propranolol	1.39	4.36	1.63	1.86	2.11	1.28
Pindolol	1.30	15.00	1.43	7.29	1.72	6.71
Bufuralol	1.93	2.79	2.50	1.43	4.08	0.73

[a]Selectivity
[b]Capacity factor for the first eluted enantiomer using methylene chloride-ethanol (19:1, v/v) containing $CH_3CO_2NH_4$ (0.5 g l^{-1}) as the mobile phase; flow rate, 2 ml min^{-1}. The detector was operating at 254 nm; column: 10 cm × 4.6 mm i.d.; particle size: 5 μm. (Reproduced with permission from Pirkle and Burke, in press, copyright 1991, Elsevier Science).

With a 10 cm, 3 μm α-Burke 1 CSP column and a mobile phase composition of $CH_3CN:CH_3CH_2OH:CH_3CO_2NH_4$ (38:06:02 v/v/w) reducing column temperature from 21 to 0C decreased collmn efficiency slightly. However, enantioselectivity was improved and retention was decreased (Fig. 13) (Perrin and Bhat, unpublished results). By changing the modifier from ammonium acetate to ammonium formate, enantioselectivity and resolution improved as column temperature decreased (Tables 9-10).

These studies to date have demonstrated that column temperature should be considered as a chromatographic parameter that may be varied to improve both resolution and enantioselectivities in FLC when conventional methods are unsuccessful. Furthermore, they suggest that sub-ambient column temperatures may be useful for selected resolution problems if retention is not already excessive at ambient temperature.

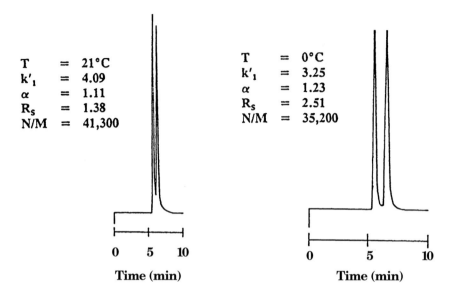

T   = 21°C
k'₁ = 4.09
α   = 1.11
Rₛ   = 1.38
N/M = 41,300

T   = 0°C
k'₁ = 3.25
α   = 1.23
Rₛ   = 2.51
N/M = 35,200

Time (min)   Time (min)

Fig. 13. Reducing the column temperature from 21 to 0°C decreased column efficiency slightly, but this is more than offset by the increase in enantioselectivity, from 1.11 to 1.23.

Table 9. The effect of temperature (21°C) upon chromatographic conditions for some β-blocking agents[a]

Analyte	k'_1	k'_2	α	R_S	N
Propranolol	3.64	4.48	1.23	2.30	3645
Metoprolol	2.22	2.39	1.07	0.70	3720
Pindolol	3.92	4.75	1.21	2.20	3760
Atenolol	3.81	4.07	1.06	0.64	3170

[a]Column size: 10 cm × 4.6 mm i.d.; particle size: 5 μm; mobile phase: methylene chrloride-ethanol (4:1, v/v) containing $CH_3CO_2NH_4$ (0.65 g l⁻¹); flow rate 1.0 ml min⁻¹; 5 μl injection; uV: 254 nm.

Table 10. The effect of subambient column temperature (0°C) upon the chromatographic conditions for some β-blocking agents[a]

Analyte	k'_1	k'_2	α	R_s	N
Propranolol	3.97	5.55	1.40	2.87	2260
Metoprolol	2.17	2.46	1.13	0.80	1520
Pindolol	4.24	5.74	1.35	2.61	2250
Atenolol	4.04	4.43	1.09	0.79	2200

[a]The same conditions as in Table 9.

5. Method Development

In attempting to optimize retention, analysis time, and resolution simultaneously, certain compromises such high efficiency are necessary. Sometimes optimization of one variables is made at the expense of one or more of the others. The effects of column temperature on the resolution of enantiomers are complex and often unpredictable. Therefore, the simplest, initial approach to method development is to keep the temperature constant and vary the flow rate and mobile phase composition. If this simple approach is unsuccessful, then other approaches should be considered such as changing the column length, varying the temperature, or changing the chiral selector.

5.1. Optimization

In discussing the influence of the mobile phase on chiral separations, Zief considered the following variables for optimization: solvent selectivity, choice of solvent strengths, and the adsorption-desorption of solvent from binding sites within the CSP (Zief, 1988). These same guidelines can be effective in the optimization of chromatographic conditions for columns of reduced length.

Several problems can be encountered in developing an FLC analysis. In one case, retention may be in the ideal range (k' = 2 - 10) but resolution inadequate; in another, retention may be outside this range.

Depending on the types of chiral compounds involved, adequate resolution can be achieved by changing either the temperature (subambient or elevated) or the amount or type of organic modifier. If changes in either column temperature or mobile phase composition do not provide adequate resolution of the peaks, column length may be increased to greater than 3 cm. In some cases, the use of conventional 25-cm columns is required to achieve the desired resolution.

Changing the chiral selector may either modify or optimize enantioselectivity, resolution or efficiency (Table 11). Changing the derivative may improve matters, as is illustrated in Table 11. Ibuprofen can be derivatized with the achiral reagent 3,5-dimethylaniline to form the 3,5-dimethylanilide (DMA) derivative, which gave an acceptable separation on the (S,S) β-GEM 1 CSP. However, the DNAn derivative of ibuprofen was resolved faster on both the (S)-N-1-N Leu and (S)-NEU CSP.

Table 11. The effect on resolution and enantioselectivity of varying the chiral selector and the derivative using 3 μm silica particles with 10 cm, 5 cm, 3 cm, and 1 cm columns

Chiral Selector	α	R_s	N	N/m	Analysis time (min)
(S)-N-1-N Leu	1.59	5.8	6060	60,600	6.3[a]
(S)-N-1-N Leu	1.45	3.2	3470	69,500	2.4[b]
(S)-N-1-N Leu	1.43	1.4	800	26,700	1.3[c]
(S)-N-1-N Leu	1.51	4.8	4520	90,400	4.6[d]
(S)-N-1-N Leu	1.82	1.3	140	14,200	2.5[e]
(S)-NEU	2.49	11.1	4650	46,500	3.3[f]
(S,S) β-GEM 1	1.75	8.8	5670	56,700	8.7[g]

[a]Ibuprofen-DNAn 2-propanol-hexane (10:90, v/v) ; 1.0 ml min⁻¹; 10 cm
[b]Ibuprofen-DNAn 2-propanol-hexane (10:90, v/v); 1.0 ml min⁻¹; 5 cm
[c]Ibuprofen-DNAn 2-propanol-hexane (10:90, v/v); 1.0 ml min⁻¹; 3 cm
[d]Ibuprofen-DNAn 2-propanol-hexane (5:95, v/v); 1.0 ml min⁻¹; 5 cm
[e]Ibuprofen-DNAn2-propanol-hexane (2/98, v/v); 1.0 ml min⁻¹; 1 cm
[f]Ibuprofen-DNAn 2-propanol-hexane-acetonitrile (10/85/5, v/v/v); 2.0 ml min⁻¹; 10 cm
[g]Ibuprofen-DNAn 2-propanol-hexane (10/90, v/v); 1.0 ml min⁻¹; 10 cm

To optimize retention, many operating parameters should be considered: composition of the eluent, elution mode, and flow rate. In the optimization process for very short columns, retention may be found inadequate. To increase retention with columns of reduced lengths, the polarity of the eluent should be decreased. For example, Fig. 14 shows that decreasing the concentration of the organic modifier, 2-propanol in the mobile phase, increased the retention of DNB-leucine butyl ester on a 1 cm, 5 μm (S)-N-1-N Leu column (Fig. 14), thereby improving resolution.

Separation of DNB-leucine butyl ester

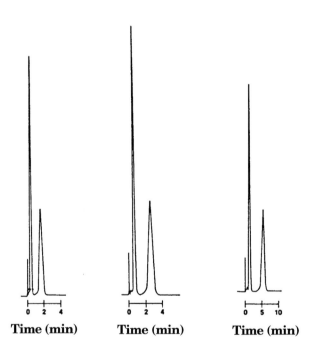

Time (min)   Time (min)   Time (min)

Fig. 14. Decreasing the mobile phase concentration of 2-propanol increases the retention of DNB-leucine butyl ester separated on the 1 cm, 5 μm (S)-N-1-N Leu column from 2.4 min at 2-propanol-hexane (30:70, v/v) to 6.8 min at 2-propanol-hexane (10:90, v/v).

5.2. Screening Process

5.2.1. Normal phase

To evaluate a chiral compound of an unknown enantioselectivity on a column of reduced length, method development should begin with a mobile phase composed of 2-propanol-hexane (20:80, v/v) at a flow rate of 2.0 ml min^{-1} at ambient temperature. If little or no enantioselectivity is observed, the flow rate should be maintained at 2.0 ml min^{-1} and the concentration of 2-propanol in the mobile phase should be decreased systematically until peak splitting is observed (Perrin, 1991). If this approach produces only partial resolution within the optimal k' range, then further improvements in resolution may be achieved by decreasing the flow rate.

For example, Fig. 15 shows the improvement in the resolution of ibuprofen derivatives that was achieved by the systematic reduction

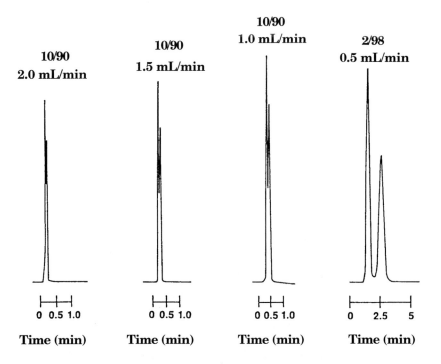

Fig. 15. At a flow rate of 2 ml min^{-1} and a mobile phase composition of 10% 2-propanol, only partial resolution of ibuprofen enantiomers was observed. However, after optimizing both flow rate and mobile phase composition, almost baseline resolution of ibuprofen was obtained.

of the 2-propanol concentration and the flow rate. Baseline resolution was attained by using a mobile phase of 2% 2-propanol at a flow rate of 0.5 ml min^{-1} (Fig. 15) (Perrin, 1991).

5.2.2. Reversed phase

In using Pirkle-Concept CSP for compounds of low polarity or non-ionic compounds, high enantioselectivities are generally obtained under normal phase conditions. Recently, however, Pirkle reported using reversed-phase conditions with achiral ion-pairing reagents in the mobile phase for the separation of N-protected amino acids (Pirkle et al., 1989). Although reversed-phase conditions can be used, they almost always afford less enantioselectivity than normal phase conditions. In this case, reversed-phase rather than normal-phase conditions improved resolution and shortened analysis times.

Pirkle studied the influence of the structure and concentration of achiral ion-pairing additives upon the retention and enantioselectivity of the N-3,5-dinitrobenzoyl derivatives of α-amino acids and 2-amino-phosphonic acids on a (R)-N-(2-naphthyl)alanine CSP. A mobile phase of methanol-phosphate buffer was used and the effect of the concentration of methanol was investigated (Pirkle et al., 1989). Increasing the concentration of the alkyltrimethylammonium ion-pairing reagent enhances retention of these derivatives. An increase in the chain length of the alkyl portion of the ion-pairing reagent gave increased retention but no significant change in enantioselectivity. The concentration of the organic modifier in the mobile phase dramatically affects the retention. Perry also studied the influence of ion-pairing reagents in a reversed-phase mode. However, his study used conventional C-18 stationary phases (Perry and Chang, 1990).

If isocratic elution with mixed solvents does not provide the desired resolution, the most obvious way to improve resolution and shorten analysis times is by gradient elution, in which the properties of the organic modifier, the amount added, and the type of modifier used in the eluent will affect both resolution and analysis times. If the capacity factors of the chiral compounds to be separated are equally affected by the modifier, then gradient elution may solve the problem. With gradient elution, both the slope and gradient time will affect the retention behavior. The result is to provide faster relative elution of the most retained enantiomer in the chromatogram compared with the case of simple isocratic elution.

Using achiral ion-pairing reagents in the mobile phase with columns of reduced lengths, Perrin developed a new gradient elution reversed-phase technique for the separation of the DNB-amino acids

(Perrin, unpublished results from this laboratory) (Table 12). Several amino acids were separated by first derivatizing with an achiral derivatization reagent (3,5-dinitrobenzoyl chloride (DNBC)). The N-DNB amino acid derivative still contains a free carboxylic acid functionality (Fig. 16). The resulting N-protected amino acid as the free carboxylic acid ion-pairs with an alkyltrimethylammonium achiral ion-pairing reagent (Fig. 17).

Table 12. Isocratic and gradient elution of some DNB-amino acids[a]

Analyte	k'_2	α	Gradient #	Column Length
DNB-leucine	19.0	3.8	no	3
DNB-leucine	12.0	2.6	1	3
DNB-leucine	9.0	3.0	2	5
DNB-phenylalanine	26.0	2.1	no	3
DNB-phenylalanine	11.0	1.8	2	3
DNB-norleucine	16.0	4.0	no	3
DNB-norleucine	12.0	2.4	1	3
DNB-norleucine	9.6	2.6	1	5
DNB-norvaline	7.5	2.5	no	3

[a]The chromatographic conditions for the isocratic elution were as follows: Columns: 3 μm (S)-N-1-N Leu; 3 cm × 4.6 i.d.; 5 cm × 4.6 mm i.d. Mobile phase: 4 mM Q_6 ion-pairing reagent in 0.1 M phosphate buffer (pH 6.9)-methanol (30/70, v/v).

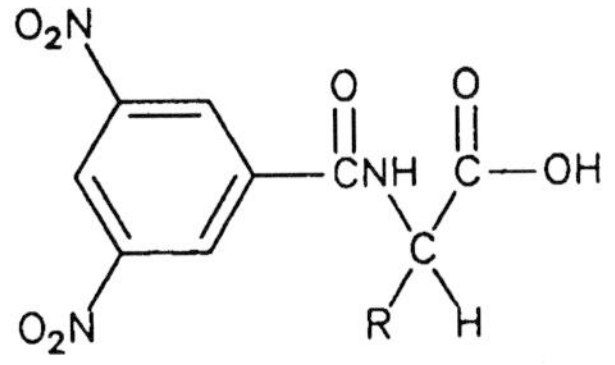

Fig. 16. A typical DNB-amino acid derivative formed from the achiral reaction with 3,5-dinitrobenzoyl chloride. Note that after this achiral reaction, the acidic moiety is still available for ion-pairing.

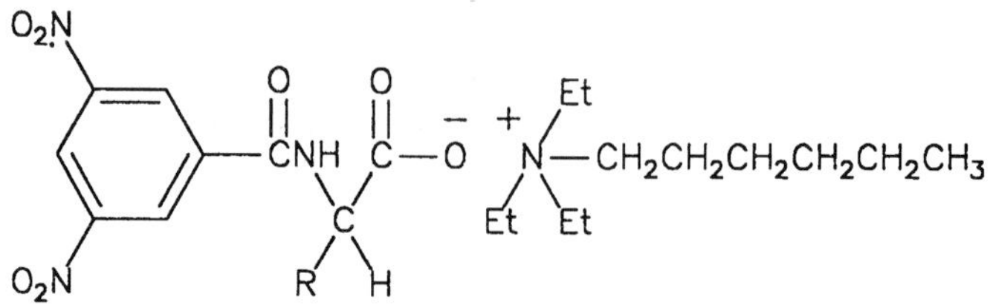

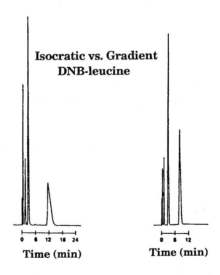

Fig. 17. Chemical structure of the *N*-DNB-amino acyl
quaternary ammonium ion pair that is formed by
the interaction with a quaternary ammonium ion-
pairing reagent present in the mobile phase.

With this approach, most of the natural amino acid *N*-DNB
derivatives were resolved on a short 3 μm (*S*)-*N*-1-*N* Leu column,
using either isocratic or gradient elution (Fig. 18). For some of the
DNB-amino acids, isocratic elution provided the needed resolution
(Fig. 19, Table 12).

Isocratic vs. Gradient
DNB-leucine

Time (min)   Time (min)

Fig. 18.   Imposing gradient elution improves peak shape
of both eluting enantiomers of DNB-leucine
separated on a short 3 cm *N*-1-*N* Leu column.

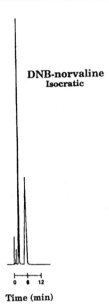

Fig. 19. Gradient elution is not always
required for the separation of some
of the DNB-amino acids.   DNB-
norvaline can be separated in less
than 6 min.

To improve resolution for the second eluting enantiomer (k'_2), a gradient system was employed. The chromatographic conditions were as follows:

Columns: 3 μm (S)-N-1-N Leu; 3 cm × 4.6 i.d.; 5 cm × 4.6 mm i.d.

Gradient #1:   **Pump A** 5 mM Q_6 ion-pairing reagent in methanol-0.1 M phosphate buffer (pH 6.9) (30/70 v/v)
                Pump B methanol;
                Gradient hold for 1.0 min at 99-1 (A:B); linear gradient over 5 min 99-1 to 80-20 (A:B)

Gradient #2:   **Pump A** 5 mM Q_6 ion-pairing reagent in methanol-0.1 M phosphate buffer (pH 6.9) (30/70 v/v)
                Pump B methanol;
                Gradient hold for 2.0 min at 99-1 (A:B); linear gradient over 5 min 99-1 to 75-25 (A:B)

6. Conclusions

With an appropriate injection valve (Fig. 20) coupled to five different chiral selectors a wide variety of potential pharmaceutical candidates can be screened in a rapid fashion. With an autosampler, the pharmaceutical candidate can be injected repeatedly and the chiral selector that shows signs of enantioselectivity identified. The separation may be re-validated with columns of lengths greater than 3 cm if necessary.

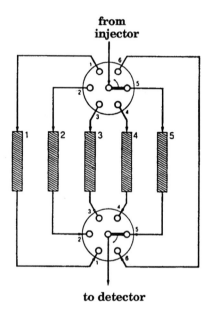

from injector

to detector

Fig. 20. With this equipment, many potential pharmaceutical candidates could be screened rapidly on
several CSP. The CSP showing enantioselectivity could be quickly identified. A model 7066
Rheodyne injector is used with five different
chiral columns.

Acknowledgment

The author gratefully acknowledges Dr. John Perry of Regis Technologies, Inc. for his helpful discussion, editing, and critical comments on the manuscript.

7. References

Caldwell, J. (1989). *Chirality*, *1*, 249–250.

Cayen, M. N. (1991). *Chirality*, *3*, 94–98.

Crosby, J. (1991). *Tetrahedron*, *47*, 4789–4846.

Danielson, N. D. and J. J. Kirkland (1987). *Anal. Chem.*, *59*, 2501–2506.

De Camp, W. H. (1989). *Chirality*, *1*, 2–6.

Diasio, R. B. and M. E. Wilburn (1979). *J. Chromatogr. Sci.*, *17*, 565–567.

Doyle, T. D. (1991). In: *Chiral Separations by LC* (S. Ahuja, Ed.), ACS Symposium Series *471*, Chap. 2. American Chemical Society, Washington DC.

Finn, J. M. (1988). In: *Chromatographic Chiral Separations*; (M. Zief and L. J. Crane, Eds.), Vol. 40, Chap. 3. Marcel Dekker, New York.

Gilpin, R. K. and W. R. Sisco (1980). *J. Chromatogr.*, *194*, 285–295.

Guiochon, G. (1980). In: *High-Performance Liquid Chromatography* (C. Horvath, Ed.), Vol. 1, Chap. 1. Academic Press, New York.

Henderson, D. E. and D. J. O'Connor (1984). In: *Advances in Chromatography* (J. C. Giddings, E. Grushka, J. Cazes and P. R. Brown, Eds.), Vol. 23, Chap. 2. Marcel Dekker, New York.

Hirukawa, M., Y. Arai and T. Hanai (1987). *J. Chromatogr.*, *395*, 481–487.

Hutt, A. J. (1991). *Chirality*, *3*, 161–164.

Jinno, K., T. Nagoshi, N. Tanaka, M. Okamoto, J. C. Fetzer and W. R. Biggs (1988). *J. Chromatogr.*, *436*, 1–10.

Macaudiere, P., M. Lienne, A. Tambute and M. Caude (1989). In: *Chiral Separations by HPLC, Applications to Pharmaceutical Compounds* (A.M. Krstulovic, Ed.), Chap. 14. Wiley, New York.

Majors, R. E. (1975). *Analysis*, *10*, 549.

Mazzo, D. J., C. J. Lindemann and G. S. Brenner (1986). *Anal. Chem.*, *58*, 636–638.

Papadopoulou-Mourkidou, E. (1989). *Anal. Chem.*, *61*, 1149–1151.

Pearson, J. D. (1986). *Anal. Biochem.*, *152*, 189.

Perrin, S. R. (1991). *Chirality*, *3*, 188–195.

Perrin, S. R. and W. H. Pirkle (1991). In: *Chiral Separations by LC* (S. Ahuja, Ed.), ACS Symposium Series *471*, Chap. 3. American Chemical Society, Washington DC.

Perry, J. A., L. G. Glunz and T. J. Szczerba (1983). *LC-GC*, *1*, 40–41.

Perry, J. A. and M. A. Chang (1990). *J. Pharm. Sci.*, *79*, 437–439.

Pirkle, W. H. *J. Chromatogr.*, *558*, 1–6 (1991).

Pirkle, W. H. and D. L. Sikkenga (1975). *J. Org. Chem.*, *40*, 3430.

Pirkle, W. H. and L. Schreiner (1981). *J. Org. Chem.*, *46*, 4988–4991.

Pirkle, W. H. and A. Tsipouras (1984). *J. Chromatogr.*, *291*, 291–298.

Pirkle, W. H. and J. M. Finn (1983). In: *Asymmetric Synthesis* (J.D. Mosher, Ed.), Vol. 1, Chap. 6. Academic Press, New York.

Pirkle, W. H. and M. H. Hyun (1985). *J. Chromatogr., 328*, 1–9.

Pirkle, W. H. and T. C. Pochapsky (1986). *J. Am. Chem. Soc., 108*, 352–354.

Pirkle, W. H. and J. E. McCune (1988a). *J. Liq. Chromatogr., 11*, 183–187.

Pirkle, W. H. and J. E. McCune (1988b). *J. Chromatogr., 441*, 311–322.

Pirkle, W. H. and J. A. Burke. *J. Chromatogr., 557*, 173–185 (1991).

Pirkle, W. H., J. M. Finn, Schreiner, J. L. and B. C. Hamper (1981). *J. Am. Chem. Soc., 103*, 3946–3966.

Pirkle, W. H., M. H. Hyun and B. Bank (1984). *J. Chromatogr., 316*, 585–604.

Pirkle, W. H., Pochapsky, T. C., G. S. Mahler and R. E. Field (1985). *J. Chromatogr., 348*, 89–96.

Pirkle, W. H., T. C. Pochapsky, G. S. Mahler, D. G. Corey, D. S. Reno and D. Alessi (1986). *J. Org. Chem., 51*, 4991–5000.

Pirkle, W. H., G. S. Mahler, T. C. Pochapsky and M. H. Hyun (1987). *J. Chromatogr., 388*, 307–314.

Pirkle, W. H., J. P. Chang and J. A. Burke (1989). *J. Chromatogr., 479*, 377–386.

Pirkle, W. H., K. C. Deming and J. A. Burke (1991). *Chirality, 3*, 183–187.

Schmitt, J. A., R. A. Henry, R. C. Williams and J. F. Dickman (1971). *J. Chromatogr. Sci., 9*, 645.

Shindo, H. and J. Caldwell (1991). *Chirality, 3*, 91–93.

Snyder, L. R. and J. J. Kirkland (1979). In: *Introduction to Modern Liquid Chromatography*; Chaps. 2 and 5. Wiley, New York.

Szczerba, T. J., D. N. Baehr, L. G. Glunz, J.A. Perry and M. J. Holdoway (1988). *J. Chromatogr., 458*, 281–286.

Tchapla, A., S. Heron, H. Colin and G. Guiochon (1988). *Anal. Chem., 60*, 1443–1448.

Testa, B. and W. F. Trager (1990). *Chirality, 2*, 129–133.

Unger, K. K., G. Jilge, J. N. Kinkel and M. T. W. Hearn (1986). *J. Chromatogr., 359*, 61.

Wainer, I. W. (1988). In: *Drug Stereochemistry, Analytical Methods and Pharmacology* (I. W. Wainer and D.E. Drayer, Eds.), Vol. 11, Chap. 6. Marcel Dekker, New York.

Wainer, I. W. and M. C. Alembik (1988). In: *Chromatographic Chiral Separations*; (M. Zief and L.J. Crane, Eds.), Vol. 40, Chap. 14. Marcel Dekker, New York.

Zief, M. (1988). In: *Chromatographic Chiral Separations* (M. Zief and L.J. Crane, Eds.), Vol. 40, Chap. 12. Marcel Dekker, New York.

Part Two: **Recent Developments in the Isolation of Compounds from Biological Matrices**

CHAPTER 4

Solid Phase Extraction for Sample Preparation

MORRIS ZIEF AND SUNIL V. KAKODKAR

Research Laboratory, J.T. Baker, Inc.
222 Red School Lane, Phillipsburg, NJ 08865 USA

1. Introduction

Sample dissolution followed by liquid-liquid extraction was a popular sample preparation procedure for many years (Zweig and Sherma, 1972). Traditional liquid-liquid extractions, performed in separatory funnels, are tedious, time-consuming and costly. These methods not only require several sample handling steps but may also present the following problems to the analyst: phase emulsions, large solvent volumes, and impure and wet extractions. Other classical sample preparation techniques include centrifugation, filtration, distillation, precipitation, open column chromatography and lyophilization.

In about 1978 a simpler alternative approach, solid phase extraction, was introduced (Poole and Schuette, 1984). This concept, similar to low pressure liquid chromatography, is the basis for the design of a practical sample preparation technique that uses small, disposable extraction columns filled with a variety of sorbents.

2. Sorbents

Table 1 lists most of the sorbents commercially available for solid phase extraction. In the early development of liquid chromatography, only polar sorbents such as silica, kieselguhr and alumina were routinely available. These adsorbents were used to isolate polar organics from relatively non-polar solutions possessing eluotropic values ($\varepsilon°$) less than 0.38 eluting solvent strength on silica (Snyder and Kirkland, 1979). Polar compounds (alcohols, esters, amines) dissolved in hexane:ethyl ether (3:1, v/v) are representative examples. These analytes can be eluted from a silica column with solvents

Table 1. Sorbents for Solid Phase Extraction

Sorbent	Structure	Particle Diameter, Shape	Pore Size (Å)
Adsorption			
Kieselguhr (Diatomaceous earth)	$-SiOH$	40 µm, irregular	--
Silica gel (precipitated)	$-SiOH$	40 µm, irregular	60
Florisil®	$-Mg_2SiO_3$	73–140 µm, irregular	--
Alumina (neutral)	Al_2O_3	50–200 µm, irregular	--
Reversed Phase			
Octadecyl (C_{18})	$-(CH_2)_{17}CH_3$	40 µm, irregular	60
Octyl (C_8)	$-(CH_2)_7CH_3$	40 µm, irregular	60
Ethyl (C_2)	$-CH_2CH_3$	40 µm, irregular	60
Cyclohexyl	$-CH_2CH_2-$⬡	40 µm, irregular	60
Phenyl	$-CH_2CH_2CH_2-$⬡	40 µm, irregular	60
Normal Phase			
Cyano (CN)	$-(CH_2)_3CN$	40 µm irregular	60
Amino (NH_2)	$-(CH_2)_3NH_2$	40 µm, irregular	60
Diol (COHCOH)	$-(CH_2)_3OCH_2CH-CH_2$ $\quad\quad\quad\quad\;\;$ HO OH	40 µm, irregular	60
Ion-Exchangers			
Amino (NH_2)	$-(CH_2)_3NH_2$	40 µm, irregular	60
1°,2°-Amino (NH)(NH_2)	$(CH_2)_3NHCH_2CH_2NH_2$	40 µm, irregular	60
Quaternary amine (N^+)	$-(CH_2)_3N^+CH_3)$	40 µm, irregular	60
Carboxylic acid (COOH)	$-(CH_2)_2COOH$	40 µm, irregular	60
Propyl sulfonic acid (SO_2OH)	$-(CH_2)_3SO_2OH$	40 µm, irregular	60
Aromatic sulfonic acid ($ArSO_2OH$)	$-(CH_2)_2-$⬡$-SO_2OH$	40 µm, irregular	60
Size Exclusion			
Sephadex® G-25	Dextran	50–150 µm, irregular	--
Wide-Pore Reversed Phase			
Butyl (C4)	$-(CH_2)_3CH_3$	40 µm, irregular	275

having an $\varepsilon°$ value greater than 0.6. When methanol is the eluting solvent, strong hydrogen bonds are formed with the silanol groups of silica, thus displacing the adsorbed polar organic compound. The classical use of polar adsorbents with non-polar solvents was termed normal phase extraction.

As early as 1950 the treatment of Kieselguhr with dimethyl-dichlorosilane afforded the first non-polar sorbent (Howard and Martin, 1950). Gradually, a variety of bonded silicas appeared in the literature. It was not until 1974, when octadecyl silica columns became commercially available from several manufacturers, that nonpolar bonded siloxanes began to dominate sample preparation (Melander and Horvath, 1980). The availability of octadecyl silica then allowed efficient removal of non-polar compounds from aqueous solutions. This new bonded phase offered attractive possibilities for analysis in biological, environmental, and industrial matrices. Over time a very large number of papers on clinical applications to drugs, pharmaceuticals, and metabolites in biological fluids such as plasma and urine have appeared in the literature.

The new system in which the sorbent was less polar than the mobile phase or sample solution was termed reversed-phase chroma-tography. Here non-polar or slightly polar analytes could be extracted from solvents with $\varepsilon°$ greater than 0.6. The analyte was then eluted from the sorbent with a solvent having a low $\varepsilon°$ value.

The octadecyl bonded phase has the highest retention for non-polar compounds. In some instances, the interaction between analyte and sorbent is too great and is difficult to disrupt with a non-polar elution solvent. When this is a problem, retention can be reduced by using a bonded phase with shorter alkyl chains such as octyl, butyl or methyl.

Although the original purpose for the preparation of siloxanes was the conversion of unbonded silica to a bonded non-polar phase, both polar and non-polar bonded silicas are now available. Bonded phases which can be used in normal phase fashion are formed when the R group of the silyl derivative is a cyano, amino, or diol group (see Table 1). These bonded siloxanes are less retentive than silica gel toward very polar analytes (*e.g.*, carbohydrates) and therefore permit extractions impossible to achieve with unbonded silicas.

The most polar bonded silicas in Table 1 contain SO_3^- and $N^+(CH_3)_3$ ionic functional groups. These phases are used for the extraction of acids and bases from aqueous solutions according to the classic theories of ion exchange. To achieve optimum extraction conditions, the ion-exchange sorbent and the analyte should be oppositely charged and the counter ion concentration of the sample solution low. For example, epinephrine, a catecholamine, is adsorbed on an aromatic sulfonic acid bonded phase at a pH of 7.0 to 7.5. Elution is achieved with a combination of high ionic strength and low pH (2 M ammonium sulfate at pH 1.5). The low pH neutralizes the sorbent while the ammonium ions compete with catecholamine for negatively charged ion-exchange sites.

The retention of ionic compounds is achieved by promoting ionization. Thus, the pH is lowered for basic analytes and increased for acidic analytes in aqueous solution. The optimum pH for 100% ionization of an ionic analyte depends on its pKa, defined as the pH at which 50% of the ionizable groups are charged and 50% are neutral. Adjustment of the pH of an acid analyte solution to 2 pH units higher than the pKa of the analyte results in approximately 99% ionization (each pH unit changes the percentage of charged or uncharged molecules by a factor of 10). The acid would then be in the proper form for retention on an anion exchange column. For elution of an acid from a strong anion exchange (SAX) column, the pH should be adjusted to two pH units below the pKa, converting the acid to the non-ionized form. In contrast, a base is most effectively extracted onto a cation exchange column from a solution two pH units below the pKa value. At this pH, the base is 99% protonated. The basic analyte is eluted from the column with the elution solvent two pH units above the pKa of the base.

Sephadex G-25 is a 1,6-glucose polymer cross-linked with epichlorohydrin to create a matrix with a controlled pore size. It is used for desalting or buffer exchange of protein solutions. Smaller molecules (<10,000 molecular weight) enter the pores of the hydrated carbohydrate polymer and are significantly retarded as the sample solution percolates through the gel.

Wide-pore bonded silicas (40 micron, 275 Å) find utility in the isolation of high molecular weight compounds (*e.g.*, proteins, peptides and nucleic acids). Compounds with molecular weights above approximately 2,000 encounter restricted mass transfer into the pores of the standard pore (60 Å) sorbents.

3. Solid Phase Extraction Columns

Configurations of disposable columns, prepacked with one of the adsorbents or bonded phases listed in Table 1, have already been described in the literature (Zief, 1985). Usually polypropylene columns are packed with 100, 200, 500 or 1000 mg of sorbent sandwiched between two 20 μm polyethylene frits. Sample volume capacities of the columns are 1, 3 or 6 ml. Large volumes of samples can be added to the columns via detachable reservoirs. Solutions are usually aspirated through the columns by vacuum. To facilitate the handling of multiple samples, manifold column processors which can be connected to a 10– to 20–inch Hg vacuum source have become standard equipment in many laboratories.

4. Column Capacity

Analyte capacity for absorbents and bonded silica has been estimated to be 10 to 20 mg of analyte g^{-1} of packing (Majors, 1986). The empirical capacities of 200 mg columns for several analytes of practical interest have been found to be ~5.0 to 10 mg. These levels are far above the analyte concentrations in practical pharmaceutical analysis. Although drugs and drug metabolites in urine and serum are usually at the ng ml^{-1} to μg ml^{-1} level, these fluids contain many potential interferences which may co-extract. Columns containing 100 mg of sorbent have been found adequate for extraction of nanogram levels of analyte from urine and serum volumes of less than 1 ml. The typical loading of bonded ion-exchangers is in the 0.4–1.4 meq g^{-1} range.

5. Chemistry and Characterization

Extraction columns from a number of suppliers are commercially available (including Analytichem International, J.T. Baker, Inc., Burdick & Jackson, Fisher Scientific, Supelco, and Waters). It is important to realize that differences in bonding chemistry and manufacturing methods yield products that are not necessarily interchangeable. The surface properties of the bonded sorbents depend primarily on the type of silica selected for the bonding reaction. The degree of loading and endcapping is also very important in determining the surface characteristics of a bonded phase.

6. Synthesis of Bonded Sorbents

The preparation of bonded phases involves the reaction of free silanol groups of silica with mono-, di- or tri-halo or alkoxy silanes. Polymeric silanes have been used by Alpert and Regnier (1979) for adsorption on silica as well as by Ramsden (1985) for covalent binding with silanols (Fig. 1).

In all the above cases one can always find free silanol groups. The presence of unbonded silanols causes the bonded phase to exhibit heterogeneous surface characteristics; those due to attached –R groups and those due to unreacted silanols. These silanols are deactivated by end-capping with trimethylchlorosilane as follows (Fig. 2).

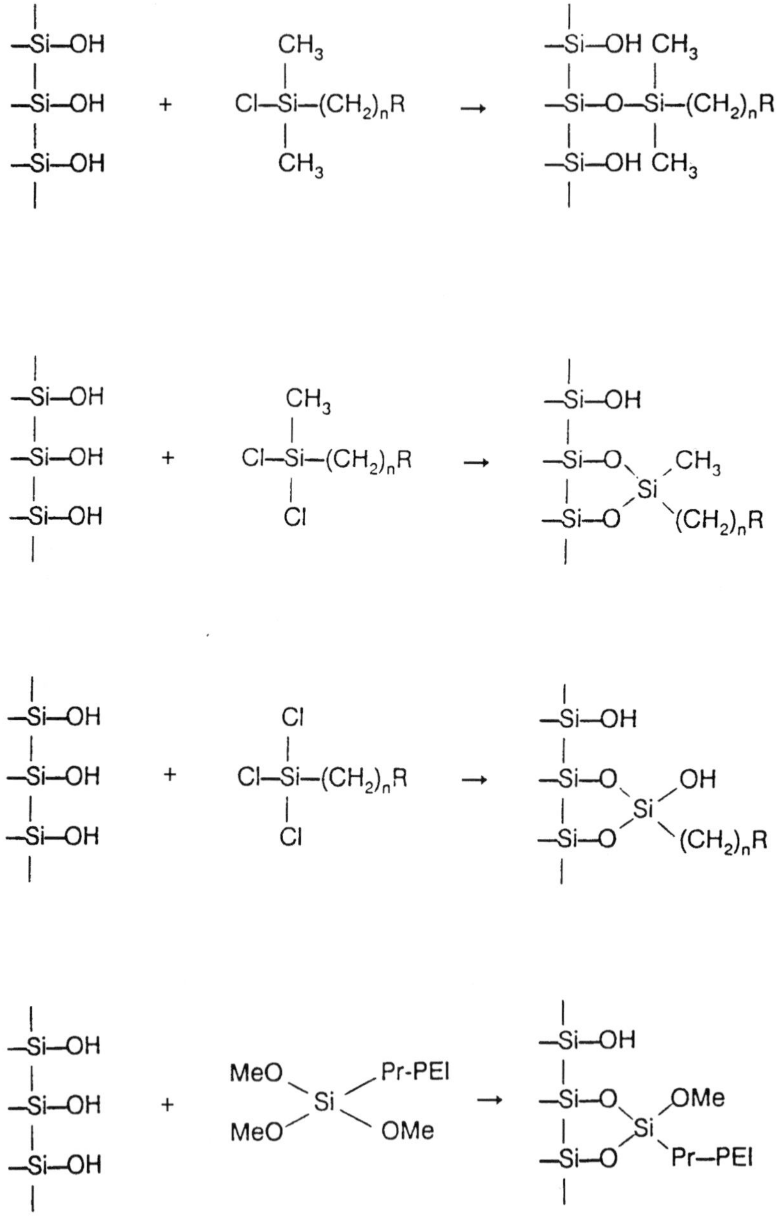

Pr = *n*-propyl group

Figure 1

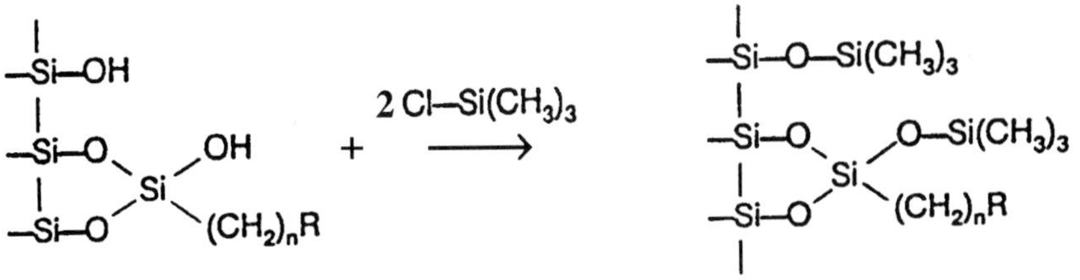

Figure 2

The potential for competitive adsorption on the hydroxy sites of an otherwise non-polar surface is thus eliminated or minimized. The bonded phases which are used in solid phase extraction mainly consist of polar, non-polar and charged bonded phases.

Polar-bonded phases are usually prepared by reaction of cyano-propylsilane or aminopropylsilane with silica. In the case of the diol bonded phase, glycidoxypropylsilane or any other silane-containing terminal epoxy group is reacted with silica and subsequently hydrolyzed to diol with dilute acids. The non-polar bonded phases are prepared by reacting the octadecyl, octyl, butyl, cyclohexyl or phenyl silanes with silica. The strong cation exchange bonded phases are usually prepared by reacting the appropriate silane, followed by some chemical reactions, e.g., 3-mercaptopropyltrimethoxysilane is reacted with silica and the resulting bonded phase is oxidized with 50% nitric acid; the aromatic sulfonic acid bonded phase is prepared by reacting 2-(4-chlorosulfonylphenyl)ethyltrimethoxysilane with silica followed by hydrolysis with aqueous THF. The carboxylic acid weak cation exchanger is prepared by reaction of 2-(carbomethoxy)ethyltrichlorosilane followed by hydrolysis of the ester to acid using dilute acids. The strong anion exchanger is prepared by reacting silica with N-trimethoxysilylpropyl-N,N,N,-trimethylammonium chloride.

7. Characterization

Initially, identification of bonded phases was carried out by C, H and N analysis. This approach failed to yield any information about functional groups on the bonded phases or differentiate the bonded phases according to their mode of preparation. There was not a clear method available which gave an exact and detailed picture of chemical species bonded to silica. With the availability of instruments, diffuse reflectance IR spectroscopy and ^{13}C cross polarization, magic angle spinning (CP-MAS) along with ^{29}Si NMR measurements were increasingly used in exploring the structure of bonded phases, including the nature of the alkyl chain, the presence of endcapping agents and the functionality of the silane used to modify the silica surface.

Diffuse reflectance FTIR spectroscopy was used to study the hydrolytic stability of cyano-bonded silica samples (Murthy and Berry, 1990). Murthy and co-workers have recently evaluated the performance of cyano bonded phases from different vendors using diffuse reflectance FTIR spectroscopy and ^{13}C CP/MAS and ^{29}Si NMR spectra (Murthy et al., 1991). The characteristic IR bonds for the cyano, amide and carboxyl groups were employed to determine the functional purity of cyano bonded phases. The authors compared the ratio of the cyano group stretch band at 2252 cm^{-1} and the silica substrate Si–O–Si combination band at 1871 cm^{-1}. This ratio afforded the relative concentration of cyano groups. Since trihalosilanes were used in the manufacture of cyano bonded phases, the by-product hydrochloric acid caused hydrolysis of cyano groups to amide, which was evidenced by bands at 1666 cm^{-1} and 1624 cm^{-1} (amide I and amide II bands, respectively). In some cases the amide group was further hydrolyzed to acid (carboxyl band at 1720 cm^{-1}), thus conferring ion exchange character on the resultant phase. The NMR spectra confirmed the above data. The bonded phases which showed the presence of amide and carboxyl groups in IR showed corresponding peaks in ^{13}C NMR. In addition, they were able to show that at least two vendors used mono- or di-functional silanes while the majority used tri-functional silanes. The above results could explain the difficulties encountered when cyano cartridges from several vendors are intermixed.

Very little data have been reported for ^{13}C NMR (Davison et al., 1981; Leydon et al., 1981; Chiang et al., 1982) and ^{29}Si NMR (Maciel and Sindorf, 1980; Engelhardt et al., 1981) on silica gels modified with organosilanes. Detailed experimental data on chemically modified silica gels were published by Bayer et al. in 1983. The authors studied eight different bonded phases prepared mainly by reaction of silica gel with trialkoxyalkylsilanes. The authors also compared the ^{13}C NMR chemical shifts of organofunctional silanes which were used in the surface modification. Jinno (1989) studied the ^{13}C NMR CP/MAS spectra of octadecylsilica from different manufacturers. He provided data on surface structure and silane loading that agreed with the chromatographic retention behavior method using polycyclic aromatic hydrocarbons.

An excellent review of octadecyldimethylsilyl derivatized silica using high power ^{1}H decoupling and ^{13}C NMR was published by Maciel et al. (1985). Caravajal and coworkers (Caravajal et al., 1988) have studied in great detail the structure of 3-(aminopropyl)triethoxy-silane-modified silica using ^{13}C and ^{29}Si NMR. They were able to estimate the percent of ethoxy groups that have reacted, the ratio of unreacted ethoxy groups remaining, and the number of siloxane attachments to the silane moiety. Akapo and Simpson (1990) studied silica-based octyl bonded phases which were prepared by the fluidized bed technique. They concluded that although the bonded phases have different structure, both polymeric and monomeric phases exhibit almost the same chromatographic properties.

8. Applications

A clear understanding of the separation mechanisms outlined in Table 2 facilitates rapid method development and successful use of solid phase extraction. The separation mechanisms in this table occur due to intermolecular interactions between the analyte molecules and the functional groups of the sorbent. Hydrogen bonding is an important force on bonded phase surfaces. Analyte or interference molecules with the ability to hydrogen bond can interact significantly with the isolated silanols.

The burgeoning literature on solid phase extraction already includes manuals (Analytichem International, 1985; Zief and Kiser, 1988), symposia (Proceedings of the Second Annual International Symposium on Sample Preparation and Isolation using Bonded Silicas, 1985; Third Annual International Symposium, 1986), and compilations of applications (Baker, 1982; Waters, 1986).

Table 3 lists a variety of solid phase extractions successfully completed with silica gel and bonded silica columns. The table illustrates applications for steroid purification using silica as well as octadecyl solid phase extraction columns. Optimum sorbent selection depends on the sample matrix and the groups attached to the steroid nucleus. When the non-polar character of the molecule predominates, as in cholesterol, the octadecyl bonded phase is the sorbent of choice. When polar groups are present on the steroid skeleton (hydrocortisone, for example), silica becomes a good candidate for adsorption of the analyte. Likewise, oil-soluble vitamins and phenols can be separated on either the octadecyl or cyano bonded phases. The retention on the octadecyl bonded phase is due to interaction with the non-polar section of the molecule.

Table 2. Summary of mechanisms for solid phase separations

Separation Mechanism	Analyte Type	Dissolving Solvents	Eluting Solvents
Normal phase (silica)	Slightly to moderately polar	Low $\varepsilon°$*, e.g., hexane, chloroform	High $\varepsilon°$, e.g., methanol
Normal phase (polar bonded phase)	Moderately to strongly polar	Low $\varepsilon°$, e.g., hexane, chloroform	High $\varepsilon°$, e.g., methanol
Reversed-phase (non-polar bonded phase) hexane, For	Nonpolar	High $\varepsilon°$, e.g., methanol/water acetonitrile/water	For non-polar analytes: low $\varepsilon°$ e.g., chloroform. polar analytes: high $\varepsilon°$, e.g., methanol
Anion exchange (SAX, WAX)	Ionic acid	Water or buffer (pH=pKa +2)†	1) Buffer (pH= pKa −2) 2) pH where sorbent or analyte is neutral 3) Solvent with high ionic strength
Cation exchange (SCX, WCX)	Ionic base	Water or buffer (pH=pKa +2)†	1) Buffer (pH= pKa −2) 2) pH where sorbent or analyte is neutral 3) Solvent with high ionic strength
Size exclusion	Proteins	Water or buffer	Water or buffer

*ε = eluotropic strength
†pKa = $-\log_{10}$Ka where Ka is a measure of the ionic activity of the analyte

Table 3. Applications for disposable silica and bonded phase columns

Sorbent	Application
Octadecyl (C_{18})	Reversed-Phase Extraction of Nonpolar Compounds--abused drugs, acetaminophen, amines, analgesics, antiarrhythmics, anticonvulsants, antiepileptics, antibiotics, aromatics, barbiturates, benzodiazepines, cannabis, carbohydrates, carboxylic acid, cholesterol esters, ethchlorvynol, ethosuximide, fatty acids, hypnotics, lidocaine, lipids, oil-soluble vitamins, phenols, sedatives, steroids, sulfonamides, tetracyclines, theophylline, tricyclic antidepressants, triglycerides, valproic acid
Octyl (C_8)	Reversed-Phase Extraction of Moderately Polar Compounds--compounds adsorbed too tightly to octadecyl (C_{18})
Phenyl (C_6H_5)	Reversed-Phase Extraction of Nonpolar Compounds--offers less retention of hydrophobic compounds
Cyano (CN)	Normal Phase Extraction of Polar Compounds--amines, benzyl alcohol, oil-soluble vitamins, phenols, sugar alcohols
Silica Gel	Adsorption of Polar Compounds--alcohols, aldehydes, alkaloids, amines, amino acids, amphetamines, antibiotics, antioxidants, barbiturates, carbohydrates, flavinoids, heterocyclic compounds, hydrocortisone, ketones, lipids, nitro compounds, organic acids, peroxides, phenols, polypeptides, steroids, vitamins
Diol ($CHOHCH_2OH$)	Normal Phase Extraction of Polar Compounds (similar to silica gel)--proteins, peptides, aqueous surfactants
Amino (NH_2)	Weak Anion Exchange Extraction--carbohydrates, nucleotides, peptides, saccharides, steroids, sugars, vitamins
Diamino	Weak Anion Exchange Extraction--amino acids
Aromatic Sulfonic	Strong Cationic Exchange Extraction--amino acids, catecholamines, acid ($C_6H_5SO_3H$) hormones, nucleic acid bases, nucleosides, purines, pyrimidines, water-soluble vitamins
Quaternary Amine	Strong Anion Exchange Extraction--antibiotics, cyclic nucleotides, nucleotides, nucleic acids

9. The Four Steps of Solid Phase Extraction

There are a variety of ways in which solid phase extraction may be carried out, but usually there are four discrete steps:

1. Condition the column by aspirating methanol followed by water or buffer through the column.

2. Aspirate the sample through the column, extracting compounds of interest from a weak solution.

3. Wash the bonded phase to remove impurities or interferences selectively.

4. Selectively elute compounds of interest with a strong solvent.

10. Method Development

The flow chart in Figure 3 illustrates the sequential development of a solid phase extraction method. Factors which should be considered during evaluation of the analytical requirements are the degree of analyte concentration, the degree of purification, and solvent constraints imposed by the analytical detection method. Information about the functional group arrangement, molecular weight, polarity, solubility and pKa of both analytes and interferences is invaluable in developing an extraction strategy which separates the analytes from the matrix and the interferences.

When the analyte is more polar than the associated impurities in the sample, normal phase conditions are advisable. The polar analyte can then be removed from the column with a more polar solvent such as methanol. When the impurities are more polar than the analyte, a reversed-phase system is more effective. The polar impurities are more strongly attracted to the polar solvent system. Therefore, they pass through the column unretained.

The optimization of a proposed extraction method should be approached systematically. The first step should be optimization of the elution solvent. This can be accomplished by adding the analyte to the sorbent and subsequently determining an elution profile from solvents with a range of eluotropic strengths.

Selection of suitable wash solvents is usually accomplished by trial and error.

EVALUATE EXTRACTION PROBLEM
↓   ↓

EVALUATE ANALYTICAL
REQUIREMENTS
                              CHARACTERIZE SAMPLE

CONCENTRATION REQUIRED   ANALYTE(S)
How dilute is sample   Molecular weight • Functional groups
How sensitive is detection method   Polarity • Solubility • pKa

PURIFICATION REQUIREMENTS   SAMPLE MATRIX
Specificity of detection method   Extract from unmodified matrix
What are potential interferences   Release from solid matrix
                           (homogenize, sonicate, digest)
                           Reduce matrix solution strength
                           (dilute, pH, ionic strength)
                           Potential interferences in matrix

SOLVENT CONSTRAINTS   INTERFERENCES
Aqueous elution acceptable for   Molecular weight • Functional groups
detection method   Polarity • Solubility • pKa
Volatile elution solvent required

↓

PROPOSE PRELIMINARY METHOD

Select extraction mode
Select extraction sorbent
Select elution solvent
Select sample volume
Select matrix modification method

OPTIMIZE METHOD
↓
Verify elution of analyte from sorbent using standard solution
YES
↓
Extract analytes from standard solution without matrix
YES
↓
Verify extraction and elution of analytes from complex matrix
YES
↓
Determine if interferences are co-eluted with analytes
NO
↓
Validate method

Fig. 3. Method development flow chart.

Published extraction methods serve as excellent guides in method development. Details of two procedures utilizing different sorbents are presented here to illustrate the concepts described above. In Example 1, the extraction of barbital from urine onto an octadecyl column, is typical of reversed-phase extraction. Here water proves to be an excellent wash solvent. Elution with acetone:chloroform (50:50, v/v) affords near quantitative removal from the sorbent.

In Example 2, the extraction of bacitracin from an oily ointment, the matrix is dissolved in methylene chloride. Upon pouring the suspension of bacitracin onto a diol column, the analyte remains on the column. Methylene chloride is a good wash solvent for removal of small traces of the non-polar ointment. The pure analyte is eluted by 0.1 N HCl. The techniques described in these two examples provide attractive alternatives to traditional sample preparation methods.

EXAMPLE 1

Rapid Extraction of Barbiturates from Urine

Discussion of Method: Barbital is representative of the barbiturate drug class. It is a diethyl-substituted barbituric acid with moderately low solubility in water. Although it is a moderately polar analyte dissolved in a polar matrix (urine), it has sufficient hydrophobic character to be easily adsorbed by an octadecyl column and eluted with acetone:chloroform (50:50, v/v).

Column: 6 ml octadecyl (C_{18}), 500 mg

Extraction Time: 10 min/10 samples

Sample Preparation: Adjust 20 ml of urine to pH 7 with 0.5 M potassium phosphate buffer.

Column Conditioning: Aspirate two column volumes of methanol followed by two column volumes of distilled water. Do not allow the columns to dry before addition of samples.

Sample Addition: Fill column two-thirds full with prepared urine sample. Attach a 15 ml reservoir and adaptor to the top of the column and add the remaining prepared urine to the reservoir. Aspirate sample through column at a rate of approximately 4 ml min^{-1}.

Column Wash: Aspirate one column volume of distilled water.

Sample Elution: Elute with two 0.5 ml aliquots of acetone: chloroform (50:50, v/v).

EXAMPLE 2

Rapid Extraction of Bacitracin from Ointments

Discussion of Method: Commercial bacitracin is a mixture of at least nine antibiotic polypeptide complexes in an oil ointment. It is very polar, soluble in water and alcohol, and insoluble in ether, chloroform and acetone. It can be easily separated from the ointment base by solubilizing the matrix in methylene chloride. The polar bacitracin is then adsorbed on a diol column. The methylene chloride wash removes traces of the non-polar ointment matrix. The bacitracin is then eluted from the polar diol column with 0.1 N HCl. Use of a more polar sorbent such as silica gel in this normal phase extraction would result in irreversible retention on the column.

Column: 3 ml diol, 500 mg

Extraction Time: 10 min/10 samples

Sample Preparation: Weigh 200 mg of bacitracin ointment into a small vial. Add 2 ml methylene chloride, stopper, and dissolve ointment base with gentle heating and shaking. (Note: bacitracins are insoluble in methylene chloride and form a suspension.)

Column Conditioning: Slowly aspirate one column volume (2.5 ml) of methylene chloride through the column.

Sample Addition: Quantitatively transfer contents of vial to diol column and aspirate through column. Wash vial and stopper with two 1 ml aliquots of methylene chloride and aspirate each aliquot through the column.

Column Wash: Aspirate two 1-ml aliquots of methylene chloride. Air-dry column under vacuum for 3 min.

Sample Elution: Using the volumetric collection rack and 2 ml volumetrics, elute bacitracin with two 1-ml aliquots of 0.1 N HCl. Dilute to volume with eluting solvent. Mix well prior to LC analysis.

11. References

Akapo, S. and C. F. Simpson (1990). *J. Chromatogr. Sci.*, *28*, 186.

Alpert, A. and F. Regnier (1979). *J. Chromatogr.*, *185*, 375.

BAKER-10 spe Applications Guide, Vol I (1982), Vol. II (1984). J.T. Baker, Inc., Phillipsburg, NJ

Bayer, E., K. Albert, J. Reiners, M. Neider and D. Muller (1983). *J. Chromatogr.*, *264*, 197.

Caravajal, G. S., D. E. Leyden, G. Quinting and G. E. Maciel (1988). *Anal. Chem.*, *60*, 1776.

Chiang, C. H., N. Liu and J. L. Koenig (1982). *J. Colloid. Interface Sci.*, *86*, 26

Davison, W. H., S. W. Kaiser, P. D. Ellis and R. R. Inners (1981). *J. Amer. Chem. Soc.*, *103*, 6780.

Engelhardt, G., H. Jancke, E. Lipmaa and A. Samoson (1981). *J. Organomet. Chem.*, *210*, 295.

Handbook of Sorbent Extraction Technology (1985). Analytichem International, Inc., Harbor City, CA.

Howard, G. A. and A. J. Martin (1950). *Biochem. J.*, *46*, 532.

Jinno, K. (1989). *J. Chromatogr. Sci.*, *27*, 729.

Leydon, D. E., D. S. Kendall and T. G. Waddell (1981). *Anal. Chim. Acta*, *126*, 207.

Maciel, G. A. and D. W. Sindorf (1980). *J. Amer. Chem. Soc.*, *102*, 7606.

Maciel, G. A., R. C. Ziegler and R. K. Taft (1985). In: *Silanes, Surfaces and Interfaces*," p. 413. Gorden & Breach Science Publishers, London.

Majors, R. E. (1986). *LC-GC*, *4*, 972.

Melander, W. R. and C. Horvath (1980). In: *High Performance Liquid Chromatography: Advances and Perspectives* (C. Horvath, Ed.), Vol. 2, p. 115. Academic Press, New York.

Murthy, R. S. S. and J. P. Berry (1990). In: *Chemically Modified Oxide Surfaces*, (D.E. Leydon and W.T. Collins, Eds.), Vol. 3, p. 151, Gordon & Breach, London.

Murthy, R. S. S., L. J. Crane and C. E. Bronnimann (1991). *J. Chromatogr.*, *542*, 205–220.

Poole, C. F. and S. A. Schuette (1984). *Contemporary Practice of Chromatography*, p. 459. Elsevier, Amsterdam.

Proceedings of the Second Annual International Symposium on Sample Preparation and isolation Using Bonded Silicas (1985). Analytical International, Harbor City, CA; also *Proceedings of the Third Annual International Symposium*.

Ramsden, H. (1985). *U.S. Pharmacopoeia*, *4*, 486.

Sindorf, D. W. and G. A. Maciel (1981). *J. Amer. Chem. Soc.*, *103*, 4263.

Snyder, L. and J. J. Kirkland (1979). *Introduction to Modern Liquid Chromatography*, 2nd ed., p. 366. John Wiley & Sons, New York.

Waters Sep-Pak Cartridge Applications Bibliography, 4th Ed. (1986). Waters Division, Millipore Corporation.

Zief, M. (1985). In: *Liquid Chromatography in Pharmaceutical Development* (I. Wainer, Ed.), Aster Publishing Company, Springfield, OR, p. 133.

Zief, M. and R. Kiser, *Solid Phase Extraction for Sample Preparation*. J.T. Baker, Inc., Phillipsburg, NJ.

Zweig, G. and J. Sherma (1972). *Handbook of Chromatography*, CRC Press, Cleveland, OH, Vol. II, pp. 191-254.

CHAPTER 5

Application of Restricted-Access Media to the Direct Analysis of Biological Samples

JOHN A. PERRY*

*Regis Chemical Company, 8210 Austin Avenue
Morton Grove, Illinois 60053 U.S.A.*

1. Introduction

The internal-surface reversed-phase particle was invented at Purdue University by Pinkerton (Pinkerton and Hagestam, 1985; Hagestam and Pinkerton, 1985; for reviews, see Pinkerton *et al.*, 1986; Westerlund, 1987; Pinkerton, 1988; Perry, 1990; Pinkerton, 1991). Internal-surface reversed-phase (ISRP) particles have different inner and outer surfaces. As suggested in Fig. 1 and described presently in more detail, proteins are neither adsorbed nor denatured by ISRP outer surfaces, and they do not penetrate the pores which contain the absorptive, hydrophobic surface. Functionally, therefore, ISRP permits the direct HPLC determination of drugs in plasma or serum without prior removal of proteins.

In this section of the chapter, ISRP concepts and applications are scanned thoroughly enough to lend insight into ISRP history, function, and synthesis; the growing range of ISRP applications; factors that affect ISRP selectivity, with application to method development; and some characteristics of a second-generation, commercially available ISRP packing.

Finally, the origin and characteristics of a class of materials conceptually related to ISRPs are explored. These materials are called semipermeable surface (SPS) particles and, like ISRPs, are subclasses of restricted-access media (RAM), materials that allow small molecules to be analyzed in the presence of large biopolymers.

*Present address: Harbitzalléen 16B, 0275 Oslo, Norway

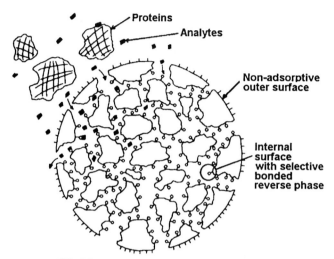

Rigid porous hydrophilic particle

Fig. 1. The ISRP particle has a hydrophilic outer surface
that does not adsorb proteins and an inner surface
that proteins cannot reach. Proteins pass right
through ISRP columns. Analytes, however, can
reach and become separated by the internal-surface
reversed phase. (Reproduced with permission from
Perry, 1990.) (Copyright, Marcel Dekker, Inc.)

2. The Rationale for ISRPs

The principal method for the analysis of drugs in serum or plasma
is liquid chromatography (LC). The number of such analyses is phe-
nomenal: in the early 1980s, for instance, a typical pharmaceutical
company was already running about 30,000 such LC analyses per
year (Pinkerton and Hagestam, 1984). Before ISRP, protein had to
be removed from each potential LC serum or plasma sample before it
could be injected.

The LC stationary phase of choice for the analysis of drugs is the
paraffinic octadecylsilylsilica (ODS). Protein-containing fluids such as
serum and plasma may not be directly injected into such a column
because proteins denature on the ODS surface, adsorb onto the
porous silica to which the ODS is bonded, and thus accumulate within
the column. Initially, such accumulation clogs the pores, inhibiting
diffusion of the analytes into the pores (within which chromatographic

separations primarily take place) (Arvidsson *et al.*, 1984; Juergens, 1984) and decreasing the column capacity (Hern, 1982). Continuing protein accumulation narrows interparticle channels, constrains the flow of mobile phase, and raises the pressure required to maintain a given flow rate of mobile phase through the column, to an increasingly unacceptable level. Eventually, the column becomes unusable.

If the LC column is not to be ruined in this way, the sample must first undergo some manner of clean-up. The analytes must first be separated and isolated from the protein matrix (Snyder and Kirkland, 1979; DeJong, 1980; Roth *et al.*, 1981; Hux *et al.*, 1982; Juergens, 1984; Nazareth *et al.*, 1984; see also a more recent reference, Puhlmann *et al.*, 1992). In one clean-up method (Snyder and Kirkland, 1979), the proteins are precipitated, the analytes are extracted from the supernatant, the extraction solvent is evaporated, and the analytes are reconstituted in the mobile phase. This approach is often slow and frequently inaccurate, because the protein-coprecipitated analytes are not fully recovered from the supernatant; nevertheless, it removes endogenous species and preconcentrates the analytes. In another method (Roth, 1983; Juergens, 1984; Arvidsson *et al.*, 1984; Nazareth *et al.*, 1984; Roth *et al.*, 1981), the analytes are first extracted onto disposable, large-particle, bonded-phase silica supports—a costly approach requiring one disposable column per sample.

In a third clean-up method, samples are injected into short packed precolumns connected to the LC analytical columns through switching valves. The precolumns are packed with either coarse (30–50 μm diameter) reversed-phase silica particles (Juergens, 1984; Roth *et al.*, 1981; Nazareth *et al.*, 1984) or ion exchange materials (Hux *et al.*, 1982). The analytes, strongly retained on the precolumn, are then backflushed onto the analytical columns. Proteins retained on the precolumns must be removed by washing, and the expensive precolumns are generally replaced after perhaps 100 injections. While the analytes are preconcentrated in this method, the backflushes, purge washes, and precolumn replacements are all time consuming.

3. ISRP Principles

In 1985, Hagestam and Pinkerton introduced an elegant new LC concept: the Internal-Surface Reversed-Phase (ISRP) (Hagestam and Pinkerton, 1985; Pinkerton and Hagestam, 1985) to obviate the difficulties invariably associated with these protein clean-up procedures. The external surfaces of ISRP supports are hydrophilic and do not adsorb proteins; moreover, these supports are chosen to have pores too small—with the ISRP in place, 5.2 nm diameter—to admit proteins. The hydrophobic partitioning phase of an ISRP support, confined exclusively to the internal surface of the porous support, is accessible to low molecular weight analytes, which are retained and separated.

3.1.   Protein Exclusion and Recovery

Albumin is the most abundant plasma protein. Therefore, it is the most important protein to consider within the context of drug analysis in plasma.   Given a mobile phase of appropriate pH and ionic strength, proteins are indeed not adsorbed onto silica surfaces or silica to which a glycerylpropyl group has been bonded (Schmidt *et al.*, 1980). Albumin has a molecular weight of 65,600 daltons, a prolate ellipsoid shape, 15 nm by 3.8 nm, and a radius of gyration of 3.11 nm (Andereg, 1955). With regard to its exclusion from pores, albumin may be considered to have an effective radius of 4.0 nm (Yau *et al.*, 1979). Because spherical silica supports are made up of agglomerated silica microspheres, the pores can be modeled as squares in cross-section (Unger *et al.*, 1984). A molecule is completely excluded from a pore when its radius equals that of the pore (Yau *et al.*, 1979). Thus albumin is excluded from spherical silica pores that have diameters less than 8.0 nm.

3.2.   Analyte Retention and Separation

Small-molecule analytes can penetrate into the small pores from which the proteins are excluded and be retained and separated within the pores if the internal surface areas or the pores are 150–400 $m^2 g^{-1}$ (the external surface area of a 5–micron spherical silica is only 0.5 to 1% as much or about 2 $m^2 g^{-1}$) (Unger *et al.*, 1984). The internal silica surface can be made suitable for retention by covalently attaching a hydrophobic partitioning phase.

4. Basics of ISRP Synthesis

For synthesis of ISRP, a glycerylpropyl phase is first bonded to a silica particle (or glass bead) of suitably small pore diameter (5.2 nm after the ISRP has been attached). A suitable polypeptide is then bonded via the amine function to this glyceryl surface. Through the action of an enzyme too large to penetrate the pores and thus affect the internal surfaces, the polypeptide is then removed from the external surface (Pinkerton and Hagestam, 1985).

The silica or glass bead is first modified with the glycerylpropyl phase (Lassen *et al.*, 1983), then activated with carbonyldiimidazole

(Bethel *et al.*, 1979), and finally coated with the tripeptide glycine-L-phenylalanine-L-phenylalanine. The external surface is then treated with the enzyme, carboxypeptidase A (Hofman and Bergman, 1940). Carboxypeptidase acts specifically on free carboxyl groups (Bethel *et al.*, 1979) especially toward aromatic amino acids. Phenylalanines but not glycine are thus removed in a sequential fashion.

Carboxypeptidase A (35,000 daltons) has a spherical radius of approximately 3.1 nm and thus is excluded from pores smaller than 6.0 nm in diameter. Because it is smaller than the 65,600–dalton, 4.0–nm-radius albumin, this enzyme has somewhat greater access to the larger pores than do the serum proteins. Therefore, serum proteins cannot reach whatever ISRP peptides may remain after enzyme treatment.

5. The Nature of the Internal Surface Reversed Phase

5.1. Surface Chemistry

5.1.1. The outer surface

The external surface functionality of the first commercially available ISRP was glycine, attached through an amide link to the rest of the ISRP stationary phase molecule with its free carboxylic acid group displayed to the mobile phase. This outer ISRP surface does not interact with or denature proteins. However, it is similar to the inner surface in retention and ion-exchange capability.

5.1.2. The inner surface

The corresponding inner surface of the ISRP is glycine-phenyl-alanine-phenylalanine (GFF), which is attached through the glycine amide to silica (Fig. 2), and, like the outer surface, displays the free carboxylic acid group of the outer phenylalanine to the mobile phase.

As will be discussed later in the treatment of GFF2, because GFF is polymeric at the silica surface whereas GFF2 is monomeric, both the retention and the chromatographic efficiency of GFF2 are greatly improved compared with those of GFF. However, because both GFF and GFF2 have the same gly-phe-phe functional group as well as the same underlying silica particle, both probably show mutually similar chromatographic selectivites and exclusion properties. Consequently, method development procedures with GFF2 should resemble closely those for GFF. The ISRP GFF is suitable for the analysis of a wide variety of drugs and small peptides (Table 1) (Pinkerton and Koeplinger, 1988).

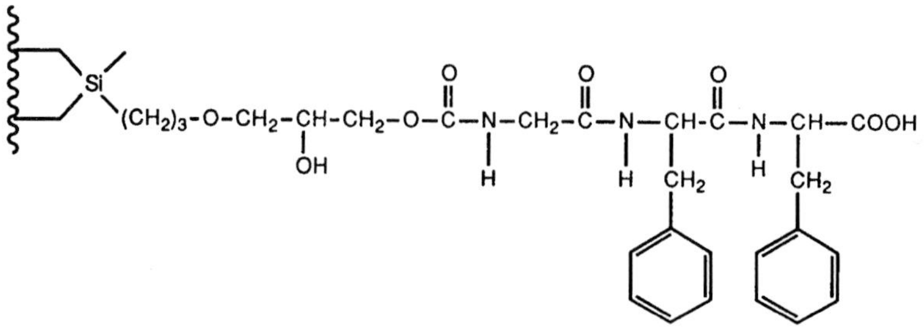

Fig. 2. The structure of the complete GFF internal surface stationary phase. (Reproduced with permission from Perry, 1990. Copyright, Marcel Dekker, Inc.)

5.2 Size Exclusion Characteristics

The size exclusion characteristics of the ISRP GFF particles, and presumably those of GFF2, are suggested in Fig. 3 (Rateike, 1987). Of six separately injected proteins of known molecular weight, two of the molecular weights below 15,000 were retained, showing penetration into the pores. However, the proteins of molecular weights of greater than 29,000 were eluted within the dead volume, showing that they were excluded from the pores. [This is as expected. In the final step in ISRP preparation (Pinkerton and Hagestam, 1985; Hagestam and Pinkerton, 1985), the 36,000–dalton enzyme carboxypeptidase A cleaves the phenylalanine residues from the GFF, leaving only glycine on the outer surface.] Apparently the molecular weight cutoff of the GFF (and probably also of GFF2) ISRP particles is between 15,000 and 25,000.

The 5.2–nm pores apparently exclude proteins heavier than about 20,000–25,000 daltons—for instance, the 29,000–dalton carbonic anhydrase. The 12,400–dalton Cytochrome C and 6,500–dalton aprotinin, however, enter the pores and are retained. (The aprotinin was retained so long that it could not be detected).

Table 1. Some retention values for drugs on ISRP-GFF columns

	Retention	
Antibacterial drugs	Time (min)	k'
Sulfapyridine	3.6	1.0
Sulfamethazole	3.8	1.1
Trimethoprim	6.0	2.4
Sulfasalazine	23	12
Anticonvulsant drugs		
Ethosuximide	2.2	0.27
Theophylline	2.4	0.43
Primidone	3.0	0.78
Phenobarbital	4.9	1.9
Carbamazepine	7.4	3.3
Phenytoin	13	6.4
Antidepressant drugs		
Imipramine	38	21
Trimipramine	45	25
Amitriptyline	47	26
Antihypertensive drugs		
Chlorthalidone	6.4	2.6
Hydrochlorothiazide	7.0	3.1
Furosemide	9.2	4.4
Antiinflammatory/analgesic drugs		
Acetylsalicylic acid	1.7	0.0
Acetaminophen	2.5	0.47
Salicylic acid	3.1	0.85
Ibuprofen	4.4	1.6
Phenylbutazone	8.2	3.9
Sulfinpyrazone	8.4	4.0
Indomethacin	14	7.0
Cardiac drugs/metabolites		
Procainamide	3.5	1.1
Lidocaine	4.3	1.5
2-Naphthoxy acetic acid	5.2	2.1
Nifedipine	15	7.9
Quinidine	18	9.5
Propranolol	22	12
α-Naphthol	22	12
Verapamil	28	14

Mobile phase: 0.1 M phosphate/isopropanol/tetrahydrofuran 84/10/6. Flow rate: 1.0 ml min^{-1}. Column: 15 cm × 4.6 mm i.d.

JOHN A. PERRY

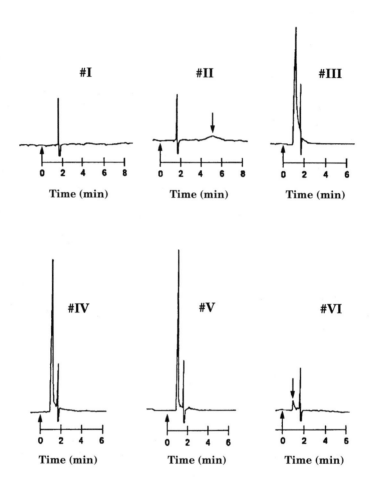

Fig. 3. Analytes: #I: Aprotinin (MW, 6,500); #II: Cytochrome
C (MW, 12,400) (top arrow); #III: Carbonic anhydrase
(MW, 29,000); #IV: Globulin (MW, 55,000); #V:
Bovine serum albumin (MW, 66,000); #VI: Blue
Dextran (MW, 2,000,000) (top arrow). Sample
matrix: 2.5 mg ml^{-1} of 0.1 M KH$_2$PO$_4$. Sample filtered through 0.2–μm membrane prior to injection.
Sample size: 20 μl. Mobile phase: 90% 0.1 M KH$_2$PO$_4$
(pH 6.8), 10% acetonitrile (LC-grade). Flow rate: 1.0
ml min^{-1}. Detection: 254 nm, 0.16 AUFS. Column
packings: 5–μm GFF ISRP. Guard column, 1 cm ×
3.0 mm i.d.; analytical column, 15 cm × 4.6 mm i.d.
(Reproduced with permission from Perry, 1990.)

6. Initial GFF Applications: Simple Analyses for Drugs in Serum

6.1. Protein Recovery: Column Life

If a protein is retained by the ISRP column, continuing serum injections quickly destroy the column. The first evidence of this is an increase in the pressure, quickly followed by a dramatic reduction in column efficiency and complete deterioration of column performance.

Adequate ISRP column life, given adequate sample filtering through 0.2–μm filters before injection, is achieved in two ways: with an adequately functional ISRP packing and ISRP guard columns. A proper ISRP packing retains essentially no protein. Whether the ISRP pressure increase eventually seen with injection of several hundred serum samples is caused by serum protein or by other serum components has never been determined; however, the ISRP guard column is essential to protect the analytical column.

Generally, a GFF ISRP guard column can accommodate over 500 10– or 20–μl injections of plasma or serum before it must be replaced (Meriluoto and Erikson, 1988). The performance of the GFF ISRP analytical column, meanwhile, remains essentially unchanged (Szczerba, 1986).

6.2. Drugs: Free, Bound, and Whole

In plasma or serum, an equilibrium exists between the free and the protein-bound drug. The sum of the bound and the free drug is the whole drug. This section deals with methods for the analysis of the total drug concentration; for the separate ISRP GFF determinations of free and bound concentrations, see section 7.2.

6.3. Analyses for Whole Drugs

6.3.1. Whole drug recoverability

Even when the analyte is tightly bound to plasma protein, if the size of the injected serum sample is 10 to 20 μl, the concentration of the whole drug is determined by ISRP (GFF) analysis. In calibrating for the GFF ISRP analysis of phenylbutazone in horse plasma, for instance, Sams (1987) found that "water and plasma samples agree," indicating that the total drug concentration was measured even though phenylbutazone is 98% bound. Similarly, phenytoin is more than 90% bound, but in a GFF ISRP analysis, recovery was 98% (Hagestam and Pinkerton, 1985). In other GFF analyses, probenecid

and lidocaine were 100% recovered from ISRP columns even though
the two drugs are 83–94% and 65–77% protein bound, respectively
(Nakagawa *et al.*, 1987).

6.3.1.1. Organic solvent effects. In the analysis of cefpiramide in
human plasma, Nakagawa *et al.* found that 2.5% organic solvent in
the buffered mobile phase improved the chromatographic efficiency of
the GFF ISRP column, the precision of the analysis, and the degree
of recovery (from 70% to 80% without the solvent to almost 100%
with it). These results suggest that the organic solvent accelerates
the rate of drug release from plasma proteins (Nakagawa *et al.*,
1987).

6.3.2. Variables that affect selectivity

6.3.2.1. Organic modifiers. As noted above (*6.3.1.1, Organic solvent
effects*), a small mobile phase concentration of organic solvent benefits ISRP analyses in several ways (Nakagawa *et al.*, 1987). Should
they be indicated, larger concentrations are also allowed. ISRP
mobile phases may contain up to 20% by volume of any of four
organic solvents: acetonitrile, isopropanol, methanol, or tetrahydrofuran. Mobile phases containing two or more organic modifiers may
also be used, but the total must not be more than 20%. As shown
below (*6.3.3. Method development*), each of the organic modifiers
affects selectivity in a different way.
      Like the 6.0–7.5 allowable pH range, the solvent concentration
maximum is set by the proteins in the sample, not by the ISRP
column. During the period that the protein is present within the
ISRP column, it must not be denatured. Once the protein has been
swept from the column, however, the mobile phase composition
restrictions become no more limiting than they are for any other
silica-based column.

6.3.2.2. pH. Figure 4 shows the structures of seven drugs chosen
because of their dissimilar acid-base properties. Figure 5 shows the
GFF retentions of these drugs at three pH values, 6.0, 6.8, and 7.5
(Perry and Chang, 1987). It is seen that GFF selectivity increases
strongly with increasing pH. [For 21 drugs studied at pH 7.5 by Sams
and Evec (1987), ISRP GFF selectivity was found to be more than
twice that of ODS.] As can also be seen, retention by GFF varies
inversely, although relatively weakly, with pH. The decrease in GFF
selectivity (and no doubt also that of GFF2) with decreasing pH can

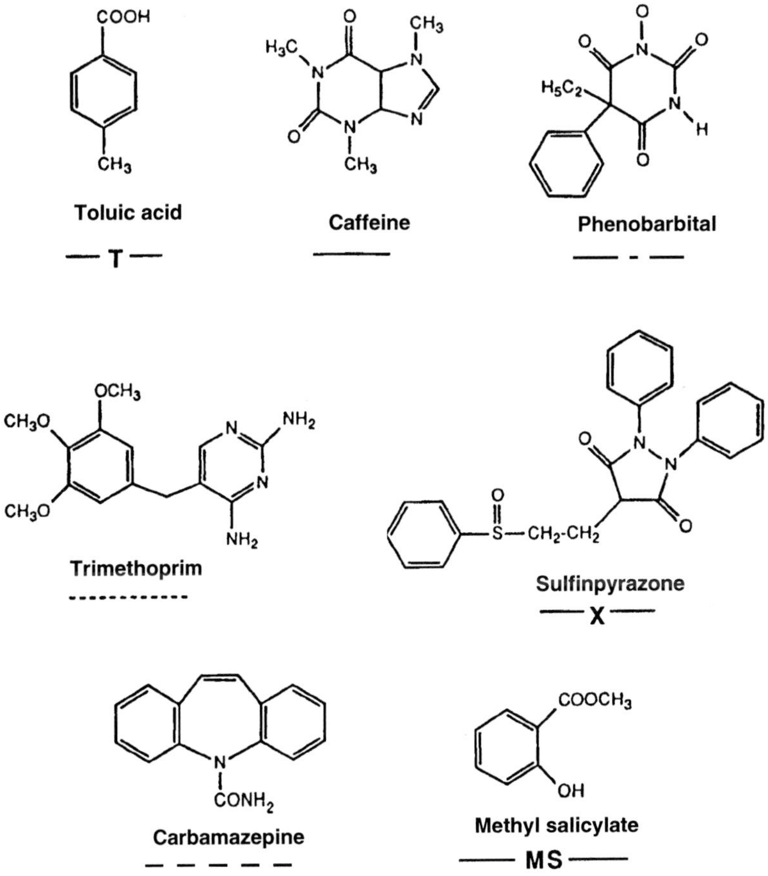

Fig. 4. Seven drugs chosen for their dissimilar acid-base properties. Their ISRP behavior is shown in Fig. 5. (Reproduced
with permission from Perry, 1990.)

be explained as follows.   The ISRPs GFF and GFF2 present a free
carboxylic acid group to the usually buffered, aqueous mobile phase.
With decreasing pH, the ionization of the carboxylic acid group
decreases, in turn decreasing the ion-exchange capacity of the GFF/-
GFF2 phase.   Achieving reproducible retentions with GFF/GFF2
requires equilibration of the system. When water is used as a mobile
phase, retention appears to reflect the pH of a buffered system previously in the column. The retention of theophylline demonstrates
these points. With buffered mobile phases, the retention of theophylline does not change as a function of pH. However, with water as the

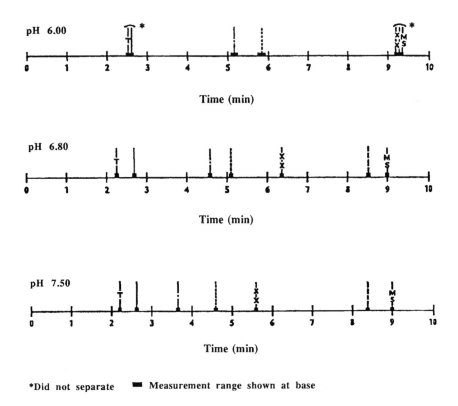

*Did not separate   ■ Measurement range shown at base

Fig. 5. The retentions of acid-base-dissimilar drugs (shown in Fig.
        4) nicely illustrate the increase of GFF selectivity with
        increasing pH. Mobile phase: 80% 0.1 M KH₂PO₄ (pH 6.0,
        6.8, 7.5), 20% acetonitrile (HPLC-grade reagents). Mobile
        phase flow rate: 1.0 ml/min. Sample size: 200 µl. Column
        dimensions: 15 cm × 4.6 mm i.d. Packing: 5–µm GFF
        ISRP. (Reproduced with permission from Perry, 1990.)

mobile phase, theophylline retention by GFF varies by a factor of 3,
depending on the pH of the previous mobile phase. This unexpected
behavior is shown in Fig. 6.

6.3.2.3. Temperature. The retention times of four drugs are shown
in Table 2 for column temperatures of 10°C and 26°C. Decreasing the
temperature tends to increase retention by GFF (and probably
GFF2). Table 2 also shows that the ratios of these retention times
increase substantially with decreases in column temperature. (This
conclusion probably holds for GFF2 as well.)

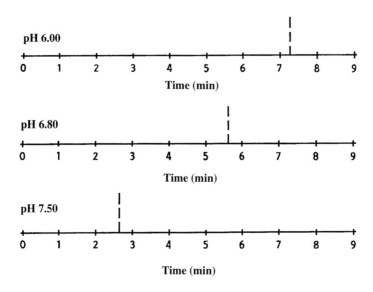

Fig. 6. Unexpected irreproducible retention behavior of theophylline in water. Mobile phase: water (HPLC grade). Mobile phase that preceded water: 80% 0.1 M KH_2PO_4 (pH 6.0, 6.8, 7.5), 20% acetonitrile (HPLC-grade reagents). Mobile phase flow rate: 1.0 ml min^{-1}. Sample size: 20 µl. Column dimensions: 15 cm × 4.6 mm i.d. Packing: 5–µm GFF ISRP. (Reproduced with permission from Perry, 1990.)

Table 2. Decreasing column temperature increases GFF ISRP retention

Drugs	Retention (min)	
	26°C	10°C
Caffeine	2.63	2.65
	(1.43)*	(1.64)
Phenobarbital	3.76	4.35
	(1.30)	(1.47)
Sulfinpyrazone	4.87	6.34
	(1.15)	(1.20)
Carbamazepine	5.58	7.58

*Separation factor (t_2/t_1)

6.3.3. Method development

The following method development study shows that a wide variety of mobile phase compositions is usable with GFF/GFF2. Figure 7 shows the structures of furosemide, phenylbutazone and oxyphenbutazone, three drugs commonly administered to racehorses. GFF ISRP selectivity was optimized for this separation by altering the composition of the mobile phase. Organic modifiers were tested first, then pH. Three organic modifiers, each up to 20% by volume, were tested: acetonitrile (MeCN), isopropanol (IPA), and tetrahydrofuran (THF). The separations that resulted from using 10% of these are shown in Fig. 8. Obviously, THF was found to be the modifier of choice. A systematic, empirical approach to method development (see Figs. 8–10) was found more successful than analysis of the structures of the solutes and the GFF stationary phase (Fig. 2).

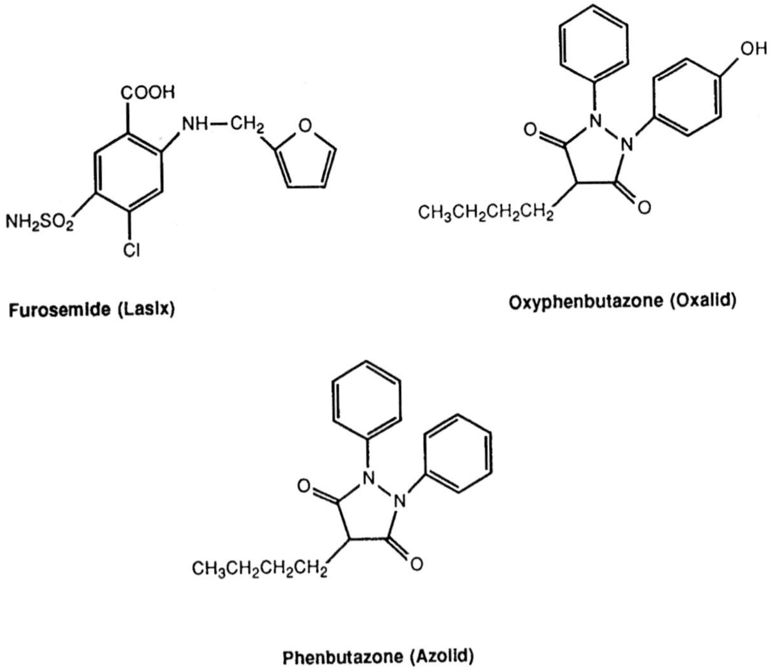

Furosemide (Lasix)

Oxyphenbutazone (Oxalid)

Phenbutazone (Azolid)

Fig. 7.   The structures of furosemide, phenylbutazone and oxyphenbutazone.

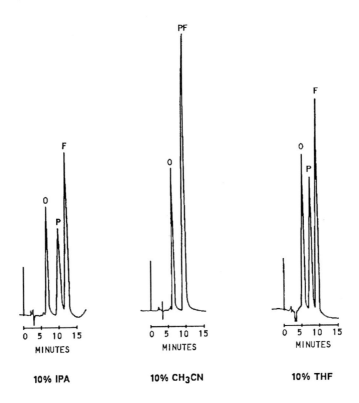

Fig. 8. GFF ISRP column length: 15 cm. Mobile phase
composition: 0.1 M KH_2PO_4 buffer/10% organic
modifier; pH 6.8. Flow rate: 1.0 ml min^{-1}.
(Reproduced with permission from Perry, 1990.)

Next, the pH was varied between 6.0 and 7.5 (Fig. 9). The general
conclusion of this study was that GFF/GFF2 selectivity increased with
increasing pH even though decreasing pH increases GFF/GFF2
retention.   In this case, the best GFF separation occurred at pH 7.5.
(Again, note that the pH range is limited not by the ISRP column but
by the protein, which must remain globular during the separation.
Once the protein has left the ISRP column, pH may be varied at will
over the full range available to any other silica-based column.) The
optimum 12–min GFF separation of furosemide, phenylbutazone, and
oxyphenbutazone from horse plasma and from each other is shown in
Fig. 10 (Perry and Rateike, 1986; Perry et al., 1987).

JOHN A. PERRY

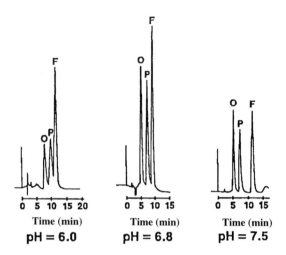

Fig. 9. Experimental conditions as in Fig. 8, with tetrahydrofuran
        as buffer. (Reproduced with permission from Perry, 1990.)

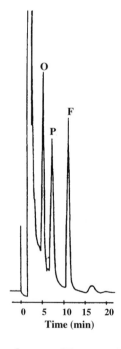

Fig. 10. GFF ISRP column, 15 cm × 4.6 mm i.d. Mobile
         phase composition: 0.1 M KH_2PO_4 buffer/10%
         organic modifier; pH 7.5. Flow rate: 1.0 ml min^{-1}.
         (Reproduced with permission from Perry, 1990.

7. Extensions of GFF Applicability

7.1. Analytes

7.1.1. Peptides

A different application of the selectivity of GFF was demonstrated by Pinkerton and Koeplinger (1988) with the separation of peptides (Fig. 11).

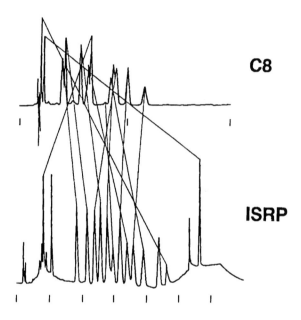

Fig. 11.   Comparison of C8 and ISRP separation of peptides. For this mixture of peptides, GFF selectivity is different from but compares favorably with that of octyl (C8). (Figure reproduced with permission from Perry, 1990.)

7.1.2. Toxins

Meriluoto and Erikson (1988) applied the combined size exclusion and peptide-selective characteristics of the GFF ISRP to analyze for 1,000–dalton toxins in environmental waters containing cyanobacteria. They stated that the use of the ISRP column drastically simplified sample cleanup and thus shortened the total analysis time. The detection limits were good, and several hundred samples could be injected before it was necessary to install a new guard column.

7.2. Techniques

7.2.1. Frontal analysis

7.2.1.1. Analyses for free and bound drugs. In a lecture presented in 1987 in Japan, Pinkerton described the discovery in his laboratories of the ISRP capability for separating the free drug forms from the bound, and its elucidation (Pinkerton *et al.*, 1987). In 1989, Pinkerton and coworkers reported that when imirestat-containing serum samples in excess of 200 µl were injected onto a 5–cm GFF ISRP column, the imirestat unexpectedly eluted as two peaks. The same phenomenon was observed with a separate, similar injection of phenytoin (see Fig. 12). It was surmised that the peaks represented the bound and unbound fractions of the drug, which had been separated on-column (Pinkerton *et al.*, 1989).

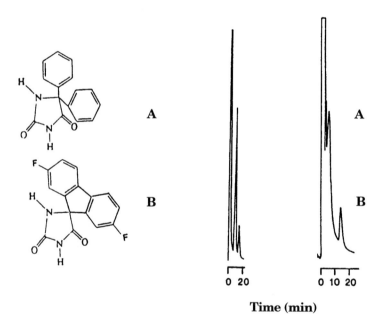

Time (min)

Fig. 12. Split-peak phenomena observed from direction injection of (A) imirestat at 20 µg/ml (200 µl injected) and (B) phenytoin at 51 µg ml^{-1} (500 µl injected) in human serum. The first peaks are serum, the second the "bound" drug—the drug that had been bound to protein at the time of injection—and the third and last peaks are the "free" drug. Column, GFF ISRP, 5 cm × 4.6 mm i.d; mobile phase, 0.01 M phosphate (pH 6.8); flow rate, 1.0 ml min^{-1}. (Reproduced with permission from Pinkerton *et al.*, 1989.)

Further study indicated that the split-peak phenomenon was uniquely dependent on sample size and required the on-column presence of drug and serum proteins in a serum matrix. The area of the first peak should approximate the drug bound to the protein prior to injection; the second peak should represent the free drug and any drug released immediately upon introduction to the column (Pinkerton *et al.*, 1989).

The splitting of the drug fractions into two chromatographic peaks is facilitated when the following conditions are met: the sample contains binding proteins, the sample size is equal to or greater than 200 µl injected into a 4.6 mm × 25 cm ISRP column, and the mobile phase is weak (Pinkerton *et al.*, 1989).

Subsequent work by investigators at Kyoto University (Shibukawa *et al.*, 1988) using warfarin, bovine serum albumin (BSA), a 4.6 mm × 25 cm ISRP column, and a 200–µl sample size has demonstrated the same type of peak-splitting phenomenon. Shibukawa *et al.* reported further investigations of this GFF ISRP capability. Extending it to frontal analysis, they were able to determine the free drug concentrations of warfarin and of indomethacin (Shibukawa *et al.*, 1989a). Later, through the two warfarin peaks eluted from a GFF column, they evaluated the warfarin-albumin interactions for both strong and weak drug-protein binding; in this case, the free warfarin (1%) was evaluated by ultrafiltration (Shibukawa *et al.*, 1989b; see also Shibukawa *et al.*, 1988). Later still, Pinkerton applied the approach to the determination, within ±10%, of the binding parameters for warfarin-human serum albumin (Pinkerton and Koeplinger, 1990).

Drugs that are strongly bound to proteins may elute as two peaks when a serum sample is directly injected onto an ISRP GFF column, under very select conditions. The drug molecules bound to a strong primary binding site experience a delayed release, whereas drugs bound to weaker secondary binding sites are released immediately upon introduction to the mobile phase. As a result of the exclusion of proteins from the packing, the drug that had been bound to the primary site elutes in a major peak ahead of a second smaller peak, which contains both the drug bound initially to secondary sites and the original free drug. The peak splitting depends markedly on column length, sample size, and mobile phase. The peak positions can be controlled with mobile phase ionic strength, pH, and organic modifier concentration. Alternatively, the peaks can easily be merged into one peak if desired. The phenomenon has been observed for phenytoin and its derivatives in human serum (Miller, 1987; Pinkerton *et al.*, 1987; 1989), and warfarin combined with bovine serum albumin (BSA) (Shibukawa *et al.*, 1987; 1988). If the drug is combined with pure protein and the free drug concentration is measured by an independent method (*e.g.*, ultrafiltration), then the concentration of the drug bound to the secondary site can be determined from the second peak by difference. With this and a knowledge of the number of moles bound per site, the binding constants for each binding site can be calculated directly (Shibukawa *et al.*, 1987). In 1990, Pinkerton reported the use of this technique to determine the parameters for the binding of warfarin to human serum albumin within 10% (Pinkerton and Koeplinger, 1990).

7.2.2. *Column switching*

The Internal-Surface Reversed Phase (ISRP) was the first example of Restricted Access Media (RAM), discussed presently in the section of the SemiPermeable Surface. For generality, we use RAM in the following treatment of the column switching technique and in Fig. 13.

7.2.2.1. *The column switching technique.*
A RAM column can be used for sample cleanup (Hagestam and Pinkerton, 1985; Puhlmann *et al.*, 1992; Haginaka, 1990; Matlin *et al.*, 1990; Pompon *et al.*, 1992; Szczerba and Perry, 1986). Valve configurations suitable for column switching are presented in Fig. 13 (Szczerba and Perry, 1986).

In this use, the sample is charged to a guarded RAM column, which here functions merely for cleanup: on this column, the small-molecule analyte of interest is separated from the bulk of the large-molecule matrix. This matrix, which contains the bulk of the large molecules, usually but not necessarily protein (see, for instance, Pompon *et al.*, 1992), is passed on to waste.

The separated small-molecule analytes are then transferred to a second column for analysis. With the second column, the stationary and mobile phases need be chosen only as required by the analysis—tolerance of the second column to protein being no longer pertinent. If required, the solute may be concentrated on the second column, then eluted by a solvent gradient.

At least in concept, solute detectability may be enhanced considerably by this approach. Given adequate separation of the low molecular weight analytes from the macromolecules, sample volume may be increased to enhance detection sensitivity.

An interesting facet of column-switching and indeed the whole field of RAM chromatography is that both <u>seem</u> to involve trace-last chromatography, and would therefore seem suspect. Trace-last chromatography refers to partition retention for both the nontrace moieties of the sample and the trace constituents. When that happens, trace retention of the trace component becomes nonreproducible. However, that is not what is happening in RAM. In RAM, trace retention

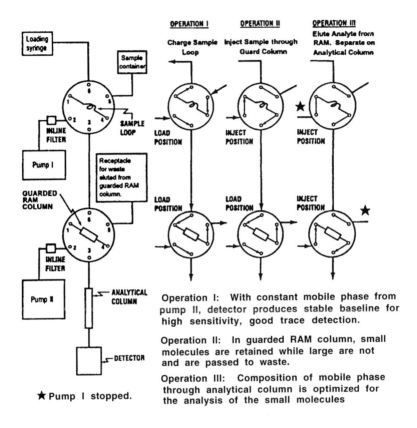

Operation I: With constant mobile phase from pump II, detector produces stable baseline for high sensitivity, good trace detection.

Operation II: In guarded RAM column, small molecules are retained while large are not and are passed to waste.

Operation III: Composition of mobile phase through analytical column is optimized for the analysis of the small molecules

★ Pump I stopped.

Fig. 13. On-line cleanup. In column switching the functions of sample-cleanup and analyte-analysis are combined. <u>Note</u>: sample prefiltering and guard columns remain necessary. Samples should be filtered through 0.2–μm filters before being injected. A guard column is placed upstream of the larger RAM column, the series-connected guard column-main RAM column combination being placed between ports 1 and 4 of the second valve. (Adapted and reprinted with permission from Szczerba and Perry, 1986.)

must be both 1) essentially unaffected by the size-excluded large molecules and 2) statistically reliable—in contrast to partition-only trace-last separations. Trace-first chromatography may be relied on (Perry *et al.*, 1987; Perry and Chang, 1990); trace-last may not (Perry and Rateike, 1990). Trace-last chromatography, when used for quantitative analysis, should be accompanied by statistical testing— as, indeed, applications of it tend to have been (Pompon *et al.*, 1992; Puhlmann *et al.*, 1992). It is the fine track record of these just-cited applications, supported by statistical analysis, that calls for further examination.

The conclusions from the trace-first—trace-last papers (Matlin *et al.*, 1990; Perry and Chang, 1990; Perry and Rateike, 1990) are firm enough, and the data withstand examination. However, the statistical tests of the "trace-last" column-switching papers (Pompon *et al.*, 1992; Puhlmann *et al.*, 1992) are also firm. The conclusions are mutually contradictory.

Realizing the differing natures of RAM operation and trace-last chromatography seems to resolve the apparent contradiction. In RAM separations, the large molecules are not separated from the small by the same mechanism. The large molecules are simply excluded from RAM packings, and are swept from the columns. In contrast, the small molecules are retained by chromatographic partition, which partition operates largely within the RAM particles. This dual-mode operation of RAM differs from the single-mode operation of trace-last chromatography.

Two column-switching papers serve to illustrate the advantages of the approach. The first, by Pompon *et al.*, also involves an unusual application, namely, separating small molecules from oligonucleotides (rather than from proteins) (Pompon *et al.*, 1992). The second, by Puhlmann *et al*, shows how detectability and reliability can be considerably improved by ISRP cleanup combined with C18 concentration and selectivity (Puhlmann *et al.*, 1992).

7.2.2.2. Application to oligonucleotides. The behavior of an antisense oligonucleotide and its degradation products was analyzed at a low concentration (10^{-5} M) in cell culture medium by LC without radiolabeling, sample preparation, or an internal standard. For this purpose a novel on-line switching technique was developed, with a precolumn containing ISRP material acting as a 'reversed-filter' of proteins. This technique was also adapted to ion-exchange LC (Pompom *et al.*, 1992). (Note: the ISRP column used for the above work was of a make different from that of GFF/GFF2.)

7.2.2.3. Application to creatinine. The continued need for ISRP columns was discussed recently by Puhlmann *et al.* (1992), who noted that different liquid chromatographic procedures for the determination of creatinine in serum have been described in recent years. The majority of these methods are associated with laborious sample preparation to separate proteins (protein precipitation, ultrafiltration, or solid-phase extraction), which require external standards or time-consuming calculations. Furthermore, calibration inaccuracies caused by matrix effects may occur in some cases. The direct injection of serum samples is possible with only a few procedures, the pre-columns used allowing only short periods of system standstill as they have not been designed for a serum protein load.

A column-switching method for determination of creatinine was described by Puhlmann *et al.* (1992). Serum is first injected directly into a 25–cm ISRP GFF precolumn. (Note: it would have been less expensive in the long run to have used a GFF-guard column-GFF 25–cm column combination, rather than the GFF 25–cm column alone. JP) Then, after elution of the bulk of the serum proteins, the creatinine-containing peak is diverted onto a C18 column, from which it is then gradient eluted. The analytical sensitivity and reliability afforded by use of the ISRP column alone (Fig. 14) are clearly improved by column-switching (Fig. 15). The column-switching method shows good long-term stability, simple sample handling without pretreatment, high selectivity, broad linearity (0.3–30 mg dl^{-1} creatinine), good reproducibility (interassay RSD <3%) and high recovery (97–100%) relative to values obtained with gas chromatography-mass spectrometry (Puhlmann *et al.*, 1992).

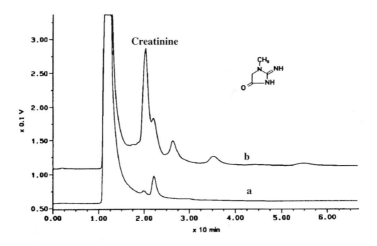

Fig. 14. Serum chromatograms from a GFF column alone, for quantification of creatinine in serum: a, normal serum; b, serum from a dialysis patient. See also Fig. 15. The structure of creatinine is shown in the diagram. (Reproduced with permission from Puhlmann *et al.*, 1992.)

7.2.2.4. Application to chiral analysis: warfarin. Column switching has been used (a singular application) to determine the distribution in serum of warfarin enantiomers (Chu and Wainer, 1988). The serum

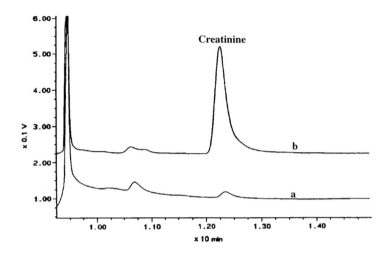

Fig. 15. Chromatograms from GFF-cleanup—ODS-analyzing
         column switching for quantification of creatinine in a)
         normal serum and b) serum from a dialysis patient. See
         also Fig. 14. (Reproduced with permission from Puhlmann *et
         al.*, 1992.)

sample was injected on an ISRP column to free the mixture of warfarin enantiomers from serum protein and from the major warfarin
metabolites. The concentration of the combined warfarin enantiomers was determined from the effluent eluted from the ISRP
column. The warfarin peak from the ISRP column was diverted to a
chiral column (bovine serum albumin immobilized on 10–μm spherical
silica). The distribution of the warfarin enantiomers was then determined from the effluent of the BSA column.

As mentioned earlier, Shibukawa combined this column-switching
technique with both chiral separations and column frontal analysis to
determine the concentrations of the free forms of the warfarin
enantiomers (Shibukawa *et al.*, 1990).

8. GFF2

8.1. Review

This section is based largely on the material in Perry (1992), a
description that was published less than three months before the
preparation of this chapter was begun. At the time of this writing, no
other reviews of the nature and applications of this material existed.

8.2. Improving GFF

8.2.1. Background: GFF deficiencies.

8.2.1.1. Limited applicability. Inadequate retention precluded the application of the original GFF to the analysis of many hydrophilic drugs. The poor retention of hydrophilic substances by GFF prevented not only the straightforward application of GFF to the analysis of these materials but also any substantial increase in the sensitivity concomitant with GFF use, for instance by the injection of larger-than-normal amounts. Correspondingly, a general increase in GFF ISRP retentivity would be expected to expand both GFF ISRP applicability and the sensitivity attainable with GFF ISRP columns.

8.2.1.2. Mediocre chromatographic efficiency. The original GFF phases always displayed column efficiencies of only 25,000 to 35,000 plates m^{-1}. GFF has a complex structure that underlies not only its highly useful selectivity but also, it has been suggested, its low efficiency. That several concurrently acting GFF retention mechanisms inevitably cause poor GFF efficiency is a suggestion worth examining, for it has also to do with the general theory of chromatography. That concurrent mechanisms prevent high efficiency has long been part of conventional wisdom in chromatography. The suggestion could be tested—and possibly disproven—only by the development of the much higher efficiencies that are to be expected in modern LC. With the original GFF, this was never possible, so the suggestion could not be gainsaid. Low GFF column efficiency, however, was not the only problem. Quality control had also been found inadequate.

8.2.1.3. Poor reproducibility. For quality control in the synthesis of the original GFF, standard multicomponent mixtures were devised. However, large and apparently mutually unrelated retention differences between successive lots of the original GFF were found. This was unacceptable. For some analyses, off-the-shelf ISRP columns must be exact replicates of one another, to be used as is, without inspection or adjustment of analytical conditions. The inability of the original GFF to satisfy this criterion was considered the most necessary to correct.

8.3. GFF-Improvement Targets

Achievement of much improved GFF reproducibility was the prime objective. The extreme reproducibility that had recently been achieved at Regis in synthesizing other monomeric stationary phases suggested that a similar approach to GFF synthesis might best improve GFF reproducibility. Accordingly, this was made the top objective in the overall ISRP research and development program.

Among subsidiary targets of the research, once achievement of the primary objective was assured, the GFF-related ISRP materials glycine-phenylalanine (GF) and glycine-phenylalanine-phenylalanine-phenylalanine-phenylalanine (GFFFF) were also synthesized and characterized, each in monomeric form.

8.4. Research Materials and Methodology

The 3-glycidyloxypropyldimethylethoxysilane used for the revised ISRP synthesis was purchased from Huls America (Piscataway, NJ, USA); the 70% perchloric acid, A.C.S. reagent grade, from Fisher Scientific Company (Fairlawn, NJ, USA).

Except for the change to be described, the synthesis of the monomeric glycine-phenylalanine-phenylalanine was performed essentially as described earlier (Pinkerton and Hagestam, 1985). In this change, the monoethoxysilane was substituted for the trimethoxysilane and bonded to silica by refluxing in heptane for 48 hours.

8.5. Results and Discussion

8.5.1. Achieving GFF monomericity

Analysis of GFF chemical synthesis suggested possible sources of variability. Several enter with the bonding of 3-glycidoxypropyltri-methoxysilane (I) to silica. With reagent I, the first methoxy-silanol reaction leaves two reactive $-OCH_3$ leaving groups still available. Reaction of either with water produces a silanol on the bound reagent and leads to a polymerization that is difficult to control. The resultant stationary phase is not reproducible, nor are its properties. However, polymerization can be completely avoided by using a reagent that is monofunctional, rather than polyfunctional.

The commercially available monofunctional 3-glycidoxypropyl-di-methylethyl-oxysilane serves this purpose. Having the same type of alkoxy leaving group as the trifunctional, it was found to have similar bonding conditions, and it was adopted. The consequent chromatographic behavior and coverage then provided a reliable basis for optimizing the various parameters associated with the overall synthesis.

8.5.2. Surface coverage

8.5.2.1. Consistent coverage. Coverage is measured by elemental analysis and is reported as the percent carbon per unit area of the silica surface. Although the coverage produced by the new mono-functional reagent was lower (4.9–6.0%) than that of the preceding trimethoxy reagent, it was much more consistent. Paradoxically, the reduced surface coverage resulted in increased analyte retention.

8.5.2.2. Efficiency. Despite its lower carbon content, GFF2 shows greater retentivity than GFF as well as greater column efficiency. This improved retention—and thus wider applicability—is strikingly evident in Figure 16, which shows the concomitant improved separa-

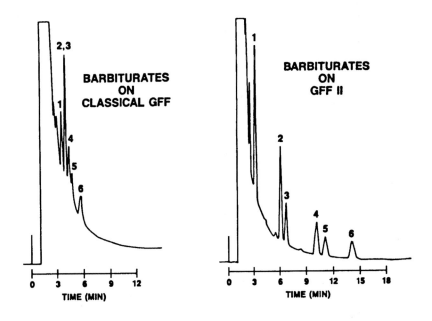

Fig. 16. Improved performance of the GFF2 phase for the separation of barbiturates. Analytes: a standard mix of six barbiturates in human serum. 1: Barbital; 2. Pheno-barbital; 3: Butabarbital; 4: Amobarbital; 5: Pento-barbital; 6: Secobarbital. Sample volume: 10 μl. Mobile phase: 5/95 methanol/0.1M KH_2PO_4 (pH 7.5) v/v, at 1.0 ml min^{-1}. Columns: 5 μm ISRP, 15 cm × 4.6 mm i.d. Detection: 240 nm. (Reproduced with permission from Perry, 1992.)

tion of six barbiturates (Perry, 1992; Rateike, 1990). Using carbamazepine as a test solute, GFF2 not only produces twice the retention of GFF but also yields 63,000 plates m^{-1} rather than the 35,000 found with GFF (Fig. 17) (Perry, 1992; Rateike, 1990).

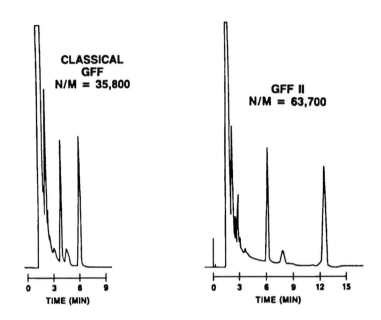

Fig. 17.   Increased column efficiencies and retention with GFF2 for the analysis of phenobarbital and carbamazepine (the second peak). Analytes: phenobarbital and carbamazepine in human serum. Sample volume: 10 µl. Mobile phase: 20/80 acetonitrile/0.1M KH_2PO_4 (pH 6.8) v/v, at 0.6 ml min^{-1}. Columns: 5 µm ISRP, 15 cm × 4.6 mm i.d. Detection: 254 nm. (Reproduced with permission from Perry, 1992.)

Possibly the improved retention and efficiency result from the constructive participation, in the monomeric GFF2, of <u>each</u> stationary phase molecule. In contrast, some of the molecules in the partially polymeric GFF on the one hand may hinder mass transfer and thus decrease chromatographic efficiency, and on the other cannot participate fully in solute retention.

8.5.3. Chain length vs. performance

The retentions of the monomeric GF, GFF2, and GF$_4$ increase in order of chain length—the number of phenylalanines—and also all exceed that of the original GFF. The chromatographic efficiency of the monomeric GF is similar to that of the monomeric GFF, *i.e.*, GFF2; but that of the GF$_4$ is mediocre, only about 40,000 plates m^{-1}.

In Fig. 18 the chromatographic efficiencies and retentions of "classical" GFF, monomeric GF, and monomeric GFF are presented for intercomparison. On the left are shown the chromatograms; on the right, the retentions, with the corresponding chromatographic efficiencies given in thousands of plates m^{-1}.

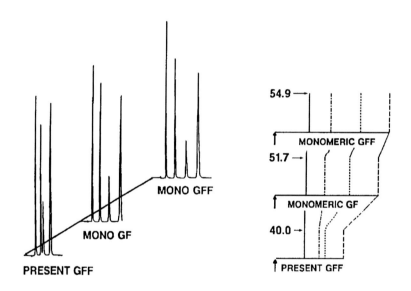

Fig. 18.   Retention, selectivity, and efficiency of "classical" GFF, monomeric GF, and monomeric GFF compared in chromatographic and diagrammatic forms.   In the diagrams, the numbers show thousands of theoretical plates m^{-1}.   Column: 15 cm × 4.6 mm i.d.   Sample volumes: 10 μl. Solutes, in order of appearance: caffeine, phenobarbital, trimethoprim, and carbamazepine. Mobile phase composition: 0.1 M KH$_2$PO$_4$/acetonitrile, 80/20 (v/v). Flow rates: 1.0 ml min^{-1}. Detection: 254 nm, 0.1 AUFS. (Reproduced with permission from Perry, 1990.)

Both the GF and GFF2 monomeric materials show greater retentions and efficiencies than the "classical" GFF. Indeed, the monomeric GF and GFF efficiencies reach 55,000 plates m^{-1}, reasonably efficient by current LC standards. The relatively low efficiency of "classical" GFF was apparently imposed by its physical form, not its multiplicity of retention mechanisms. The monomeric GFFFF shows retentions even greater than the monomeric GFF, but its efficiency is a relatively low 25,000 plates m^{-1}.

9. Restricted Access Media (RAM)

9.1. The Internal Surface Reversed Phase (ISRP)

The term Restricted Access Media was devised by Regnier (Desilets et al., 1991). The Internal Surface Reversed Phase (ISRP), invented by Pinkerton and Hagestam (1985), set a benchmark for performance of Restricted Access Media (RAM), of which it is an example. RAM packings permit chromatography of small molecules in the presence of large biomolecules such as proteins. By one means or another, RAM packings prevent access of large biomolecules to some inner surface. RAMs that function similarly to the ISRPs include the Shielded Hydrophobic Phase (SHP) (Gisch et al., 1988) and the Semi-Permeable Surface (SPS) (Glunz et al., 1992).

9.2. The Shielded Hydrophobic Phase (SHP)

The SHP consists of a hydrophilic, polymeric, water-solvated phase, covalently bonded to silica. Small analytes such as drugs can reach hydrophobic regions embedded in the SHP (Gisch et al., 1988), but proteins cannot. An acceptable column life of over 1,000 injections has been reported (Gisch et al., 1988).

10. The Semipermeable Surface (SPS)

10.1. Genesis

10.1.1. The adsorbed-surfactant RAMs of Desilets.

10.1.1.1. Direct serum injection. The concept of the SemiPermeable Surface (SPS) stems from Desilets and Regnier at Purdue University. In their work, Desilets showed that monolayers of oxyethylene-based nonionic surfactants are readily adsorbed on conventional reversed phase packings (Desilets, 1988; Desilets *et al.*, 1991). Such a packing displays RAM function. When serum is injected, the layer of hydrophilic polymer prevents serum proteins from making contact with hydrophobic sites. The history and characterization of SPSs has also been reviewed by Glunz *et al.* (1992).

Figure 19 shows a chromatogram resulting from a direct injection of human serum onto a C8 column that Desilets had coated with Brij-700 surfactant (Code: P-100-AE-18). In the chromatogram, diuretics are seen to be separated from the proteins and from each other.

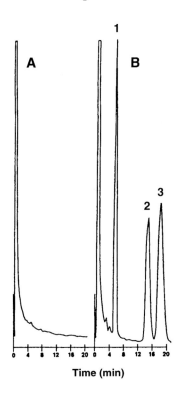

Time (min)

Fig. 19. An oxyethylene-surfactant-coated conventional C8 stationary phase functioning as a Restricted Access Medium. A, immediate excellent protein recovery from an injection of human serum; B, separation of diuretics from serum proteins and each other. 1, hydrochlorothiazide (50 µg ml⁻¹); 2, chlorthalidone (100 µg ml⁻¹); 3, furosemide (100 µg ml⁻¹). Column dimensions: 5 cm × 4.6. mm i.d. Mobile phase: 97/3 0.05 M phosphate buffer, pH 6.5/isopropanol. Flow rate: 1.0 ml min⁻¹. Sample volume: 25 µl. Detection: 254 nm, 0.04 a.u.f.s. (Reproduced with permission from Glunz *et al.*, 1992.)

10.1.1.2. Protein recovery. Protein recoveries from that material were excellent: 97 ± 3% as determined by the method of Bradford (Bradford, 1976). Similar results were obtained using several other detergents on C8 and C18 packings. Although there was some loss of chromatographic efficiency with these coated columns, the major disadvantage was the leaching of the detergent layer from the column.

10.2. Commercial Development

10.2.1. Hydrophilic polymers, chemically bound to silica

Proceeding from the work of Desilets, methods were developed at Regis Chemical Company for bonding hydrophilic polymers to the silica surface. With this bonding, the polymer forms an outer semipermeable surface, at the same time leaving an inner layer for the chromatography of smaller molecules.

10.2.2. Experimental

10.2.2.1. SPS packings, and SPS and SHP columns. SPS packings and columns were manufactured at Regis. The SHP column was purchased from Supelco (Bellefonte, PA, USA).

10.2.2.2. Chromatography. Mobile phases (pH 6.8) were prepared with HPLC-grade water, KH_2PO_4 (0.1 M) and acetonitrile (EM Science, Cherry Hill, NJ, USA). Carbamazepine and phenobarbital were purchased from Sigma (St. Louis, MO, USA).

Chromatographic experiments were carried out with standard commercial equipment. The Kratos (Ramsey, NJ, USA) system used included a Model 783G detector (flow-cell, 12 μl internal volume) set at 254 nm and a Kratos Spectroflow 400 pump set to produce 0.6 ml min^{-1}. The internal diameter of the capillary tubing throughout the system was 0.01 in. Samples, all 10 μl in volume, were injected with a Model 7010 injection valve (Rheodyne, Calabasas, CA, USA).

10.3. Nature

As shown in Fig. 20, SemiPermeable Surface (SPS) media consist of particles that bear two phases: an outer semipermeable surface that is also a hydrophilic phase, and an inner hydrophobic phase. The outer semipermeable surface prevents large molecules such as proteins from reaching the inner phase. Small molecules interact with the outer phase/surface, particularly if they are hydrophilic; small molecules also can and do penetrate the outer surface, reach the inner phase, and interact with it, particularly if they are hydrophobic. The chemical nature of both the outer SPS phase/ surface and the SPS inner phase can be varied independently.

HYDROPHILIC   HYDROPHOBIC
OUTER PHASE/SURFACE   INNER PHASE
$[-O-CH_2-CH_2-O-]$ $[CH_2-CH_2-CH_2]$

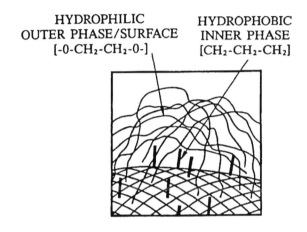

Fig. 20.   SemiPermeable Surface (SPS) media. (Reproduced with permission from Glunz et al., 1992.)

10.4. Characteristics

10.4.1. Versatility

A major difference of SPS from other restricted access packings is its inherent versatility. With the SPS, the outer and inner phases are synthesized independently, and thus can be controlled individually. The density of the outer phase/surface can be varied to change the penetration threshold; the inner phase can be varied from that of the particle surface itself to any reversed phase commonly used in liquid chromatography. Thus, for a given separation with SPS media, the inner or outer phases can be optimized independently.

10.4.2. Retention mechanisms

Another distinguishing characteristic of SPS is its unique retention mechanism. As Desilets and Regnier first suggested (Desilets, 1988; Desilets et al. (1991) and confirmed later by Wang et al., 1992), small molecules appear to be retained by a combination of hydrogen bonding at the polyoxyethylene phase/surface and hydrophobic interaction at the inner reversed phase. Depending on its hydrophobicity and hydrogen-bonding capability, a small molecule can penetrate into and partition with different regions of the hydrophilic layer and the reversed phase below.

10.4.2.1. Inner surfaces: hydrophobic interaction. Table 3 presents retention times found with SPS C1, C4, C8, and C18 inner surfaces for a group of aromatic hydrocarbons. As might be expected, retentions increase from C1 through C18. SPS silica packings have also been made with the hydrophilic polymer used for the commercial SPS for the outer surface, but only the bare support for the inner surface. This makes possible with SPS the use of silica for normal-phase separation of smaller molecules in the presence of globular biopolymers such as proteins.

Table 3. Aromatic Hydrocarbons on 15 cm SPS Columns

	C1	C4	C8	C18
		Retention Time (min)*		
Benzene 2.51	2.60	2.93	3.24	
Naphthalene	3.56	3.94	4.76	6.32
Phenanthrene	5.56	6.71	8.8	14.38
Chrysene	9.45	12.24	17.98	37.76
		k'		
Benzene	0.6	0.7	1.0	1.3
Naphthalene	1.2	1.6	2.2	3.5
Phenanthrene	2.5	3.5	4.9	9.3
Chrysene	4.9	7.2	11.0	26.0

*Mobile phase, at 1.0 ml min^{-1}: 75/25 methanol/water. Void time is time from injection to injection-caused deflection from base line.

10.4.2.2. Outer surfaces: hydrogen bonding. Hydrophilic solutes tend neither to penetrate deeply into the reversed phase nor therefore to distinguish between the successively longer paraffinic chains of the C1, C4, C8, and C18 reversed phases. The retentions shown in Table 4 differ among the hydrophilic solutes, but for a given hydrophilic solute, the retentions do not differ from one SPS alkyl reversed phase to the next. As mentioned, both patterns agree with earlier comments of Desilets and Regnier concerning the overall mechanism (Desilets, 1988; Desilets *et al.*, 1991).

Table 4. Selected Compounds on 15 cm SPS Columns

	C1	C4	C8	C18
		Retention Time (min)		
Tartaric acid	2.40	2.50	2.29	1.93
Citric acid	3.19	3.61	3.83	2.73
Barbital	2.11	2.08	2.03	1.89
Secobarbital	3.66	4.33	4.56	4.65
Methyl paraben	2.57	2.65	2.67	2.83
Propyl paraben	3.18	3.56	3.91	4.51

In Table 5, the retention times of seven hydrophilic drugs on two conventional C8 and C18 columns and four SPS phases (C8, C18, CN, and phenyl) are compared. In general, the conventional alkane phases retain the drugs longer than the SPS phases. Notice too that the SPS C18 does not retain either sulfinpyrazone or carbamazepine any longer than the SPS C8. Table 6 shows SPS column reproducibility and efficiency.

Table 5. Comparison of Retention of Selected Drugs
on Conventional SPS Phases

	Conventional		SPS 5PM			
	C8	C18	C8	C18	CN	Phenyl
Toluic acid	2.6	3.0	2.2	1.8	2.3	2.2
Caffeine	2.6	2.4	1.8	1.6	1.8	1.8
Trimethoprim	4.9	4.6	3.9	3.2	4.2	3.9
Phenobarbital	11.7	10.9	10.0	9.9	10.1	9.8
Sulfinpyrazone	26.0	27.7	15.1	13.4	15.4	15.2
Carbamazepine	36.7	41.8	15.1	13.6	15.2	14.9
Methyl salicylate	62.6	92.9	29.4	32.6	20.5	20.9

Table 6. Lot-to-lot reproducibility, SPS C8

| | Retention Time (min) | | | |
| | Lots | | | |
	A	B	C	D
Toluic acid	2.2	2.1	2.2	2.2
Caffeine	1.8	1.9	1.8	1.7
Trimethoprim	3.9	5.0	3.9	3.8
Phenobarbital	10.0	10.4	10.8	10.6
Sulfinpyrazone	15.1	14.2	16.3	15.7
Carbamazepine	15.1	18.3	16.6	15.9
Methyl salicylate	29.4	33.2	35.7	36.5
	Plates (10^3 m^{-1})			
	61.1	67.8	66.6	61.3

Column dimensions: 15 cm $\times$ 4.6 mm i.d.; mobile phase flow rate: 1.0 ml min^{-1}; mobile phase composition: 80/20 0.1 M KH_2PO_4, pH 6.8/ CH_3CN; detection, 254 nm; sample volume: 10 μl. Column efficiency measured separately on dibutyl phthalate peak, using a 10 μl volume of a solution containing 75 mg dibutyl phthalate per 100 ml of mobile phase; the mobile phase was methanol/water 70/30, used at 0.5 ml min^{-1}; the dibutyl phthalate capacity factor was 3.7.

A characteristic of SPS columns that follows from this retention mechanism is the excellent peak shapes obtained for basic drugs. In Fig. 21, for instance, excellent peak shape is obtained for the basic drug sulfapyridine on the SPS-5PM-C8 SPS column.

10.5. Direct injection

10.5.1. Serum

A sample of human serum containing phenobarbital and carbamazepine was injected into a 15–cm SPS C8 column. The packing pore diameter was 100 Å; the particle diameter, 5 μm. In the chromatogram shown in Fig. 22, the efficiency measured on the carbamazepine peak was over 60,000 plates m^{-1}, which is about equal to that of the improved GFF, GFF2 (Perry, 1992). If not outstanding,

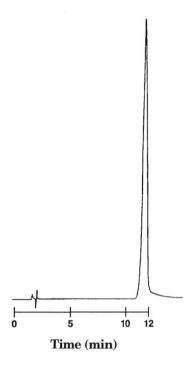

Time (min)

Fig. 21. The hydrophilic SPS outer phase is well suited for the chromatography of basic drugs, as demonstrated here by a
chromatogram of sulfapyridine. Column: 15 cm × 4.6 mm i.d.
SPS-5PM-S5-100-C8. Mobile phase composition: 95/5 0.1 M
KH_2PO_4 buffer, pH 6.8/acetonitrile; and flow rate: 1.0 ml
min^{-1}. (Reproduced with permission from Glunz *et al.*, 1992.)

such efficiency is nevertheless acceptable (a Regis Rexchrom 15–cm
C18 column provides 85,000 plates m^{-1}). Note also the carbamazepine retention: approximately 27 min (in comparison with 6.0/12.5
min with GFF/GFF2, respectively—see Fig. 17; and 15 min with the
SHP). Over 400 serum injections were made on this column without
any increase in pressure or change in performance.

Figures 23 and 24 show other chromatograms of drugs in human
serum at therapeutic levels. In Fig. 23, theophylline is shown well
separated from serum proteins, as well as from caffeine and acetaminophen. In Fig. 24, five barbiturates are shown well separated
from each other; the early-eluting barbital is adequately separated
from the serum proteins for analysis in plasma (compare with Fig. 16
for barbiturate separations over GFF/GFF2).

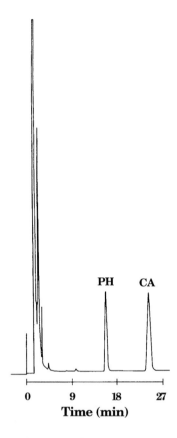

Fig. 22.   Phenobarbital and carbamazepine are separated from each other and from serum proteins
by injection onto an SPS C8 column. Column
dimensions: 15 cm × 4.6 mm i.d. Packing: 100
A pore diameter, 5 μm particle diameter.
Mobile phase: 80/20 phosphate buffer, pH
6.8/acetonitrile; at 0.6 ml min^{-1}. Sample
volume: 10 μl. Concentrations in μg/μl: phenobarbital, 160; carbamazepine, 60. (Reproduced
with permission from Glunz et al., 1992.)

He et al. (1992) have used an SPS ODS-inner-phase column for the
direct injection analysis of the anticonvulsants carbamazepine (CBZ)
and its active 10,11-epoxide metabolite (EPO) in plasma. The two
were separated from each other and from commonly coadministered
drugs such as phenobarbital (PB) and phenytoin. The sample volume, 5 μl, was large enough to afford satisfactory sensitivity, yet not

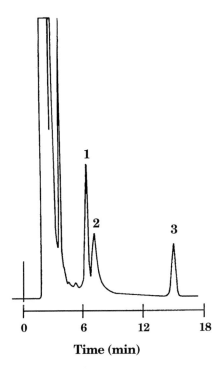

Time (min)

Fig. 23. The 15 cm SPS C8 separates theophylline (1) from serum very cleanly, and could be used to separate it from caffeine (2) and acetaminophen (3) as well. Column: 15 cm × 4.6 mm i.d. SPS-5PM-S5-100-C8. Mobile phase: 0.01 M KH_2PO_4 buffer, pH 7.0, at 0.5 ml min^{-1}. Sample volume: 10 µl. Concentrations in µg ml^{-1}: theophylline, 15; caffeine, 15; acetaminophen, 10. Detection: 254 nm. (Reproduced with permission from Glunz et al., 1992.)

so large as to bring about peak splitting from involvement of protein-binding equilibria. Having 20% acetonitrile in the mobile phase did not precipitate plasma protein, but did effect immediate release from the protein of the drugs CBZ and EPO and also quick elution of CBZ, EPO, PB from the column. Chromatograms reflecting optimized conditions are shown in Fig. 25 (He et al., 1992).

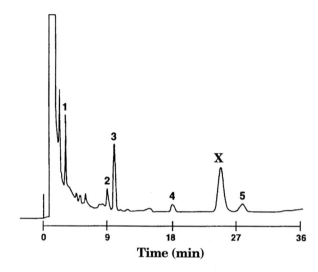

Fig. 24. Barbiturates at therapeutic levels in serum are resolved from
each other and from serum proteins on an SPS C8 column.
Column: 15 cm × 4.6 mm i.d. SPS-5PM-S5-100-C8. Mobile
phase: 0.1 M KH_2PO_4 buffer, pH 6.8, at 1.0 ml min^{-1}. Sample
volume: 10 µl. Concentrations: 15 µg ml^{-1}. Detection: 254
nm. 1, Barbital; 2, Butabarbital; 3, Phenobarbital; 4, Amobarbital; 5, Secobarbital. The substance causing peak "X"
comes from the particular lot of serum used and is not
present in all lots. (Reproduced with permission from Glunz
et al., 1992.)

10.5.2. Natural materials—a new capability

Materials of natural origin present difficulties to chromatographers. Each such material possesses far too many components to
be resolved by one method applied to the sample in one injection. But
pretreatment to render the sample less complex may alter the
original composition of the sample, giving rise to artifactual results.
Thus, successful chromatographic examination of natural materials
requires both the highest possible resolution, at some given
selectivity, and direct injection. SPS columns offer potential in this
area.

SPS columns can receive a wide variety of natural materials by
direct injection, and show a surprising resolution of the many components in each. The examples shown here—human urine, red wine,
grape juice, and caffeine-containing coffee (caffeine identified by arrow)
(Figs. 26–29)—merely illustrate the many types of materials that may
successfully be injected and resolved.

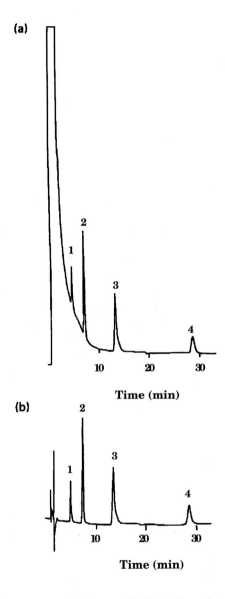

Fig. 25. Carbamazepine 10,11-epoxide [1], phenobarbital [2], carbamazepine [3], and phenytoin [4], are shown separated from
serum (a) and from each other (b) on an SPS C18 column.
Column dimensions: 15 cm × 4.6 mm i.d.; mobile phase,
phosphate buffer (pH 7.1, ionic strength 0.1)-acetonitrile (4:1,
v/v); flow rate, 1.0 ml min^{-1} (but 1.5 ml min^{-1} after 17 min.);
UV 214 nm; injection volume, 5 μl; column temperature,
30°C. (Reproduced with permission from He *et al.*, 1992.)

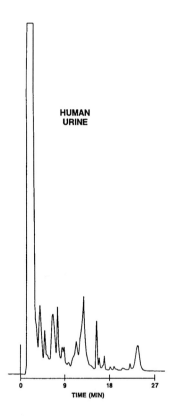

Fig. 26. Human urine. Column: 15 cm × 4.6 mm i.d.
SPS-5PM-S5-100-C8. Mobile phase: A: H_2O,
pH 2.5; B: acetonitrile. Gradient: 99% A to
50% A in 30 min. Flow rate: 1.0 ml min^{-1}.
Sample volume: 50 µl. Detection: 214 nm.
(Reproduced with permission from Glunz *et
al.*, 1992.)

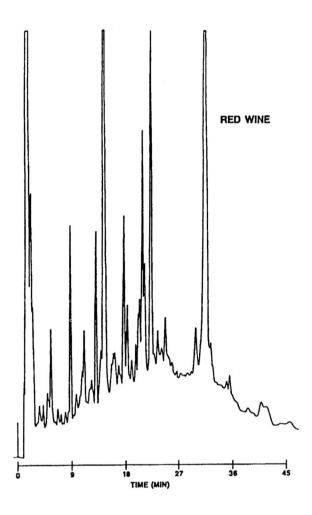

RED WINE

TIME (MIN)

Fig. 27.   Red wine.   Column: 15 cm × 4.6 mm i.d.
          SPS-5PM-S5-100-C8. Mobile phase: A: H_2O,
          pH 2.5; B: acetonitrile. Gradient: 99% A to
          50% A in 30 min. Flow rate: 1.0 ml min^{-1}.
          Sample volume: 50 µl. Detection: 214 nm.
          (Reproduced with permission from Glunz *et
          al.*, 1992.)

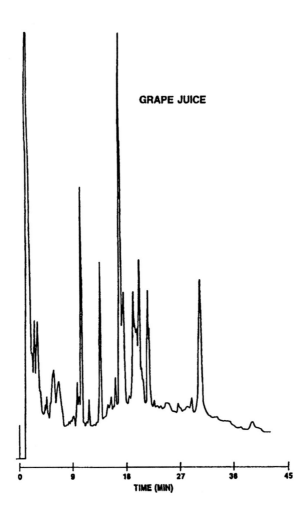

Fig. 28. Grape juice. Column: 15 cm × 4.6 mm i.d.
SPS-5PM-S5-100-C8. Mobile phase: A: H_2O,
pH 2.5; B: acetonitrile. Gradient: 99% A to
50% A in 30 min. Flow rate: 1.0 ml min^{-1}.
Sample volume: 50 μl. Detection: 214 nm.
(Reproduced with permission from Glunz *et
al.*, 1992.)

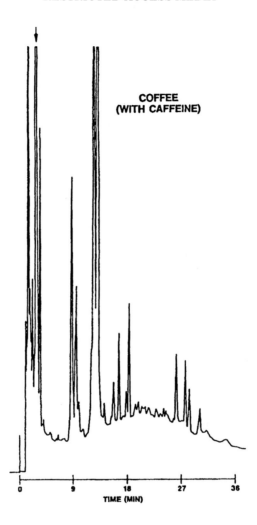

Fig. 29. Coffee with caffeine. Column: 15 cm × 4.6 mm i.d.
SPS-5PM-S5-100-C8. Mobile phase: A: H_2O, pH
2.5; B: acetonitrile. Gradient: 99% A to 50% A in
30 min. Flow rate: 1.0 ml min^{-1}. Sample volume:
50 μl. Detection: 214 nm. (Reproduced with permission from Glunz *et al.*, 1992.)

Acknowledgments

The author acknowledges with sincere gratitude the manifold assistance of Mrs. Carolyn Gillespie of the Regis Chemical Company, in preparing this manuscript; and also the immediate, generous, and effective response of Dr. Thomas C. Pinkerton of the Upjohn Company, in providing the author with his ongoing compilation of GFF literature citations.

The research programs that led to both the GFF2 and the SPS were funded by Small Business Innovation Research grant GM36215-03, from the National Institute of General Medical Sciences.

The author extends his thanks to Dr. Steven Wong, editor of the *Journal of Liquid Chromatography*, and to its publisher, Marcel Dekker, Inc., for allowing the reproduction of much of the three papers on which this chapter was based.

11. References

Andereg, J. W. (1955). *J. Am. Chem. Soc.*, 77, 2927.

Arvidsson, T., K. G. Wahlund and N. Daoud (1984). *J. Chromatogr.*, *317*, 213–226.

Bethel, G. S., J. S. Ayerer, W. S. Hancock and M. T. Hearn (1979). *J. Biol. Chem.*, *254*, 2572–2574.

Bradford, M. M. (1976) *Anal. Biochem.*, *72*, 248-254.

Chu, Y-Q. and I. W. Wainer (1988). *Pharm. Res.*, *5*, 680–683.

DeJong, G. J. (1980). *J. Chromatogr.*, *183*, 203–211.

Desilets, Carla, Thesis, Purdue University, West Lafayette, IN 43707.

Desilets, C. P., M. A. Rounds and F. E. Regnier, (1991). *J. Chromatogr.*, *554*, 25–39.

Gisch, D. J., B. T. Hunter and B. Feibush (1988). *J. Chromatogr.*, *433*, 264–268.

Glunz, L. J., J. A. Perry, B. Invergo, H. Wagner, T. J. Szczerba, J. D. Rateike and P. W. Glunz (1992). *J. Liq. Chromatogr.*, *15*, 1361–1379.

Hagestam, I. H. and T. C. Pinkerton (1985). *Anal. Chem.*, *57*, 1757–1763.

Haginaka, J. (1990). *J. Chromatogr.* *529*, 455–461.

He, J., A. Shibukawa and T. Nakagawa (1992). *J. Pharm. Biomed. Anal.*, *10*, 289–294.

Hearn, M. T. W. (1982). *Adv. Chromatogr.*, *20*, 4–82.

Hofman, K. and M. Bergman (1940). *J. Biol. Chem.*, *134*, 225.

Hux, R. A., H. Y. Mohammed and F. F. Cantwell (1982). *Anal. Chem.*, *54*, 113–117.

Juergens, U. (1984). *J. Chromatogr.*, *310*, 97–106.

Lassen, P. O., M. Glad, L. Hasson, M. O. Mannsson, S. Ohlson and U. Mosbach (1983). *Adv. Chromatogr.*, *21*, 41–85.

Matlin, S. A., C. Thomas and P. M. Vince (1990). *J. Liq. Chromatogr.*, *13*, 2253–2260.

Meriluoto, J. A. O. and J. E. Erikson (1988). *J. Chromatogr.*, *438*, 93–99.

Miller, T. D. Ph.D. Thesis, Department of Chemistry, Purdue University, West Lafayette, Indiana, August 1987.

Nakagawa, T., A. Shibukawa, N. Shimono, T. Kawashima and H. Tanaka (1987). *J. Chromatogr.*, *420*, 297–311.

Nazareth, A., L. Jaramillo, B. L. Karger, R. W. Giese and L. R. Snyder (1984). *J. Chromatogr.*, *309*, 357–368.

Perry, J. A. (1990). *J. Liq. Chromatogr.*, *13*, 1047–1074.

Perry, J. A. (1992). *J. Liq. Chromatogr.*, *15*, 3343–3352.

Perry, J. A. M. and Chang, Pinkerton Application Note No. 22, July 14, 1987. Regis Chemical Company, Morton Grove, Illinois.

Perry, J. A. and J. D. Rateike (1990). *J. Chromatogr.*, *503*, 403–409.

Perry, J. A. and M. Chang (1990). *J. Pharm. Sci.*, *79*, 437–439.

Perry, J. A., J. D. Rateike and T. J. Szczerba (1987). *J. Chromatogr.*, *389*, 57–64.

Perry, J. A., J. D. Rateike and T. J. Szczerba, Pinkerton Application Note No. 24, July 23, 1987. Regis Chemical Company, Morton Grove, Illinois.

Perry, J. A. and J. D. Rateike, Pinkerton Application Note No. 4, May 27, 1986. Regis Chemical Company, Morton Grove, Illinois.

Pinkerton, T. C. (1988). *Amer. Lab.*, *20*, 70–76.

Pinkerton, T. C. Pinkerton Application Note No. 32, June 15, 1989. Regis Chemical Company, Morton Grove, Illinois.

Pinkerton, T. C. (1991). *J. Chromatogr.*, *544*, 13–23.

Pinkerton, T. C. and I. H. Hagestam (1985). U.S. Patent 4,544,485.

Pinkerton, T. C. and K. A. Koeplinger (1988). *J. Chromatogr.*, *458*, 129–145.

Pinkerton, T. C. and K. A. Koeplinger (1990). *Anal. Chem.*, *62*, 2114–2122.

Pinkerton, T. C. and H. I. Hagestam (1984). Invention Record and Disclosure, Purdue Research Foundation, May 10.

Pinkerton, T. C., T. D. Miller, S. E. Cook, J. A. Perry, J. R. Rateike and T. Szczerba (1986). *J. BioChromatogr.*, *1*, 96–105.

Pinkerton, T. C., S. E. Cook, C. P., Desilets and T. D., Miller, Plenary Lecture, 30th Ann. Symp. on Liq. Chromatog., Kyoto, Japan, Jan. 27–28, 1987.

Pinkerton, T. C., T. D. Miller and L. J. Janis (1989). *Anal. Chem.*, *61*, 1171–1174.

Pompon, A., I. Lefebvre and J–L. Imbach (1992). *Biochem. Pharmacol.*, *43*, 1769–1775.

Puhlmann, A., T. Duelffer and U. Kobold (1992). *J. Chromatogr., 581*, 129–133.

Rateike, J. D. Pinkerton Application Note No, 34, March 26, 1990. Regis Chemical Company, Morton Grove, Illinois.

Rateike, J. D. Pinkerton Application Note No. 21, June 30, 1987. Regis Chemical Company, Morton Grove, Illinois.

Roth, W. (1983). *J. Chromatogr., 278*, 347–357.

Roth, W., K. Beschke, R. Jauch, A., Zimmer and F. W. Koss (1981). *J. Chromatogr., 222*, 13–22.

Sams, R. A. Personal communication 1987.

Sams, R. A. and L. L. Evec, in Pinkerton Application Note No. 25, Aug. 31, 1987. Regis Chemical Company, Morton Grove, Illinois.

Schmidt, D. E., Jr., R. W. Giese, D. Conron and B. L. Karger (1980). *Anal. Chem., 52*, 177–182.

Shibukawa, A., T. Nakagawa, H. Tanaka and J. Haginaka. 8th Conference of Liquid Chromatography, Tokyo, Japan, Oct. 27–29, 1987 (abstract).

Shibukawa, A., T. Nakagawa, M. Miyake and H. Tanaka (1988). *Chem. Pharm. Bull., 36*, 1930–1933.

Shibukawa, A., T. Nakagawa, N. Nishimura, M. Miyake and H. Tanaka (1989). *Chem. Pharm. Bull., 37*, 702–706.

Shibukawa, A., T. Nakagawa, M. Miyake, N. Nishimura and H. Tanaka (1989). *Chem. Pharm. Bull., 37*, 1311–1315.

Shibukawa, A., M. Nagao, Y. Kuroda and T. Nakagawa (1990). *Anal. Chem., 62*, 712–716.

Snyder, L. R. and J. J. Kirkland (1979). *Introduction to Modern Liquid Chromatography*, 2nd ed.; Wiley, New York, Chap. 17.

Szczerba, T. J. Pinkerton Application Note No. 10, June 5, 1986. Regis Chemical Company, Morton Grove, Illinois.

Szczerba, T. J. and J. A. Perry, Pinkerton Application Note No. 14, August 23, 1986. Regis Chemical Company, Morton Grove, Illinois.

Szczerba, T. J. and J. A. Perry, Pinkerton Application Note No. 31, May 1, 1989. Regis Chemical Company, Morton Grove, Illinois.

Unger, K. K., J. N. Kinkel, B. Anspach and B. Gieshe (1984). *J. Chromatogr., 296*, 3–14.

Unger, K. K. (1979). *Porous Silica: Its Properties and Use as Support in Column Liquid Chromatography*; Elsevier; Amsterdam, p. 17.

Wang, H., C. Desilets and F. Regnier (1992). *Anal. Chem., 64*, 2821–2825.

Westerlund, D. (1987). *Chromatographia, 24*, 155.

Yau, W. W., J. J. Kirkland and D. D. Bly (1979). *Modern Size-Exclusion Liquid Chromatography*; Wiley-Interscience; New York, Chapter 2.

CHAPTER 6

On-Line Microdialysis Sampling

CHRISTOPHER M. RILEY, JOSEPH M. AULT, JR.
and CRAIG E. LUNTE

*Center for BioAnalytical Research and Department of Pharmaceutical
Chemistry, University of Kansas, Lawrence, KS 66045, U.S.A.*

1. Introduction

Microdialysis sampling is the use of a hollow dialysis fiber 180–850 μm in diameter to dialyze a biological matrix or tissue. The fiber is slowly perfused with a sampling medium, typically chosen to closely match the extracellular fluid of the tissues being sampled. The driving force in microdialysis sampling is diffusion of a compound down a concentration gradient through a membrane into the perfusate. Recovery depends on the perfusate flow rate, the diffusion characteristics of the analyte, the nature of the matrix, and the properties and dimensions of the dialysis membrane (Westerink, 1992).

The microdialysis sampling technique is suitable for a wide range of matrices. While the initial use of microdialysis was in studies of neurotransmitters in the brain, the technique has been applied to virtually every major organ of the body, including the liver, heart, lungs, and muscles (Scott *et al.*, 1989; Scott and Lunte, 1993; Lonnroth *et al.*, 1987a,b; Lonnroth *et al.*, 1991; Larsson, 1991; Deleu *et al.*, 1991). The nature of the tissue in which microdialysis probes have been implanted varies from fluid systems such as blood and cerebrospinal fluid to solid tissues such as muscle and skin. Several reviews on the use of microdialysis in general and brain neurotransmitter studies in particular have been published (Ungerstedt, 1991; Westerink, 1992; Westerink *et al.*, 1987a; Kendrick, 1989; Benveniste, 1989; Lonnroth and Smith, 1990; Benveniste and Hüttemeier, 1990).

Microdialysis sampling has several advantages compared with other methods of sampling biological fluids and tissues. In particular, microdialysis may be used in awake, freely moving animals and does not remove fluid from or add fluid to the tissues being sampled. Dialysis samples are free from protein and typically require no clean-up prior to analysis. Once the analyte crosses the dialysis membrane enzymatic degradation is stopped because the analyte is separated from proteins and other high molecular weight components (Westerink, 1992; Lunte *et al.*, 1991). Furthermore, the microdialysis probe may also be used to deliver compounds to a specific site in the biological system.

2. Quantitative Considerations

2.1. Analyte Recovery

The relative recovery (R) of a compound by microdialysis is defined by:

$$R = \frac{C_{p,f}}{C_{e,f}} \tag{1}$$

where $C_{p,f}$ is the concentration in the perfused sample and $C_{e,f}$ is the concentration in the external solution (Wages *et al.*, 1986) (Fig. 1).

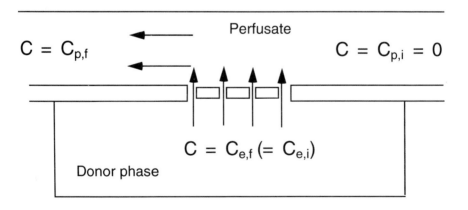

Fig. 1.   Illustration of microdialysis recovery with respect to drug concentrations in the donor phase and perfusate. $C_{p,i}$, initial drug concentration in the perfusate; $C_{p,f}$, final drug concentration in the perfusate; $C_{e,i}$, initial drug concentration in the donor; final $C_{e,f}$, drug concentration in the donor.

Thus, it follows that the measured concentration of the analyte in the microdialysis sample is always less than the actual extracellular concentration of the compound in the sampled tissue. The relative recovery is high at low flow rates, approaching 100% as the flow rate approaches zero (Fig. 2) (equilibrium dialysis) (Wages *et al.*, 1986). As the flow rate increases the relative recovery decreases resulting in a decrease in the concentration of a sample for a given volume.

The absolute recovery (A) (Eq. 2) is the mass of analyte in the perfusion medium per unit time (Wages *et al.*, 1986). This is the product of the concentration of the analyte in the external solution (C_e), the perfusion flow rate (F_v), and the relative recovery (R):

$$A = C_e \cdot F_v \cdot R \tag{2}$$

The absolute recovery may be used as an indication of the perturbation of the system undergoing sampling. The higher the absolute recovery, the more material removed per unit time and the greater the perturbation of the system. Thus, at slow perfusion rates less material is removed from the tissue increasing the reliability of the physiological experiment (Wages *et al.*, 1986). The absolute recovery is zero at zero flow and increases with flow rate. When higher flow rates are attained, the absolute recovery reaches a plateau (Fig. 2) (Wages *et al.*, 1986). In some cases, the absolute recovery may decrease with increasing flow rate due to ultrafiltration. In this situation, drug that diffuses into the membrane is forced back out due to higher pressures inside the membrane at faster perfusate flow rates. From an analytical perspective, the relative recovery is more important than the absolute recovery because it has a direct effect on the required sensitivity of the analytical method.

2.2. *Analyte Delivery*

Delivery experiments, also known as "retrodialysis" or inverse recovery experiments, are performed in the same manner as recovery experiments except that the compound of interest or an internal standard is added to the perfusion medium. Substances have been added to the perfusion medium to calibrate recovery, to prevent chemical degradation during sampling and to elicit a pharmacological response (Wong *et al.*, 1992; Wang *et al.*, 1993; Damsma *et al.*, 1987a; Westerink, 1986; Lonnroth *et al.*, 1987a,b; Menacherry *et al.*, 1992; Scott and Lunte, 1993). The relative loss or delivery of a compound (D) by microdialysis is given by Eq. 3:

$$D = \frac{C_{p,i} - C_{p,f}}{C_{p,i}} \tag{3}$$

where Cp,f is the final and Cp,i is the initial concentration of analyte in the perfused sample (Scheller and Kolb, 1991) (Fig. 3).

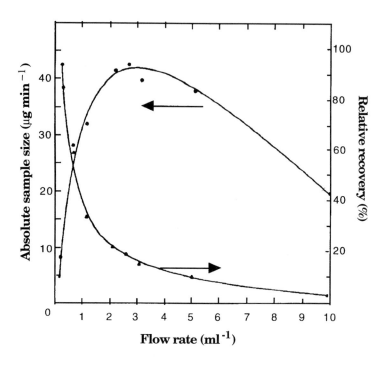

Fig. 2. Amount of sample collected per minute as a function of perfusion flow rate in terms of relative (relative recovery) and absolute (absolute sample size) recovery calculated using Equations 2 and 3, respectively. Reprinted with permission from Wages *et al.*, (1986).

2.3. *Calibration*

In order to calculate the extracellular concentration of a compound from the concentration determined in the dialysates it is necessary to know the relative recovery. The recovery is a function of the dialysis probe, the perfusion rate, the analyte, and the sample matrix. Dependence on the sample matrix means that recoveries determined *in vitro* are typically not valid *in vivo*. Accurate calibration for sampling *in vivo* requires either that the biological matrix be simulated accurately *in vitro* or that calibration be performed *in vivo*. Several approaches to calibration *in vivo* have been reported.

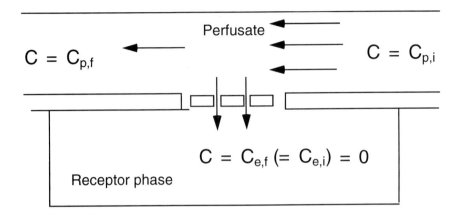

Fig. 3. Illustration of microdialysis delivery with respect to drug concentrations in the receptor phase and perfusate. initial$C_{p,i}$, drug concentration in the perfusate; $C_{p,f}$, final drug concentration in the perfusate; $C_{e,i}$, initial drug concentration in the donor; $C_{e,f}$, final drug concentration in the donor.

2.3.1. Extrapolation to zero flow

One method of calibrating a microdialysis probe *in vivo* is to vary the perfusate flow rate (Jacobson *et al.*, 1985). The change in the analyte concentration is then plotted as a function of perfusate flow rate with the actual sample concentration determined by extrapolating to zero flow. While determining probe recovery by adjusting perfusate flow rate is relatively easy, this method suffers from the need for the maintenance of steady state concentrations of the analyte in the tissue during the calibration period.

2.3.2. External calibration

Monitoring the concentration of a second analyte such as tritiated water or antipyrine, which evenly distributes throughout the interstitial fluid, both directly and by microdialysis, is another means of calibrating probe recovery *in vivo* (Larsson, 1991; Yokel *et al.*, 1992). Although this method does not require maintenance of steady-state tissue concentrations, it does necessitate even distribution of the compound throughout the body, direct sampling for calibration, and additional sample analysis.

2.3.3. Zero-net flux

The zero-net-flux method uses delivery of the analyte to determine the recovery of steady state levels of a compound *in vivo* (Lonnroth *et al.*, 1987a,b). The compound of interest is added to the perfusate at sequential concentrations greater than and less than those anticipated in the tissue. The difference in the amount of an analyte in the perfusion medium before and after dialysis is plotted against the initial concentration in the perfusate. The point at which no drug is lost or gained from the probe corresponds to the tissue concentration *in vivo* (Fig. 4). Although the zero-net-flux method uses delivery of the analyte of interest, a major disadvantage is that it requires the maintenance of steady state drug levels by peripheral infusion of the analyte over extended periods of time, limiting the practicality of the technique (Olson and Justice, 1993).

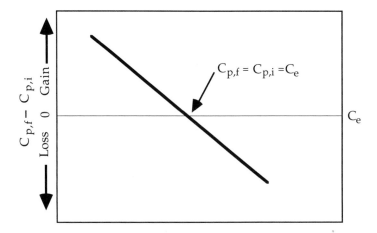

Fig. 4. Plot of the difference in analyte concentration in the perfusion medium before ($C_{p,i}$) and after ($C_{p,f}$) dialysis vs. the concentration initially in the medium ($C_{p,i}$) used to determine probe recovery *in vivo* by the zero-net-flux method.

2.3.4. Retrodialysis

Another method of probe calibration, retrodialysis, uses an internal standard included in the perfusion medium to determine probe recovery *in vivo* (Larsson, 1991; Wong *et al.*, 1992; Wang *et al*, 1993). Recovery and delivery of both the standard and the analyte are assumed to be equivalent, with the delivery of the internal standard used to establish the recovery of the analyte. The internal standard is assumed to behave like the parent molecule with regard to both physical properties (*i.e.*, diffusion characteristics, method of analysis) and biological behavior (*i.e.* metabolism, protein binding, receptor uptake and release, etc.). Thus, results can be confounded if the internal standard alters the biological activity or kinetics of the parent molecule or if the microdialysis characteristics of the analyte and the internal standard differ significantly (Ungerstedt, 1991).

2.4. Sampling interval

The sampling interval, dialysate flow rate and limit of detection of the analytical method are critical, interrelated parameters in microdialysis. Normally, the sampling interval in a kinetic or temporal study is determined by the concentration-time profile of the compound of interest; that is, the sample interval should be short enough to generate at least four data points per half-life. The choice of the sampling interval in microdialysis takes this requirement into account, but the flow rate and the limit of quantitation must also be considered. Wages *et al.* (1986) chose to classify microdialysis samples into four categories based on sampling frequency and analyte concentration, which are summarized in Table 1. For on-line analysis the speed of the analysis step will often determine the temporal resolution that can be achieved because the samples are analyzed in a serial fashion.

For on-line analysis, an LC injection valve fitted to a suitable actuator is usually used to transfer the sample to the analytical system. The rate of switching of the valve must be sufficiently rapid to minimize back pressure from the dialysate flow which would otherwise cause the perfusate to cross the membrane into the sample matrix (Linhares and Kissinger, 1992). Although the LC injection loop may be considered a sample collection loop, the concepts of sample overfilling and the length of the injection period cannot be ignored due to their potential impact on the precision and accuracy of the analytical system (Dolan and Snyder, 1989).

3. Instrumentation

3.1. Microdialysis Configurations

The basic microdialysis system consists of a probe, connective tubing, perfusate and a perfusion pump (Fig. 5). Syringe pumps are more accurate than piston-driven or peristaltic pumps. An on-line injection valve can also be added for chromatographic or flow injection analysis. For experiments with small animals, a liquid swivel allows free movement of the animal without breaking the liquid connection.

Table 1. Summary of sampling and analytical considerations of on-line microdialysis sampling coupled to high performance liquid chromatography. (Adapted from Wages *et al.,* 1986)

Sample Concentration	Sampling Frequency	Analytical Considerations
High	High	Sample interval limited by analysis time; high efficiency short column to reduce analysis time
High	Low	Sample interval limited by analysis time; conventional chromatography
Low	High	Most difficult; sampling frequency limited by limit of quantitation and analysis time; microbore chromatography
Low	Low	Larger sample volumes; either conventional of microbore chromatography; lower perfusate rates possible (higher recoveries) with microbore chromatography

For larger animals and man, small portable pumps, often attached to the subject, afford mobility but add another level of complexity to on-line analysis. Four basic probe designs have been used for microdialysis sampling: linear, loop, side-by-side and concentric (Fig. 6). The loop, side-by-side and concentric probes require only a single point of entry into a tissue. While the linear probe requires an entry and an exit, its diameter is much smaller than that of other designs, which results in smaller entry punctures or incisions.

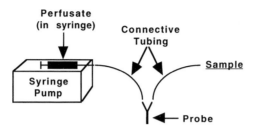

Fig. 5.   Illustration of a basic microdialysis system consisting of a
syringe pump, perfusate (in syringe), microdialysis probe,
and connective tubing.

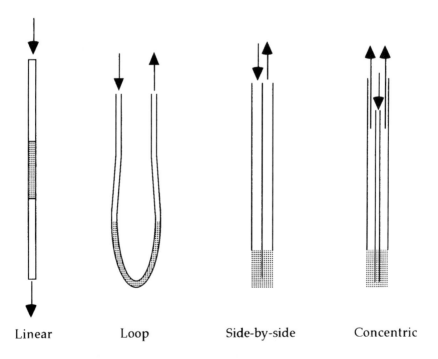

Fig. 6. Illustration of the four types of microdialysis probes.
Arrows indicate direction of flow of perfusion medium.

Table 2. Probe and microdialysis parameters for microdialysis
(see Table 3 for analytical conditions)

Analyte[1]	Probe	Volumecollected (μl)/ volume injected (μl)/ time interval (min)
DA, DOPAC	U-shape	38/20/19
DA, DOPAC, HVA	U-shape (10,000 M_r cutoff)	110–120/20/20
ACh, DA	U-shape	40/100/10.5
ACh, DA	U-shape	45 (ACh), 95 (DA)/ 40/10 (ACh) 20 (DA)
DA, DOPAC, 5-HIAA, HVA	U-shape (10,000 M_r cutoff)	54/40/19
DA	U-shaped, trans-striatal	82.5–90/50/15
GABA	Trans-striatal (10,000 M_r cutoff, 0.27 mm o.d.)	100/120/15
DA, DOPAC, HVA	Trans-striatal	120–140/20/20
DA, DOPAC, HVA, 5-HIAA	U-shape, trans-striatal, I-shaped, concentric[4]	42–455; 82.5–906/40/15
DA	U-shaped, trans-striatal, (10,000 M_r cutoff, 0.27 mm o.d.)	82.5–90/50/15
DA, DOPAC, 5-HIAA, HVA, 3-MT, Serotonin	Trans-striatal (10,000 M_r cutoff, 0.27 mm o.d.)	100/50/10

sampling coupled with chromatographic separation.

Perfusate flow rate (μl/min)	Perfusate	Reference
2	Ringer	Westerink and Tuinte, 1986
5.5–6.0	Ringer	Westerink et al., 1987b
4	Ringer	Damsma et al., 1987a
5	Ringer	Damsma et al., 1988a
2.7	Ringer	Damsma et al., 1988b
5.5–6.0	Ringer[2]	Westerink et al., 1987a
6.0	Ringer[3]	Westerink et al., 1987b
6–7	Ringer	Timmerman et al., 1989
2.8–3.05 5.5–6.06	Ringer[2]	Santiago and Westerink, 1990
5.5–6.0	Ringer[7]	Westerink et al., 1990
5	Ringer	Damsma et al., 1990

Table 2 (continued)

Analyte[1]	Probe	Volume collected (μl)/ volume injected (μl)/ time interval (min)
ACh	Concentric (3 mm membrane length)	40.5/50/15
DA, DOPAC, 5-HIAA, HVA	Side-by-side (5 mm membrane length)	1.05/0. 5/5
DA	Side by side (320 mm, 4 mm)	1.05/0.5/5
DA[8]	Side by side (5,000 M_r cutoff, 0.300 mm o.d., 2 mm membrane length)	1.05/0.5/5
APAP[8]	Concentric (0.85 mm o.d., 4 mm membrane length)	15.8/5/10
Melatonin[8], 5-MT, 5-MIAA, 5-MTOH	Side by side (5,000 M_r cutoff, 0.216 mm o.d., 1.5 mm membrane length)	43.5/100/29
DA, DOPAC, DFMD, HVA, 5-HIAA, NSD 1015, Benserazide	Concentric (6,000 M_r cutoff, 0.225 mm o.d., 2 mm membrane length)	15/15/15
Theophylline Tetracycline	Linear (2–55 mm membrane length)	na/na/10

Perfusate flow rate (μl/min)	Perfusate	Reference
2.7	Physiological salt solution	Durkin et al., 1992
0.42	Artificial cerebro-spinal fluid	Wages et al., 1986
0.21	Artificial cerebro-spinal fluid	Church and Justice, 1987
0.21	Artificial cerebro-spinal fluid	Pettit and Justice, 1989
1.58	Ringer[9]	Sabol and Freed, 1988
1.5	Ringer	Azekawa et al., 1990
1	Artificial cerebro-spinal fluid	Robert et al., 1993
1.6–4.0	Water or Krebs-Ringer buffer	Barrett et al., 1992

Table 2 (continued)

Analyte[1]	Probe	Volume collected (μl)/ volume injected (μl)/ time interval (min)
PBN, POBN	Concentric (20,000 M_r cutoff, 500 mm o.d., 2 mm membrane length)	20/20/10

[1]Abbreviations: APAP, acetaminophen; ACh, acetylcholine; DA, dopamine; DFMD, difluoromethyldopamine; DOPAC, 3,4-dihydroxyphenylacetic acid; 5-HIAA, 5-hydroxyindoleacetic acid; HVA, homovanillic acid; 3-MT, 3-methoxytyramine; GABA, γ-aminobutyric acid; 5-MT, 5-methoxytryptamine; 5-MIAA, 5-methoxyindoleacetic acid; 5-MTOH, 5-methoxytryptophol; NSD 1015, 3-hydoxybenzylhydrazine; PBN, α-phenyl-N-tert-butyl nitrone; POBN, α-4-pyridyl-N-oxide N-tert-butyl nitrone
[2]Also used calcium-free Ringer
[3]Also used calcium-free and high potassium Ringer
[4]Concentric probe investigated only in anesthetized rodents

Table 3.   Summary of the analytical methods used with on-line

Analyte	Column	Mobile phase
DA, DOPAC	Nucleosil 5 C18 (250 × 4.7 mm)	0.1 mol/l acetic acid adjusted with sodium acetate to pH 4.1: 1.8 mmol/l 1-heptane sulfonic acid, 0.01 mmol/l Sodium EDTA, 60–80 ml/l methanol
DA, DOPAC, HVA	Nucleosil 5 C18 (250 × 4.7 mm)	2 mmol/l 1-heptane sulfonic acid, 0.01 mmol/l Sodium EDTA, 60–80 ml/l methanol
ACh	Nucleosil 5 SA (75 × 2.1 mm)	0.2 M phosphate, pH = 7.4, 6–12 mM tetramethyl-ammonium chloride, flow rate 0.6 ml/min

Perfusate flow rate (μl min^{-1})	Perfusate	Reference
2	Sodium chloride solution, 0.9%	Westerink and Tuinte, 1986

[5]Ringer
[6]Calcium-free Ringer
[7]Ringer with the addition of 0.01–0.1 mmol l^{-1} of 3-hydoxybenzyl-hydrazine HCl
[8]Anesthetized rodents
[9]Ringer with 10 mM p-aminophenol and 1 mM glutathione

microdialysis sampling coupled with chromatographic separation.

	Notes	Reference
Electrochemical (rotating disc, +550 mV, Ag/AgCl)	Sodium chloride solution, 0.9%	Westerink and Tuinte, 1986
Electrochemical (rotating disc, +650 mV, Ag/AgCl)		Westerink *et al.*, 1987b
Electrochemical[2] (rotating disc, H$_2$O$_2$, +500 mV, Ag/AgCl)	Infused TTX[1] and MPP^{+}[1] while monitoring DA	Damsma *et al.*, 1987a

Table 3 (continued)

Analyte	Column	Mobile phase
ACh, DA	Nucleosil 5 SA (75 × 2.1 mm)	0.2 M phosphate, pH = 7.4, 6–12 mM tetramethyl-ammonium chloride, flow rate 0.6 ml min^{-1}
DA, DOPAC, 5-HIAA, HVA	Nucleosil 5 C18 (150 × 4.6 mm)	0.1 M acetic acid adjusted with sodium acetate to pH 4.1, 1.8 mM 1-heptane sulfonic acid, 10 μM, sodium EDTA, 60–80 ml l^{-1} methanol
DA	Nucleosil 5 C18 (250 × 4.7 mm)	0.1 mol l^{-1} acetic acid adjusted with sodium acetate to pH 4.1, 2 mmol l^{-1} 1-heptane sulfonic acid, 0.01 mmol l^{-1}, sodium EDTA, 60–80 ml l^{-1} methanol, flow rate 1 ml min^{-1}
GABA	Nucleosil 5 C18 (250 × 4.7 mm)	0.1 mol l^{-1} phosphate buffer, pH 5.5, 0.01 mmol l^{-1} sodium EDTA, 45% methanol, (final pH 6.2)
DA, DOPAC, HVA	Cp-Spher C8 (250 × 4.6 mm)	0.1 M sodium acetate adjusted with acetic acid to pH 4.1, 1.8 mM 1-heptanesulfonic acid, 0.3 mM Sodium EDTA, 125 ml l^{-1} methanol
DOPAC, 5-HIAA, HVA	Altech RSL (5 μm, 150 × 4.6 mm)	0.1 mol l^{-1} sodium acetate adjusted with acetic acid to pH 4.1, 1.8 mmol l^{-1} 1-heptane sulfonic acid, 0.3 mmol l^{-1}, sodium EDTA, 95 ml methanol, flow rate 1 ml min^{-1}
DA	Nucleosil 5 C18 (250 × 4.7 mm)	0.1 mol l^{-1} trichloroacetic acid adjusted with sodium acetate to pH 3.2, 0.01 mol l^{-1} sodium EDTA

Detection	Notes	Reference
Electrochemical[2] (rotating disc, H_2O_2, +500 mV, Ag/AgCl)	Neostigmine bromide (2 µM) added to perfusate	Damsma *et al.*, 1988a
Electrochemical (rotating disc, +650 mV, Ag/AgCl)	Neostigmine bromide (2 µM) added to ACH perfusate	Damsma et al., 1988b
Electrochemical (rotating disc, +550 mV, Ag/AgCl)		Westerink *et al.*, 1987a
Electrochemical (+400 mV), followed by fluorometric analysis		Westerink *et al.*, 1987b
Electrochemical (rotating disc, +700 mV, Ag/AgCl)	Derivatized pre-column with OPA[1]	Timmerman *et al.*, 1989
Electrochemical (glassy carbon electrode -780 mV, Ag/AgCl)		Santiago and Westerink, 1990
Electrochemical (rotating disc, +600 mV, Ag/AgCl)		Westerink *et al.*, 1990

Table 3 (continued)

Analyte	Column	Mobile phase
DA, DOPAC, 5-HIAA, HVA, Serotonin, 3-MT	Nucleosil 5 C18 (150 × 4.8 mm), Precolum : Nucleosil 5 C18 (50 × 3 mm);	0.1 mol l^{-1} acetic acid adjusted with sodium acetate to pH 4.1, 0.5 mM octanesulfonic acid, 0.01 mM sodium EDTA, 150 ml l^{-1} methanol, flow rate 1.8 ml min^{-1}
ACh	Chromospher C18 (5 μm, 75 × 2.1 mm, pretreated with laurel sulphate)	0.15 M potassium phosphate (pH 8.0), 0.27 mM sodium EDTA, 0.23 mM octyl sulphonic acid, 1 mM tetramethylammonium, flow rate 0.4 ml min^{-1}
DA, DOPAC, 5-HIAA, HVA	C18 (3 μm, 100 × 10 mm)	0.1 M citric acid, 0.17 mM sodium hexyl sulfate, 0.06% diethylamine, 0.05 mM EDTA, 7% acetonitrile, pH = 3, flow rate 80–100 μl min^{-1}
DA	Phase Sep C18 (5 μm, 100 × 10 mm)	0.05 M sodium phosphate buffer, 0.1 mM EDTA, 2.2 mM sodium octyl sulfate, 5.0 mM triethylamine with 15 % methanol
DA	(0.5 mm id)	0.1 M citric acid, 0.17 mM sodium hexyl sulfate, 0.06% diethylamine, 0.05 mM EDTA, 7% acetonitrile, pH = 3, flow rate 80–100 μl min^{-1}
APAP	Hewlett Packard (100 × 2.1 mm, 5 μm C18 ODS Hypersil)	0.05 M trichloroacetic acid, 0.2 M H_3PO_4, 0.1 mM EDTA, 2% methanol, pH = 4.48

Detection	Notes	Reference
Electrochemical (glassy carbon electrode + 700 mV, Ag/AgCl)	3-hydoxybenzyl-hydrazine HCl added to perfuste	Damsma *et al.*, 1990
Electrochemical[3] (H_2O_2, + 500 mV, Ag/AgCl)		Durkin *et al.*, 1992
Electrochemical (glassy carbon electrode, + 0.75 V, Ag/AgCl)	Neostigmine bromide (1 μM) added to perfusate	Wages *et al.*, 1986
Electrochemical (glassy carbon electrode, + 0.65 V, Ag/AgCl)		Church and Justice, 1987
Electrochemical (glassy carbon electrode, + 0.75 V, Ag/AgCl)		Pettit and Justice, 1989
Electrochemical (glassy carbon electrode, 0.69 V)		Sabol and Freed, 1988

Table 3 (continued)

Analyte	Column	Mobile phase
Melatonin, 5-MT, 5-MIAA, 5-MTOH	Eicompak (MA-ODS, 250×4.6 mm, 7μm) Precolumn: Eicom-Prepak (AC-ODS, 4×5 mm)	34% methanol, 0.1 M KH_2PO_4, 0.05 M H_3PO_4, pH 3.1, 4 μm sodium EDTA
DA DOPAC, 5-HIAA, HVA DFMD NSD 1015, Ro 4-4602	Spheri 5 RP18 (100×2.1 mm)	0.085 mol l^{-1} sodium phosphate, 1.34 mmol/l EDTA, 0.93 mmol l^{-1} octyl sulfonic acid (sodium salt), 2% methanol, flow rate 0.4 ml min^{-1}
Theophylline, Tetracycline, PBN, POBN	Hypersil ODS (50×4.6 mm, 3 μm) ODS Ultrsphere (150×4.6 mm, 5 μm) Precolumn : Spheri-10 RP-18	9% acetonitrile in 0.05 M potassium phosphate buffer (pH = 2.5) PBN:acetonitrile:water (45:55), 0.05 M sodium acetate adjusted with $HClO_4$, 0.5 ml min^{-1}; POBN: 10% acetonitrile, 2.5 ml l^{-1} $HClO_4$, 200 mg l^{-1} 1-heptanesulfonic acid sodium salt, 0.8 ml min^{-1}

[1]Abbreviations: ACh: acetylcholine, DA: dopamine, DOPAC: 3,4-dihydroxyphenylacetic acid, 5-HIAA: 5-hydroxyindolacetic acid, HVA: homovanillic acid, 3-MT: 3-methoxytyramine, GABA: γ -aminobutyric acid, APAP: acetaminophen, 5MT: 5-methoxytryptamine, 5-MIAA: 5-methoxyindolacetic acid, 5-MTOH: 5-methoxytryptophol; DFMD: difluoromethyldopamine, NSD 1015: 3-hydoxybenzylhydrazine, Ro 4-4602: benserazide, TTX: tetrodotoxin, MPP+: 1-methyl-4-phenyl-pyridinium, OPA: o-phthaldialdehyde; PBN: α-phenyl-N-tert-butyl nitrone, POBN: α-4-pyridyl-N-oxide N-tert-butyl nitrone, ACN: acetonitrile.

Detection	Notes	Reference
Electrochemical (graphite electrode, + 0.85 V, Ag/AgCl)	Perfusate contained p-aminophenol and glutathione.	Azekawa et al., 1990
Electrochemical (glassy carbon electrode, +0.65 V, Ag/AgCl)		Robert et al., 1993
Ultraviolet, 280 nm		Barrett et al., 1992
Ultraviolet, PBN: 292nm POBN: 334 nm		Westerink and Tuinte, 1986

[2]Post-column reactor containing acetylcholinesterase and choline oxidase.

[3]Enzyme reactor column: Lichrosorb-NH2 (10 μm) activate with glutaraldehyde to which acetylcholinesterase and choline oxidase were bound.

3.2. *Analytical Systems*

Analytical techniques that have been used with on-line micro-dialysis may be conveniently divided into chromatographic and non-chromatographic methods. Liquid chromatography is usually the method of choice because it provides the degree of selectivity necessary for the analysis of complex mixtures. Certain non-chromatographic methods such as flow injection analysis represent a viable alternative to LC when very rapid analyses (short sampling intervals) are required.

3.2.1. *Chromatographic methods of analysis*

Reversed-phase or ion-exchange are the modes of liquid chroma-tography that are most compatible with direct injection of aqueous microdialysis samples (Tables 2 and 3). The mode of chromatography is chosen in the usual way according to the physicochemical proper-ties of the analyte and the column configuration (length, particle size and internal diameter) is determined by the sampling interval and the required sensitivity. The sampling interval is determined by the chromatographic run time, which, in turn, is determined by the flow rate, the length of the column and the capacity ratio of the last detectable peak to elute from the column. The sensitivity of the method is determined by the injection volume and the peak disper-sion. Any LC system is constrained by the pressure drop, which limits the maximum flow rate that can be used, and the inverse relationship between linear velocity and column efficiency (Snyder and Kirkland, 1979a). The latter relationship adversely affects reso-lution at high flow rates (Snyder and Kirkland, 1979b).

In principle, equivalent separations can be achieved with columns having the same length but different internal diameters because column efficiency and analysis time are dependent on the linear velocity of the mobile phase. Consequently, no resolution advantage is gained by the use of narrow bore columns instead of conventional LC columns of greater internal diameter. However, the use of micro-bore columns results in a tremendous increase in sensitivity because the peak dispersion is proportional to the square of the column diameter. Therefore, the use of short (L=5 cm), microbore (dc $\leq$ 1 mm) columns has been advocated for analysis of microdialysis samples because they provide the optimum combination of high sensi-tivity and rapid analysis (Wages *et al.*, 1986; Kissinger and Shoup, 1990).

An important consideration in the use of microbore columns is the reduction of extracolumn contributions to band-broadening. The use of microbore columns can reduce the band spreading up to 50-fold leaving extracolumn effects as the major contribution (Kissinger and Shoup, 1990). Reductions in the length of the connection tubing and dead volume in the injector and flow cell are necessary to take full advantage of the benefits afforded by microbore chromatography (Kissinger and Shoup, 1990).

Despite the obvious theoretical advantage of short microbore columns, a review of the literature revealed that only 13% of LC methods for the analysis of on-line microdialysis samples used microbore columns ($d_c \leq 1$ mm) and 27% used narrow bore columns ($d_c = 2$ mm) (Table 3). Therefore it may be concluded that microbore chromatography is not essential to microdialysis sampling and need only be used where high sensitivity or rapid sample analysis are necessary. Conventional LC columns ($d_c = 4.6$ mm) are more convenient and do not require special instrumentation such as low flow-rate pumps, low dead-volume injectors or reduced volume detection cells. In addition, the stringent reduction in extracolumn effects required by microbore chromatography is not necessary with conventional LC. Furthermore, if adequate sensitivity exists and a reduction in analysis time is desired, a column of a shorter length and conventional diameter can be used, provided an adequate separation is maintained.

3.2.2. Non-chromatographic methods of analysis

Non-chromatographic methods of analysis of microdialysis samples have the advantage of faster analysis times than chromatographic methods. As with chromatographic analysis, flow injection analysis uses an injection valve to collect the sample and to introduce it into the analytical system (Tables 4 and 5). An alternative to flow injection analysis for the determination of microdialysis samples is continuous flow analysis (Tables 6 and 7). In contrast to chromatographic analysis, flow injection or continuous flow analysis rely on a specific method of detection such as enzymatic reaction with electrochemical detection and with mass spectrometry for analyte selectivity. Since there is no separation step, the sampling interval is determined by the response time of the detector or enzymatic reaction time.

Table 4. Probe and microdialysis parameters for microdialysis sampling coupled to flow injection systems (see Table 5 for analytical conditions)

Analyte	Probe	Volume collected ((μl)/ volume injected ((μl)/ time interval (min)
Glucose Glutamate Lactate	Concentric (0.32 mm o.d., 4 mm membrane length)	38/20/19
Glucose	Concentric (0.32 mm o.d., 4 mm membrane length)	110-120/20/20
Glucose	Concentric (0.32 mm o.d., 4 mm membrane length)	40/100/10.5
Penicillin G		45(ACh), 95(DA)/ 40/10(ACh), 20(DA)
Tris (2-chloroethyl) phosphate	(4 mm membrane length)	54/40/19

Perfusate flow rate (μl min^{-1})	Perfusate	Reference
2	Artificial cerebrospinal fluid	Boutelle *et al., 1992*
2	Artificial cerebrospinal fluid	Fellows and Boutelle, 1993
2	Artificial cerebrospinal fluid	Fellows *et al., 1993*b
5	5% (v/v) glycerol in water	Caprioli and Lin, 1990
0.8	Water	Deterding *et al., 1992*

Table 5. Analytical conditions for flow injection analysis systems coupled to microdialysis sampling (see Table 4 for microdialysis conditions)

Analyte	Analysis Conditions
Glucose Glutamate Lactate	Immobilized glucose, lactate or glutamate oxidase and horse radish peroxidase, 100 mM Na_2HPO_4, 1 mM EDTA, 2 mM ferrocene monocarboxylic acid, 0.05% Kathon CG, pH = 7.0
Glucose	Immobilzed glucose oxidase and horse radish peroxidase, 100 mM Na_2HPO_4, 1 mM EDTA, 2 mM ferrocene monocarboxylic acid, 0.05 % Kathon CG, pH = 7.0
Glucose	Immobilized lactate oxidase and horse radish peroxidase, 50 mM Na_2HPO_4, 1 mM EDTA, 0.5 mM ferrocene monocarboxylic acid, 0.05 % Kathon CG, pH = 7.0
Penicillin G	Monitored major fragment ion m/z 192 (penicillin G molecular ion m/z 330)
Tris (2-chloroethyl) phosphate	Monitored fragment ion m/z 223 (tris (2-chloroethyl) phosphate molecular ion m/z 285)

Detection	Notes	Reference
Electrochemical, ferricinium, 0.0 V (Ag/AgCl)	Dialysate samples injected onto packed bed enzyme reactor	Boutelle *et al.*, 1992
Electrochemical, ferricinium, 0.0 V (Ag/AgCl)		Fellows and Boutelle, 1993
Electrochemical, ferricinium, 0.0 V (Ag/AgCl)		Fellows *et al.*, 1993b
Tandem mass spectrometry	Anesthetized rat	Caprioli and Lin, 1990
Tandem mass spectrometry	Mixed dialysate with a 25% glycerol solution just before FAB probe tip	Deterding *et al.*, 1992

Table 6. Probe and microdialysis parameters for microdialysis coupled
to continuous flow analysis systems (see Table 7 for analytical
conditions)

Analyte	Probe
Lactate	U-shaped (20,000 M.W. cutoff, 0.2 mm o.d.)
Lactate	U-shaped (20,000 M.W. cutoff, 0.2 mm o.d.)
Lactate	U-shaped (0.27 mm i.d.)
Lactate	U-shaped (0.27 mm i.d.)
Lactate	Linear (0.2 mm o.d., 10-15 mm or 2 mm membrane length)
Glucose	U-shaped and trans-striatal
Choline	U-shape and trans-striatal (10,000 M.W. cutoff, 0.27 mm o.d.)
Glucose	Concentric (20,000 M.W. cutoff, 0.5 mm o.d., 3, 4, and 16 mm membrane length)
Glucose	Concentric (20,000 M.W. cutoff, 0.5 mm o.d., 16 mm membrane length)
Glucose	Concentric (20,000 M.W. cutoff)

Flow Rate (μl min^{-1})	Perfusate Solution	Reference
10	Artificial cerebrospinal fluid	Kuhr and Korf, 1988
10	Artificial cerebrospinal fluid	Kuhr *et al.*, 1988
10	Artificial cerebrospinal fluid	Schasfoort *et al.*, 1988
	Artificial cerebrospinal fluid	Krugers *et al.*, 1992
10	Artificial extracellular fluid	De Boer *et al.*, 1991
3.5	Artificial cerebrospinal fluid	Van der Kuil, 1991
1	Hepes buffer	Flentge *et al.*, 1992
1 to 25	Saline, ringer and others	Keck *et al.*, 1991 Keck *et al.*, 1992
4 or 7	Phosphate buffered saline (0.9%)	Meyerhoff *et al.*, 1992
4.5	Phosphate buffered saline or Ringer	Pfeiffer *et al.*, 1993

Table 6 (continued)

Analyte	Probe
Lactate	Concentric (4 mm membrane length)
Glucose	Linear (35,000 M.W. cutoff, 10 mm membrane length)
Glucose	Linear (6,000 M.W. cutoff, 0.251 mm o.d., 9,000 M.W. cutoff, 0.168 mm o.d., 10 to 20 mm membrane length)
Glucose	Linear (5,000 M.W. cutoff, 2, 9, 18, 28, and 30 mm membrane length)

[1]Studies *in vitro*
[2]Anesthetized rats

Table 7. Analysis conditions for continuous analysis systems with direct on-line reaction systems (see Table 6 for microdialysis conditions)

Analyte[1]	Analysis Conditions
Lactate	LDH and NAD+ in 6.25 mmol/l carbonate buffer, pH = 9.5
Lactate	LDH and NAD+ in 6.25 mmol/l carbonate buffer, pH = 9.5
Lactate	LDH and NAD+ in 6.25 mmol/l carbonate buffer, pH = 9.5

Flow Rate (μl min⁻¹)	Perfusate Solution	Reference
10	Normal saline	Okuda *et al., 1992*
30	Dulbecco's phosphate buffered saline	Moscone and Mascini *et al., 1992*
	Dulbecco's phosphate bufferd saline	Moscone *et al., 1992*
12 and 25	Dulbecco's phosphate buffered saline	Laurell, 1992[4]

[3]Monitored glucose levels in freely moving subjects for up to 27 h
[4]Probe encased in slotted cannula

Detection	Notes	Reference
Fluorometric, NADH	TTX was added to dialysate in some experiments.	Kuhr and Korf, 1988
Fluorometric, NADH	Probenecid was added to dialysate in some experiments.	Kuhr *et al.*, 1988
Fluorometric, NADH	NMDA added to dialysate in some experiments.	Schasfoort *et al.*, 1988

Table 7 (continued)

Analyte[1]	Analysis Conditions
Lactate	LDH and NAD+ in 6.25 mmol l-1 carbonate buffer, pH = 9.5
Lactate	LDH and NAD+ in 6.25 mmol l-1 carbonate buffer, pH = 9.5
Glucose	HK, G-6-PDH, ATP and NADP in 200 mM Tris, 2 mM MgCl, 0.2 mM EDTA, pH = 8.0
Choline	K/Na phosphate buffer, 3/1, 0.2 M; 5 mM NaCl; 0.1% Kathon CG; pH = 8.0; 60 μl min-1 split from 4 ml min-1
Glucose	Glucose oxidase
Glucose	Flow chamber with a glucose oxidase membrane
Glucose	Glucose oxidase-H_2O_2
Lactate	Reagent: 6 units ml-1 LDH and 0.6 mM NAD in glycine-hydrazine buffer (pH = 9.4)

Detection	Notes	Reference
Fluorometric, NADH		Krugers et al., 1992
Flow-through fluorimeter, NADH		De Boer et al., 1991
Fluorometric, NADPH	Reaction required 6.5 min for mixing of reagents	van der Kuil and Korf, 1991
Electrochemical, carbon composite, H_2O_2, +250 mV, (Ag/AgCl)	2 μM neostigmine bromide added to perfusate; on-line preoxidator (+750 mV); and split flow for analysis	Flentge et al., 1992
		Keck et al., 1991[1],1992[2]
Electrochemical, H_2O_2, platinum /AgCl, 700 mV	Increased flow rate in long term studies due to biosensor drift	Meyerhoff et al., 1992
		Pfeiffer et al., 1993
Ultraviolet, 340 nm	Used a z-shaped flow cell to reduce detector volume; reaction time, $\approx$ 7 min	Okuda et al., 1992

Table 7 (continued)	
Analyte[1]	**Analysis Conditions**
Glucose	Glucose oxidase on a nylon net membrane
Glucose	Glucose oxidase on a nylon net membrane, wall-jet type cell
Glucose	Glucose oxidase

[1]Abbreviations: LDH: lactate dehydrogenase, TTX: tetrodotoxin, 6-phosphate dehydrogenase

4. Applications

Microdialysis has its origins in neurochemistry where it was developed in the 1970s as a rapid method for sampling neurotransmitters in the brain causing minimal disruption to the sampled environment (Ungerstedt and Pycock, 1974). More recently, this technique has been further applied to the sampling of xenobiotics *in vivo*. With microdialysis sampling, a kinetic profile of a xenobiotic can be obtained in a single subject, resulting in a reduction in the intersubject variation and the total number of experiments required.

Microdialysis sampling has been used to determine the pharmacokinetics and metabolism of several drugs *in vivo*, most notably, acetaminophen in blood and liver (Scott *et al.*, 1990; Scott *et al.*, 1991), and phenol in blood, liver and bile (Scott *et al.*, 1989; Scott and Lunte, 1993). Caffeine and theophylline, two structurally similar compounds, have also been studied extensively by microdialysis sampling due to their differing pharmacokinetics, distribution profiles and microdialysis characteristics (Stahle, 1991; Stahle *et al.*, 1991a-c; Sjoberg *et al.*, 1992; Terasaki *et al.*, 1992; Stahle, 1992). The pharmacokinetics of the antiviral nucleoside, zidovudine in blood and brain have also been studied recently using microdialysis sampling (Wang *et al.*, 1993).

Detection	Notes	Reference
Electrochemical, H_2O_2, Platinum (Ag/AgCl)	Calibrated bio-sensor as a function of membrane length and dialysate flow rate	Moscone and Mascini, 1992
Electrochemical, H_2O_2, Platinum (Ag/AgCl)		Moscone et al., 1992
Oxygen electrode, H_2O_2, –0.8 V (Ag/AgCl)		Laurell, 1992

NMDA : N-methyl-d-aspartate, HK : hexokinase, G-6-PDH : glucose-

Steele et al. (1991) demonstrated the ability of microdialysis sampling to prevent further enzymatic degradation of aspirin once it crossed the dialysis membrane in the examination of the pharmacokinetics of aspirin. Microdialysis sampling has also been used to determine the protein binding of drugs (Herrera et al., 1990; Ekblom et al., 1992).

In a unique application, the pharmacokinetics of hydrocortisone has been investigated using microdialysis sampling linked to radioimmunoassay (RIA) (Miller and Geary, 1991). Most recently, microdialysis sampling has been employed to study the kinetics of 9-amino-1,2,3,4-tetrahydroacridine (THA) in the brain, avoiding the tissue disruption and anesthesia interferences encountered with classical techniques (Telting-Diaz and Lunte, 1993).

4.1. Dopamine and its metabolites

One of the first reported uses of microdialysis sampling in conjunction with on-line analysis was the system described by Westerink and Tuinte (1986) to investigate neurotransmitters in the brains of freely moving rats. The system consisted of a microdialysis probe, a peristaltic pump and an injection valve. For the majority of studies using this system, a reversed-phase LC column of conventional dimensions (250 × 4.7 mm i.d.) and an electrochemical detector were used for analysis (Fig. 7) (Westerink *et al.*, 1987a). This system has proven useful in monitoring a wide range of neurotransmitters and their metabolites in a number of pharmacological studies. The system in Fig. 7 is shown as a combination of several possible configurations, notably the use of a reactor column and the addition of a reagent line to the dialysate line downstream of the probe before the injection valve (dotted OPA reagent line).

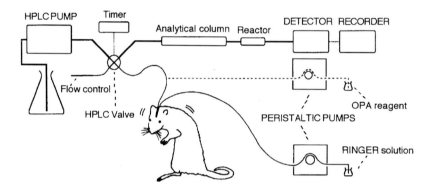

Fig. 7   Diagram of the on-line brain dialysis system developed by Westerink *et al.* The reactor, used for acetylcholine analysis, is filled with silica to which acetylcholine esterase and choline oxidase are covalently bound. The detector is an electrochemical electrode (Pt for acetylcholine and carbon paste for monoamines) or a fluorometer (amino acids). The amino acids are derivatized precolumn with o-phthaldialdehyde (OPA). (Reproduced with permission from Westerink *et al.*, 1987a)

   The analysis of dopamine and its metabolites, 3,4-dihydroxyphenylacetic acid and homovanillic acid, as well as 5-hydroxyindolacetic acid, a metabolite of serotonin, has been a major focus of research using on-line microdialysis. In some of the earliest studies, Westerink and co-workers used on-line microdialysis sampling to examine the chronic use (seven day period) of intercerebral microdialysis for monitoring dopamine in the rat striatum. Based on the dopamine response to potassium and (+)-amphetamine inclusion in the perfusate, and systemic administration of (+)-amphetamine, it was concluded that the first 48 h following implantation was the most useful time period for analytical measurements due to tissue reactions which developed with longer periods of implantation (Westerink and Tuinte, 1986). In subsequent investigations, the effect of adding two pharmacologically active agents, TTX and MPP+, in the perfusion medium has been investigated by microdialysis sampling of dopamine (Westerink *et al.*, 1987a,b).

Timmerman *et al.* used a similar system to investigate the effect of two enantiomers of a potent and selective dopamine agonist on dopamine receptors in rat striatum assessed by dopamine release (Timmerman *et al.*, 1989). The observation of stereotypical behavior during microdialysis sampling was used to evaluate the effects of each enantiomer in this study. The comparison of the effect of several routes of administration, interperitoneal, oral and transdermal, on the potency and duration of action of both enantiomers was correlated with dopamine levels sampled by microdialysis.

Westerink *et al.* (1990) used the system shown in Fig. 7 to monitor the effects of adding sub-μM concentrations of an L-aromatic decarboxylase inhibitor to the perfusate on brain tyrosine hydroxylase activity. The levels of dopamine in rat striatum were used as an index of tyrosine hydroxylase activity in response to intraperitoneal administration of apomorphine, haloperidol and γ–butyrolactone. The authors noted that the type of probe used, trans-striatal or U-shaped, appeared to determine the pharmacological outcome of the haloperidol experiments. Dopamine release was used as a measure of neuronal damage to compare the extent of tissue disruption caused by the implantation of the four different types of microdialysis probes studied (Santiago and Westerink, 1990). The microdialysis samples from the probes were analyzed by LC with electrochemical detection. The authors found a difference between probes 2 h after implantation, but no difference was evident after 24 h. They concluded that microdialysis sampling should be performed at least 24 h after probe implantation.

Another early use of microdialysis coupled to on-line LC analysis was the evaluation of dopamine, DOPAC, 5-HIAA, and HVA from the anterior striatum of anesthetized rats (Wages *et al.*, 1986). This system was comprised of a push-pull syringe system attached to a side-by-side probe whose flow ran through an LC injection valve. The dialysate was periodically switched onto a microbore column and was detected electrochemically. The conclusion of this study was that the mode of chromatography selected (conventional, fast or microbore) should be determined by the concentration of the analyte and the frequency of sampling. Further monitoring of dopamine in the anterior striatum of anesthetized rats was done following injections of haloperidol, saline and apomorphine with emphasis placed on the use of low microdialysis perfusion rates (Church, 1987).

Dopamine levels were measured on-line following regular systemic administration of cocaine by the investigator in anesthetized rats and off-line by rodent self-administration in freely moving rats (Pettit and Justice, 1989). This research highlighted some of the problems associated with low dialysate perfusion rates, such as system dead volume and animal movement.

4.2. *Acetylcholine and Choline*

For the analysis of acetylcholine (ACh) Damsma *et al.* (1987a) filled the reactor column shown in Fig. 7 with silica to which acetylcholine esterase and choline oxidase were bound. A platinum electrode was substituted for the carbon paste electrode and neostigmine bromide was added to the perfusate. An esterase inhibitor, neostigmine bromide, was added to the perfusion medium to limit the degradation of acetylcholine prior to analysis. A further modification of the system, the use of an electrochemical detector with a glassy carbon electrode, was done by Damsma *et al.* (1990) to study the effects of ischemia, death and pargyline administration on dopamine, serotonin and their metabolites in rat striatum.

Damsma *et al.* (1987b) later replaced silica with Sepharose 4B in the post-column reactor to examine the effects of tetrodotoxin, atropine and oxotremorine on the release of acetylcholine in rat striatum. This system was also used to study the central action of 4-aminopyridine and 2,4-diaminopyridine, on the release of ACh and DA (Damsma *et al.*, 1988b). Most recently, Lichrosorb NH_2 has been used instead of Sepharose 4B and the U-shaped probe has been replaced by a probe of concentric design for investigations geared toward the effect of glucose levels on memory (Durkin *et al.*, 1992).

Analysis of acetylcholine and choline in rat striatum using microdialysis, a preoxidation cell, and a sandwich-type enzyme reactor containing physically immobilized acetylcholine esterase or choline oxidase was developed by Flentge *et al.* (1992). The preoxidator was essentially an electrochemical detection cell with a large oxidative surface area and a potential set at +750 mV which was used to inactivate interfering endogenous substances. The use of the preoxidator allowed the differentiation of acetylcholine and choline without the use of a chromatographic separation. The system also incorporated a sandwich type enzyme reactor which has the advantage of a low dead volume with high enzyme concentration.

4.3. γ-Amino butyric acid

Westerink and De Vries (1989) have combined the o-phthaldialdehyde reagent system in Fig. 7 with fluorescence detection for the
analysis of γ-aminobutyric acid (GABA) in microdialysis samples.
Westerink used this precolumn derivatization system to monitor
basal release of GABA following infusion of elevated levels of potassium or nipecotic acid, a GABA uptake inhibitor (Westerink, 1989).
The effects of cadmium and tetrodotoxin on GABA dialysate concentrations were also examined.

4.4. Acetaminophen

Comparison of the striatal drug levels of acetaminophen by three
separate techniques, microdialysis, electrochemistry and tissue assay
was performed by Sabol and Freed (1988). The microdialysis system
was comprised of an infusion pump, a concentric microdialysis probe
and a pneumatically activated injection valve along with an LC
system for acetaminophen analysis. Using this system Sabol et al.
observed that acetaminophen levels monitored by microdialysis and
electrochemistry were identical, although both lagged behind the
peak tissue concentration (Sabol and Freed, 1988). The authors also
noted a difference between the distribution of acetaminophen in the
extracellular fluid (microdialysis and electrochemistry results) and
inside the cell (tissue assay).

4.5. Melatonin

Measurement of the dynamic changes of melatonin in dialysate
from the pineal glands of freely moving rodents was conducted by
Azekawa et al. (1990) using microdialysis sampling with on-line LC
and electrochemical detection. The system consisted of an infusion
pump, an autoinjector and a concentric probe described by Nakahara
et al. (1990). This system was used to observe a dramatic increase in
the extracellular melatonin levels during the light to dark transi-
tional phase.

4.6. Theophylline and tetracycline

Barrett et al. (1992) measured theophylline and tetracycline in
aqueous and plasma solutions in vitro using on-line LC with UV
detection. Since the problems associated with probe implantation

were not associated with biological fluid analysis *in vitro*, the authors utilized a home-made probe of a linear design in which the parameters under investigation were easily changed. The authors investigated various factors which influence microdialysis probe recovery, such as solution temperature, perfusate flow rate, protein binding, and membrane length and material. From the results of these studies, the authors concluded that recovery and reproducibility increased with longer membrane length and that careful control of perfusate flow rate and temperature improved accuracy and reproducibility.

4.7. α-Phenyl-N-tert-butyl nitrone and α-4-pyridyl-N-oxide N-tert-butyl nitrone

Cheng *et al.* (1993) used microdialysis sampling *in vivo* on-line with LC and UV detection to determine α–phenyl-N-*tert*-butyl nitrone (PBN) and α-4-pyridyl-N-oxide N-*tert*-butyl nitrone (POBN) in the blood and brain of intact rats. The study evaluated distribution of the nitrones between blood and brain, comparing results from both microdialysis sampling and direct tissue and blood analysis. Although the ratio of PBN/POBN was similar in both blood and brain in the two methods, the concentrations of the two compounds were higher in the samples obtained directly than in those obtained by microdialysis. The difference in PBN and POBN concentrations between the sampling methods was attributed to the difference in the two techniques; microdialysis reflects primarily the unbound fraction of a compound while direct sampling represents the total (bound and unbound) amount of a compound.

4.8. Lactate

Lactography, the continuous on-line measurement of lactate using microdialysis, was initially developed by Kuhr *et al.*, who used the technique to measure lactic acid in the extracellular fluid in the striatum of freely moving rats (Kuhr and Korf, 1988a; Kuhr *et al.*, 1988b; Korf, 1989; Korf and De Boer, 1990). In these studies, the extracellular levels of lactate in both the rat striatum and hippocampus in response to electroconvulsive shock or ischemia were investigated. The levels of lactate in the dialysate were used as a measure of brain metabolism, specifically glycolysis, or neuronal activity. Schasfoort *et al.* later used the technique to study the effect of mild stress on rat hippocampal glucose (Schasfoort *et al.*, 1988) while Krugers *et al.* used the technique to investigate the role of the entorhinal cortex and adrenal glands on hippocampal lactate (Krugers *et al.*, 1992).

De Boer *et al.* (1991) used lactography to monitor metabolism in muscle tissue. This research determined lactate concentrations at rest and under several physiological (glycosis, isometric contraction and swimming) and pathological states (partial or total ischemia). The concentration of lactate in muscle was found to be approximately twice that in arterial whole blood, with the difference being attributed to the contribution of intracellular lactate.

A method for the continuous determination of lactate in dialysates from muscle and liver following on-line reaction monitored by UV detection has been developed by Okuda *et al.* (1992). In this system a reagent stream is mixed with the dialysis stream for the reaction of lactate prior to detection. This system also incorporates a z-shaped flow cell for a reduction in volume. Using the system, levels of lactate were monitored for 30 min following acute hemorrhage (loss of 30% of total blood volume) with an observed increase in lactate of 210 and 270% in liver and muscle respectively.

4.9. *Glucose, glutamate and lactate*

The incorporation of an on-line glucose biosensor with micro-dialysis sampling was first reported by Keck *et al.* (1991, 1992). Initial studies dealt with the development of the system with emphasis on the relationship between perfusate flow rate, membrane length and device response time *in vitro* (Keck *et al.*, 1991). Subsequent application of the technique for the monitoring of glucose in the sub-cutaneous extracellular fluid of 6 rats was performed, comparing the dialysate levels to blood levels sampled from an in-dwelling catheter (Keck *et al.*, 1992). Further application of the system described by Keck *et al.* was performed by Pfeiffer *et al.* in the study of subcutaneous glucose levels in 10 healthy and 10 diabetic volunteers using microdialysis and an implanted glucose sensor (Pfeiffer, 1992). The observed dialysate glucose concentrations showed a good correlation with concentrations observed using a continuous intravenous glucose sensor. Glucose concentrations were measured continuously for up to 27 h in an individual patient.

Meyerhoff *et al.* used the combination of a concentric probe and sensitive glucose sensor to monitor subcutaneous glucose levels in healthy and diabetic patients (Meyerhoff *et al.*, 1992). In studies extending beyond 5 h, the perfusate flow rate had to be increased from 4 to 7 µl min^{-1} to prevent biosensor drift. The profiles of subcutaneous glucose dialysate concentrations closely reflected the patterns observed with glucose biosensors implanted intravenously.

van der Kuil and Korf (1991) modified the lactography technique of Kuhr and coworkers (Kuhr and Korf, 1988; Kuhr et al., 1988) to monitor extracellular glucose levels using a trans-striatal probe and a U-shaped probe placed in the hippocampus. Glucose levels were monitored during acute hyperglycemia, seizures and stress, and in a repetitive hypoxic, ischemia model. The conclusion from this investigation was that while microdialysis coupled to an enzymatic assay was inexpensive and simple, the extracellular glucose level reflected not only glucose metabolism but glucose transport as well.

The measurement of subcutaneous glucose levels by coupling a hollow microdialysis fiber with a glucose biosensor has also been accomplished with a system devised by Moscone et al. (1992). Using this system, subcutaneous glucose levels were monitored for several hours before a sharp decrease in sensitivity was noticed. Since the same phenomenon was not observed in microdialysis sampling of the blood, the decrease in sensitivity was attributed to an inflammatory response due to probe implantation, which resulted in an interfering species affecting the enzyme or electrode reaction. It was later demonstrated that by changing the molecular-weight cut-off of the membrane and altering the design of the biosensor flow-through cell, previous problems of biosensor variability could be partially overcome (Moscone et al., 1992).

Laurell et al. (1992) have developed a system for continuous glucose monitoring based on microdialysis sampling, a glucose oxidase micro-enzyme reactor and an oxygen electrode for continuous glucose monitoring.

Research demonstrating the feasibility of microdialysis sampling with an on-line packed bed, enzyme reactor with electrochemical detection was conducted by Boutelle et al. (1992). By changing the substrate oxidase, the enzyme reaction was used for the analysis of glucose, glutamate, and lactate in microdialysate samples from probes implanted in the striatum of freely moving rats.

Fellows et al. (1993) used the microdialysis system for determining extracellular glucose developed by Boutelle et al. (1992) along with striatal blood flow to evaluate the response of rats to mild stresses: tail pinch and restraint. This research found differences in the concentrations and time profiles of glucose in the dialysate between the different forms of stress. An extension of this research was later conducted by the same group who examined extracellular lactate in the striatum and hippocampus of freely moving rats. This research studied the influx and clearance of lactate from the extracellular fluid in the presence and absence of drugs (tetrodotoxin and MK801) and under two forms of mild stress (restraint and tail pinch) (Fellows and Boutelle, 1993).

4.10. Penicillin G

The combination of tandem mass spectrometry and on-line microdialysis sampling of a biological system was first reported by Caprioli and Lin (1990). In this study, microdialysis sampling and fast atom bombardment mass spectroscopy (FAB-MS) were combined to monitor the pharmacokinetics of penicillin G in a live rat. To accommodate FAB-MS analysis, glycerol was used as the perfusion medium. The resultant pharmacokinetic profile in the anesthetized rat was similar to those observed previously for dog and man by direct plasma analysis. Use of a similar system in conjunction with concurrent ear-vein sampling has appeared in an article describing recent advances and applications in continuous-flow fast atom bombardment (Caprioli and Suter, 1992).

4.11. Tris (2-chloroethyl) phosphate

Deterding et al. (1992) have used microdialysis sampling with continuous FAB-MS to monitor the pharmacokinetic profile of tris (2-chloroethyl) phosphate (an organophosphate flame retardant) in three anesthetized rats (Deterding et al., 1992). Deterding et al. used a vessel pressurized with helium to supply perfusate to the probe. Elimination of glycerol solution as the perfusate (Caprioli and Lin, 1990) was also accomplished by mixing the dialysate with a 25% glycerol solution just before the FAB probe tip. Although the microdialysis results differed significantly from results of a separate study involving venous sampling in freely moving rats, a good correlation was observed between the pharmacokinetic profiles for the three rats sampled using microdialysis and FAB spectrometry.

5. Conclusions

Microdialysis is a powerful technique for the sampling of biological systems. It is relatively non-destructive, and the biocompatibility of modern microdialysis probes is good, allowing continuous monitoring of biological fluids for as long as 72 h after implantation. The main limitations of the technique arise from the low flow rates that must be used (typically 1–10 μl min^{-1}), which limit the injection volume that can be used if high temporal resolution of the samples is required. Adsorption of analytes to dialysis membranes and/or connective tubing may limit the application of microdialysis to ionic or relatively polar analytes; however, there is a wide variety of membranes available with a range of molecular weight cut-offs. Because the samples are presented to the analytical device in a completely aqueous medium, reversed-phase liquid chromatography is the method of choice for on-line analysis. In this case, real-time data describing the disposition of endogenous and exogenous substances may be obtained.

Acknowledgments

Parts of this chapter have previously been published in the Ph.D. dissertation of J.M. Ault, Jr. This work was supported by the a training grant from the National Cancer Institute (CA-09242), a fellowship for JMA from the United States Pharmacopoeia and a grant from Hoffman-La Roche.

6. References

Azekawa, T., A. Sano, K. Aoi, H. Sei and Y. Morita (1990). *J. Chromatogr.*, *530*, 47–55.

Barrett, D. A., J. Adler, R. W. D. Nickalls and P. N. Shaw (1992). *Pharm. Pharmacol. Lett.*, *2*, 139–42.

Benveniste, H. (1989). *J. Neurochem. 52*, 1667–1679.

Benveniste, H. and P. C. Hüttemeier (1990). *Prog. Neurobiol.*, *35*, 195–215.

Boutelle, M. G., L. K. Fellows and C. Cook (1992). *Anal. Chem.*, *64*, 1790–1794.

Caprioli, R. M. and S. N. Lin (1990). *Proc. Natl. Acad. Sci. USA*, *87*, 240–243.

Caprioli, R. M. and M. J. F. Suter (1992). *Int. J. Mass. Spectrom. Ion Processes*, *118/119*, 449–476.

Cheng, H.-Y., T. Liu, G. Feuerstein and F. C. Barone (1993). *Free Radical Biol. Med.*, *14*, 243–250.

Church, W. H. and J. B. Justice, Jr. (1987). *Anal. Chem.*,*59*, 712–716.

Damsma, G., B. H. C. Westerink, A. Imperato, H. Rollema, J.B. De Vries and A. S. Horn (1987a). *Life Sci.*, *41*, 873–876.

Damsma, G., B. H. C. Westerink, J. B. De Vries, C. J. Van den Berg and A. S. Horn (1987b). *J. Neurochem.*, *48*, 1523–1528.

Damsma, G., B. H. C. Westerink, J. B. De Vries and A. S. Horn (1988a). *Neurosci. Lett.*, *89*, 349–354.

Damsma, G., P. T. Biessels, B. H. C, Westerink, J. B. De Vries and A. S. Horn (1988b). *Eur. J. Pharmacol.*,*145*, 15–20.

Damsma, G., D. P. Boisvert, L. A. Mudrick, D. Wenkstern and H. C. Fibiger (1990) *J. Neurochem.*, *54*, 801–808.

De Boer, J., F. Postema, G.H. Plijter and J. Korf (1991). *Pfluegers Arch. Eur. J. Physiol.*, *419*, 1–6.

Deleu, D., S. Sarre, G. Ebinger and Y. Michotte (1991). *Naunyn-Schmied. Arch. Pharmacol.*, *344*, 514–519.

Deterding, L. J., K. Dix, L. T. Burka and K. B. Tomer (1992). *Anal. Chem.*, *64*, 2636–2641.

Dolan, J. W. and L. R. Snyder (1989). *Troubleshooting LC Systems.* Humana Press, Clifton, NJ.

Durkin, T. P., C. Messier, J. B. De Vries and B. H. C. Westerink (1992). *Behav. Brain Res.*, *49*, 181–188.

Ekblom, M., U. M. Hammarlund, T. Lundqvist, and P. Sjoberg (1992). *Pharm. Res.*, *9*, 155–158.

Fellows, L. K. and M. G. Boutelle (1993). *Brain Res.*, *604*, 225–231.

Fellows, L. K., M. G. Boutelle and M. Fillenz (1993). *J. Neurochem.*, *60*, 1258–1263.

Flentge, F., K. Venema, T. Koch and J. Korf (1992). *Anal. Biochem.*, *204*, 305–310.

Herrera, A. M., D. O. Scott, and C. E. Lunte (1990). *Pharm. Res.*, *7*, 1077–1081.

Jacobson, I. M. Sandberg, and A. Hamberger (1985). *J. Neurosci. Meth.*, *15*, 263–268.

Keck, F. S., W. Kerner, C. Meyerhoff, H. Zier and E. F. Pfeiffer (1991). *Horm. Metab. Res.*, *23*, 617–618.

Keck, F. S., C. Meyerhoff, W. Kerner, T. Siegmund, H. Zier and E. F. Pfeiffer (1992). *Horm Metab Res.*, *24*, 492–493.

Kendrick, K. M. (1989). *Meth. Enzymol. 168*, 182–205.

Kissinger, P. T. and R. E. Shoup (1990). *J. Neurosci. Meth.*, *34*, 3–10.

Korf, J. (1989). *Neurosci. Res. Commun.*, *4*, 129–138.

Korf, J. and J. DeBoer (1990). *Int. J. Biochem.*, *22*, 1371–1378.

Krugers, H. J., D. Jaarsma and J. Korf (1992). *J. Neurochem.*, *58*, 826–830.

Kuhr, W. G. and J. Korf (1988). *J. Cereb. Blood Flow Metab.*, *8*, 130–137.

Kuhr, W. G., C. G. Van den Berg and J. Korf (1988). *J. Cereb. Blood Flow Metab.*, *8*, 848–856.

Larsson, C. I. (1991). *Life Sci.*, *49*, PL73–PL78.

Laurell, T. (1992). *J. Med. Eng. Technol.*, *16*, 187–193.

Linhares, M. C. and P. T. Kissinger (1992). *J Chromatogr.*, *578*, 157–163.

Lonnroth, P., P. A. Jansson and U. Smith (1987a). *Am. J. Physiol.*, *253*, E228-E231.

Lonnroth, P., P. A. Jansson, B. B. Fredholm, and U. Smith. (1987b). *Am. J. Physiol.*, *256*, E250–E255.

Lonnroth, P. and U. Smith (1990). *J. Intern. Med.*, *227*, 295–300.

Lonnroth, P., J. Carlsten, L. Johnson and U. Smith (1991). *J. Chromatogr.—Biomed. App.*, *568*, 419–426.

Lunte, C. E., D. O. Scott and P. T. Kissinger (1991). *Anal. Chem.*, *63*, 773A–780A.

Menacherry, S., W. Hubert and J. B. Justice, Jr. (1992). *Anal Chem.*, *64*, 577–583.

Meyerhoff, C., F. Bischof, F. Sternberg, H. Zier and E. F. Pfeiffer (1992). *Diabetologia*, *35*, 1087–1092.

Miller, M. A., and R. S. Geary (1991). *J. Pharm. Biomed. Anal.*, *9*, 901–910.

Moscone, D. and M. Mascini (1992). *Ann. Biol. Clin. (Paris)*, *50*, 323–327.

Moscone, D., M. Pasini and M. Mascini (1992). *Talanta* ,*39*, 1039–1044.

Nakahara, D., N. Ozaki and T. Nagatsu (1990). *Yakubutsu Seishin. Kodo*, *10*, 393–399.

Okuda, C., T. Sawa, M. Harada, T. Murakami and Y. Tanaka (1992). *Circ. Shock*, *37*, 230–235.

Olson, R. J. and J. B. Justice, Jr. (1993) *Anal. Chem.*, *65*, 1017–1022.

Pettit, H.O. and J. B. Justice, Jr. (1989). *Pharmacol. Biochem. Behav.*, *34*, 899–904.

Pfeiffer, E. F., C. Meyerhoff, F. Bischof, F. S. Keck and W. Kerner (1993). *Horm. Metab. Res.*, *25*, 121–124.

Robert, F., S. L. Lambás-Señas, C. Ortemann, J.-F. Pujol and B. Renaud. (1993). *J. Neurochem.*, *60*, 721–729.

Sabol, K. E. and C. R. Freed (1988). *J. Neurosci. Meth.*, *24*, 163–168.

Santiago, M. and B. H. C. Westerink (1990). *Naunyn-Schmied. Arch. Pharmacol.*, *342*, 407–414.

Schasfoort, E. M. C., L. A. De Bruin and J. Korf (1988). *Brain Res.*, *475*, 58–63.

Scheller, D. and J. Kolb. *J. Neurosci. Meth.*, *40*, 31–38 (1991).

Scott, D. O. and C. E. Lunte (1993). *Pharm. Res.*, *10*, 335–342.

Scott, D. O., M. A. Bell and C. E. Lunte (1989). *J. Pharm. Biomed. Anal.*, *7*, 1249–1259.

Scott, D. O., L. R. Sorensen, and C. E. Lunte (1990). *J. Chromatogr.*, *506*, 461–469.

Scott, D. O., L. R. Sorensen, K. L. Steele, D. L. Puckett, and C. E. Lunte (1991). *Pharm. Res.*, *8*, 389–392.

Sjoberg, P., I. M. Olofsson, and T. Lundqvist (1992). *Pharm. Res.*, *9*, 1592–1598.

Snyder, L. R. and J. J. Kirkland (1979a). *Introduction to Modern Liquid Chromatography*, 2nd edtion. Wiley, New York, pp. 56–58.

Snyder, L. R. and J. J. Kirkland (1979b). *Introduction to Modern Liquid Chromatography*, 2nd edtion. Wiley, New York, p. 240.

Stahle, L. (1991). *Life Sci.*, *49*, 1835–1842.

Stahle, L. (1992). *Eur. J. Clin. Pharmacol*, *43*, 289–294.

Stahle, L., S. Segersvard, and U. Ungerstedt (1990). *Eur. J. Pharmacol.*, *185*, 187–193.

Stahle, L., P. Arner, and U. Ungerstedt (1991a). *Life Sci.*, *49*, 1853–1858.

Stahle, L., S. Segersvard, and U. Ungerstedt (1991b). *Life Sci.*, *49*, 1843–1852.

Stahle, L., S. Segersvard, and U. Ungerstedt (1991c). *J. Pharmacol. Meth.*, *25*, 41–52.

Steele, K. L., D. O. Scott, and C. E. Lunte (1991). *Anal. Chim. Acta*, *246*, 181–186.

Telting-Diaz, M., and C. E. Lunte (1993). *Pharm. Res.*, *10*, 44–48.

Terasaki, T., Y. Deguchi, Y. Kasama, W. M. Pardridge, and A. Tsuji (1992). *Int. J. Pharm.*, *81*, 143–152.

Timmerman, W., B. H. C. Westerink, V.J. De, P. G. Tepper and A. S. Horn (1989). *Eur. J. Pharmacol.*, *162*, 143–150.

Ungerstedt, U. and C. Pycock (1974). *Bull. Schweiz. Akad. Med. Wiss.*, *1278*, 1–5.

Ungerstedt, U. (1991). *J. Intern. Med.*, *230*, 365–373.

van der Kuil, J. H. and J. Korf (1991). *J. Neurochem.*, *57*, 648–654.

Wages, S. A., C. W. H. and J.B. Justice, Jr. (1986). *Anal. Chem.*, *58*, 1649–1656.

Wang, Y., S. L. Wong and R. J. Sawchuck (1993) *Pharm. Res.*, *10*, 1411–1419.

Westerink, B., H. C. (1992). *Trends Anal. Chem.*, *11*, 176–182.

Westerink, B. H. C. and Tuinte, M. J. H. (1986). *J. Neurochem.*, *46*, 181–185

Westerink, B. H. C. and J. B. De Vries (1989). *Naunyn-Schmied. Arch. Pharmacol.*,*339*, 603–607.

Westerink, B. H. C., J. B. De Vries and R. Duran (1990). *J. Neurochem.*, *54*, 381–387.

Westerink, B. H. C., G. Damsma, H. Rollema, J. B. De Vries and A. S. Horn (1987a). *Life Sci.*, *41*, 1763–1776.

Westerink, B. H. C., J. Tuntler, G. Damsma, H. Rollema and J. B. De Vries (1987b). *Naunyn-Schmied. Arch. Pharmacol.*, *336*, 502–507.

Wong, S. L., Y. Wang, and R. J. Sawchuck. (1992). *Pharm. Res. 9*, 332–338.

Yokel, R. A., D. D. Allen, D. E. Burgio, and P. J. McNamara (1992). *J. Pharmacol. Toxicol. Meth.*, *27*, 135–142.

CHAPTER 7

Multi-column Approaches to Chiral Bioanalysis
by Liquid Chromatography

W. JOHN LOUGH and TERRENCE A.G. NOCTOR

School of Health Sciences, University of Sunderland,
Sunderland, SR1 3SD, U. K.

1. Introduction

In order to fully appreciate the increasing use of multicolumn approaches in chiral LC it is instructive to consider the position with respect to (i) the regulatory framework governing the development and marketing of chiral drugs, and (ii) recent advances in methods for chiral resolution.

Such is the importance of chirality that for at least a decade pharmaceutical companies have been hanging on every word of spokespersons of regulatory bodies such as the Food and Drug Administration (FDA) to try to pick up even a hint of a policy statement on the development and marketing of chiral drugs. Clearly for some time to come the policies of the regulators will continue to have considerable impact on the internal research strategies of all pharmaceutical companies, however large or small. Thankfully, though, the picture is clearer now than it was a few years ago, with the development of single enantiomer drugs being very much the norm. This is due in part to the critical appraisal over the years by many commentators, for example, Brown (1990) and de Camp (1989), of the need to treat each drug enantiomer as a different compound. In light of the many processes within the body that are likely to be mediated by naturally chiral biomacromolecules, it would be surprising if each enantiomer of a pair had identical pharmacological and toxicological properties. One of the enantiomers (the distomer) is likely to be less active in terms of the desired therapeutic effect than the other, the eutomer, and accordingly ought to be viewed as an impurity in the eutomer. Ideally, then, the eutomer should be developed as a single enantiomer drug. This is even more so the case given that matters have probably reached a stage where so much data would be required to prove that the enantiomers acted identically *in vivo*, or alternatively, that a 50:50 ratio was the optimum to obtain a synergistic effect, that development of a single enantiomer might always be the most financially viable route irrespective of any difficult synthetic or resolution issues.

It follows from the above that there is a great need to examine chiral aspects of drugs which have already been marketed as racemates. Much of this work remains to be done and, in particular, there will be a need for chiral bioanalysis of this nature for some time to come. With respect to chiral drugs of the future, the needs in pharmaceutical research will be for small amounts of chiral bioanalysis in association with initial testing of racemates, but more so for semipreparative chiral work to allow testing of single enantiomers. With homochiral syntheses being used to prepare the single enantiomers in the development phase, chiral analysis will be needed to determine trace impurities of distomer in eutomeric bulk drug and, to a lesser extent, in biological fluids, to look for any conversion of eutomer to distomer and to examine the enantiomeric nature of drug metabolites. This applies not only to new drugs in development but also to old compounds for which new uses are found or in which there is renewed interest because they had previously dropped out of development as a result of an undesirable property which can now be demonstrated to be attributable to only one of the enantiomers.

The real problems then in chiral analysis that still remain to be tackled adequately are the determination of trace distomer in eutomeric bulk drug, semipreparative enantiomer separations and the determination of enantiomers in biological fluids. While the balance may change in the future, the latter problem has acquired prominence. Although much of the current research in chiral analysis is still being directed toward the study of chiral recognition mechanisms in order to be able to develop systems with higher enantioselectivity and toward producing broader spectrum (but unfortunately less predictable) chiral systems, it does have some relevance to overcoming difficulties in these three problem areas. More importantly though, there is increasing evidence that these problems are being met head on, and it is quite striking that very often multicolumn approaches are finding favour.

For the determination of trace distomer in eutomer, it is normally sufficient to obtain good enantioresolution with the distomer eluting before the eutomer in order to achieve limits of detection <0.10% w/w. In principle, if limits of detection significantly lower than this were required, then a multicolumn approach could be adopted to give trace enrichment. However, in practice, such low limits of detection would only be required if the distomer were highly toxic, and in such cases it is unlikely that the eutomer would be developed as a drug candidate.

With respect to preparative isolation of enantiomers, significant factors to be borne in mind are the high cost of chiral stationary phases and the likely future trend towards semipreparative applications in drug discovery rather than production-scale applications. These factors point to a preference for smaller scale, highly automated preparative LC systems rather than ones employing very large columns. A good illustration of this is the work on chiral displacement chromatography carried out by Vigh *et al.* (1990). While this did not actually involve column-switching or a multicolumn approach, it required a high degree of automation with computer control of switching between mobile phase and displacer flow streams. A classic example of a multicolumn approach is the work of Cooper and Jefferies (1990) on the preparative isolation of enantiomers of trimeprazine. This elegant piece of work involved not two but three columns. The initial chiral separation was carried out on an octyl-silica column using β-cyclodextrin at a concentration of 9 mg/ml as a mobile phase additive. The separated enantiomers were then each diverted to separate "recovery" columns containing Lichroprep RP18. Once a sufficiently high load of pure enantiomer had been collected, the recovery columns were subjected to a stream of mobile phase to displace the cyclodextrin. Then, in turn, the enantiomers were displaced from the recovery columns using a stronger eluent to yield (+)- and (-)-enantiomers as their trifluoroacetate salts by virtue of 0.1% v/v trifluoroacetate being used in the mobile phase. Using this method the throughput for 4 mm i.d. columns was 1 mg/h for each enantiomer, produced free from cyclodextrin and with optical purity greater than 95%.

The main application area, however, for column-switching or multicolumn approaches to chiral separations is in chiral bioanalysis. This topic, therefore, provides the principal theme of this chapter.

2. The Need for Multicolumn Approaches to Chiral Bioanalysis by Liquid Chromatography

Column-switching techniques have application in pharmaceutical analysis in general (Chow, 1991). With the advances that have been made in computer-controlled LC instrumentation and the availability of high precision, low dead volume switching valves, the technology for column switching is now readily accessible. It is used primarily for automated sample preparation, particularly when there is a need for trace enrichment or fraction-cutting. Method development times for this on-line approach to sample preparation may be longer than for manual off-line methods, but they are much less time consuming when actually in routine operation (see also Chapter 2). They therefore allow a high sample throughput, and accordingly are well suited to use in pharmaceutical laboratories since applications involving high sample numbers are quite common. In addition, they allow greater operator safety, are not so subject to sample loss during transfer stages as off-line methods and, given the use of two or more high performance columns, have great resolving power.

Multicolumn approaches have been adopted even more in chiral bioanalysis than in bioanalysis in general. This goes further than the obvious need to protect the expensive chiral columns, which may often not be physically or chemically robust, from exposure to large quantities of biological material. It is fairly common that a method developed for the enantiomers of an analyte in standard solution turns out to be unsuitable when the same analyte is presented in a biological sample. Specifically, metabolites, or endogenous compounds, are found to co-elute with the peaks of interest. It is then almost as common to discover that fine-tuning of the chromatographic conditions results in the loss of the critical chiral resolution, rather than solving the interference problem. This is because the factors which influence enantioselectivity and structural selectivity on a particular chiral stationary phase (CSP) are often at odds.

A widely used solution to such problems is the use of two or more different chromatographic systems, with different chromatographic selectivities, in the same assay. One (achiral) system is used to separate the compound of interest from interfering materials, then a chiral separation is carried out (on a CSP) to resolve the enantiomers of the solute (or solutes) of interest.

The two stages of the analysis, achiral and chiral, may be combined in one of three ways. The two modes may be carried out sequentially, in two different chromatographic systems, which can employ entirely different eluents, if necessary. Alternatively, the achiral and chiral columns may be coupled together directly in series, so that the same eluent flows through both. Finally, the two systems may be essentially separate, with transfer of solutes between the two stages being effected using switching valves. These three approaches are discussed in more detail below.

2.1. Sequential Achiral-Chiral Chromatography in Isolated Systems

As stated above, drug metabolites and endogenous materials may interfere in a chiral assay. A relatively simple, yet versatile, solution to this problem is the use of two different chromatographic systems.

Firstly, the extracted sample is injected into an achiral chromatographic system. This chromatograph is optimised to give the best possible separation of the solutes of interest from compounds that would otherwise interfere in the chiral separation. The peak(s) containing the compound(s) of interest are collected as they elute from this first system, usually by means of an automated fraction collector. If an internal standard is included, achiral quantification can be carried out at this stage (Fig. 1).

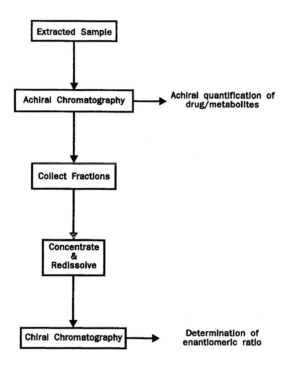

Fig. 1.   Multidimensional achiral-chiral chromatography carried out sequentially. In separation chromatographic systems. Quantification is usually carried out at the achiral stage.

When normal-phase eluents are used, the fractions collected from the achiral system may be evaporated to dryness. Alternatively, if the mobile phase contains a non-volatile buffer, the solute may either be extracted from the fraction into an immiscible solvent, which is then evaporated off, or concentrated on a solid phase cartridge from which the solute is eluted with a suitable solvent. In either case, the residue is reconstituted in a small volume of a solvent suitable for injection into the chirally selective chromatographic system, ideally the appropriate mobile phase. Separation of the enantiomers of the solute of interest in the chiral system allows the enantiomeric ratio to be determined. If quantification was performed at the achiral stage, then the concentrations of the individual enantiomers present in the original sample may be calculated. This is the preferred approach, as it gives greatest overall precision to the method. Maximum precision in quantification is obtained at the achiral stage, while each enantiomer serves as the other's "internal standard" in the determination of enantiomeric ratio, which obviates the effects of any sample loss. (Peak area ratios are normally assumed to be an accurate representation of enantiomer ratios. However, if there were any suspicion that there was chiral discrimination during sample preparation, e.g., different rates of release in complete release from protein-binding, then a calibration graph must be constructed.)

The major advantage of this sequential approach to chiral bioanalysis is the ability to use the best possible achiral and chiral separations, even when the two modes of chromatography are mutually incompatible. For instance, a reversed-phase achiral separation may be used in the first stage of the assay, and a normal phase system used for the enantioseparation. The achiral and chiral separations can therefore be truly optimised, without compromise. The approach is applicable to the enantioselective analysis of several components in a complex mixture (e.g., a drug and its metabolites). Each component of interest is collected at the achiral stage, and each is subsequently analysed enantioselectively. If necessary, different conditions for the chiral separation of each compound may be used.

The major disadvantage of this strategy is that it may be highly labour intensive, and, consequently, very time consuming. Sample injection and fraction collection may be automated, but the treatment of the collected fractions usually requires manual intervention. When large numbers of samples are involved, this may become a serious difficulty.

Despite these difficulties, the great versatility of this particular approach means that it is currently the most widely adopted strategy in multidimensional achiral-chiral bioanalysis. A relatively large number of examples of this approach have been reported. The following is a representative example.

The multidimensional enantioselective analysis of the antiepileptic compound pheneturide in serum has been achieved, using a sequential achiral-chiral method (Wad, 1988). The drug was isolated from serum using a reversed-phase separation, and subsequently analysed enantioselectively on a separate normal-phase system. A C8 column was used to resolve the drug from serum constituents, co-administered phenobarbital, two metabolites and hexobarbital, which was added as internal standard. The drug was quantified at this stage, and its peak collected as it eluted from the detector. The fraction of eluent (acetonitrile-water) containing the drug was evaporated to dryness and the residue redissolved in 200 ml of ethanol. A 25 ml aliquot of this solution was injected onto a "Pirkle-type" 3,5-dinitro-benzoyl-phenylglycine chiral stationary phase, for determination of enantiomeric ratio. The mobile phase in the chiral system was hexane-propan-2-ol (90:10, v/v). The authors found the (+)-enantio-mer of pheneturide to be present in the serum samples examined at levels approximately fifteen times greater than those of its antipode. Other drugs determined using separate achiral and chiral systems include: diltiazem (achiral = RP-C18, chiral = derivatised cellulose), (Ishii *et al.*, 1991); disopyramide (achiral = NP-Si, chiral = protein-based RP), (Enquist and Hermansson, 1989); hydroxychloroquine and its metabolites (achiral = RP-cyano, chiral = protein-based RP), (Iredale and Wainer, 1992).

2.2. Serially Coupled Achiral-Chiral Chromatography

The simplest approach to multidimensional achiral-chiral chroma-tography, in logistical terms, is to couple the two columns together in series (Fig. 2). In this approach, compounds that would otherwise

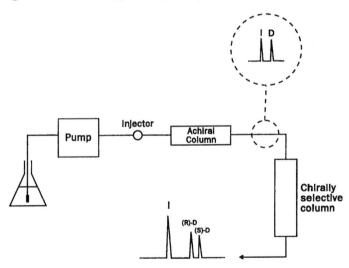

Fig. 2. Multidimensional achiral-chiral chromatography using two columns coupled in series. On the first (achiral) column, a compound 1, which would otherwise interfere in the chiral separation, is resolved from a chiral drug. The enantio-separation of D is effected on the coupled CSP.

interfere with the chiral chromatography of a solute are separated from the analyte on a short achiral precolumn. Such applications are limited by the requirement that a single mobile phase is effective in both achiral and chiral separations, and also by the fact that all constituents of the sample pass onto the "fragile" chiral column. However, this approach can be a useful method of circumventing a relatively common problem in chiral LC. Often, only a small change in selectivity is required to resolve an interfering peak, whether from a metabolite or from an endogenous compound, from the peaks of interest. However, it is usually the case that this small change is impossible to effect without losing the all-important chiral resolution at the same time. Occasionally, an achiral stationary phase can be found that, under the mobile phase conditions required for the chiral separation, is able to resolve the interfering components. If this is so, the achiral column can be coupled to the head of the chiral column. Separation of the compound of interest from the interference, as well as stereochemical resolution of the drug, can then be achieved in the same chromatographic run. However, as stated above, in practice, this may be difficult, if not impossible, to achieve, due to the unsuitability of most achiral phases.

The major problem that is often faced is the large difference in relative retentivity between achiral and chiral stationary phases, particularly those operating in reversed-phase mode. One of the most successful groups of CSP in bioanalysis are those based on immobilised proteins, which are useful in a large number of disparate applications. The protein-CSP generally require very weak eluents, compared to those used on regular RP-LC phases, which can cause problems when the two are combined. Even a short C8 or C18 column, connected at the head of a protein-CSP, can effectively trap solutes under the mobile phase conditions required by the CSP for enantioseparation. Other achiral phases may provide inadequate selectivity under the prevalent mobile phase conditions. Good compromise achiral stationary phases in these circumstances are C-2 or diol-silicas. These phases can often provide just enough selectivity to resolve solutes from interferences, while not being overly retentive when using weakly eluting mobile phases.

The direct coupling of achiral and chiral columns has provided a number of successful solutions to interference problems in the chiral LC of solutes in biological samples. For instance, disopyramide and its major metabolite, monodesisopropyl disopyramide (Fig. 3), are each stereochemically resolved on a CSP based on acid glycoprotein (AGP). However, when the two were simultaneously chromatographed on the AGP-CSP, the later eluted enantiomer of the metabolite co-eluted with the first-eluting enantiomer of the parent drug. Modification of the chromatographic conditions led only to loss of

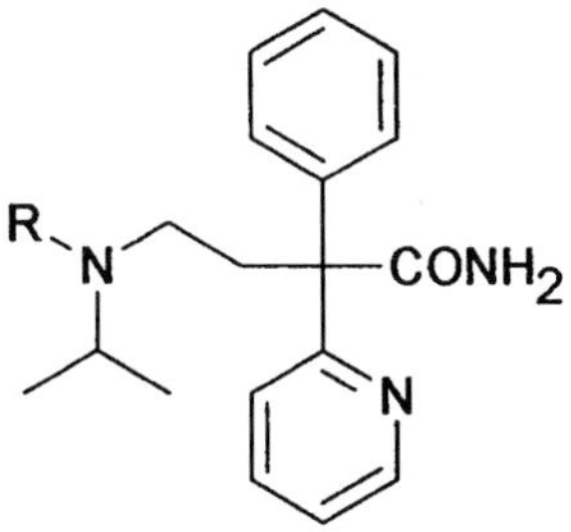

R= isopropyl: disopyramide
R = H: monodesisopropyl disopyramide

Fig. 3.   The molecular structure of disopyramide and its
major metabolite, monodesisopropyl disopyramide.

chiral resolution.   However, an achiral precolumn (either C2 or C8),
directly coupled to the head of the chiral column, separated drug and
metabolite sufficiently from each other to allow the two to be
stereochemically resolved in the same run (LeCorre et al., 1988;
Hermansson et al., 1984).

2.3. Achiral-Chiral Chromatographic Systems, Coupled by Means of Switching Valves

Frequently, it will not be possible to find an achiral and a chiral
phase that can be directly coupled to achieve direct separation of
solutes from interferences, and enantioseparation in a single injection.
It will be generally more desirable to be able to use optimal conditions for both the achiral and chiral separations.   However, as
described above, the use of a separate LC system for each chromatographic stage can be unwieldy, and the approach is not well suited to
studies involving large numbers of samples.   If the eluents required
for the different chromatographic stages are broadly compatible (e.g.,
both reversed-phase, or both normal phase), then the potential exists
to link them together by means of one or more switching valves (Fig.
4).   Compounds of interest are transferred onto the chiral column, via
a switching valve, after separation from interfering components on an
achiral phase.   The operation of the switching valves may be automated under computer control, making the whole process well suited
to large studies.   If two or more solutes are to be analysed stereochemically, these may be switched onto separate holding columns or
loops, for subsequent individual analysis.

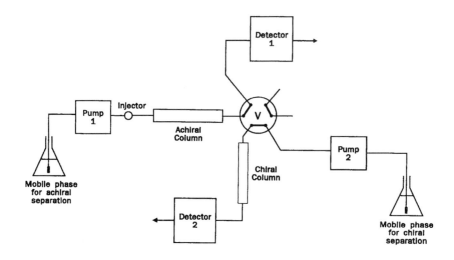

Fig. 4. Multidimensional achiral-chiral chromatography in two separate chromatographic systems linked by means of a 6-port switching valve, V. Many other configurations are possible.

Problems, particularly distortions in peak shape, may arise because small amounts of eluent from the achiral system are inevitably transferred into the chiral system. There may be some justification, therefore, in attempting to minimise differences in the eluents of the two stages, as far as is possible without compromising either separation.

The column-switching approach is attractive in bioanalysis because it is often possible to employ a less time-consuming sample work-up than would otherwise be required, as only the analyte is transferred onto the "fragile" CSP. Column-switching applications have been described for each of the major classes of CSP.

A column-switching method has recently been described for the determination of the enantiomers of the anti-malarial drug mefloquine (Fig. 5) in plasma. After a liquid-liquid extraction, samples were injected onto a cyanopropyl column which allowed separation and quantification of the drug. A double-valve system was used to back-flush the drug onto a CSP based on (S)-α-naphthylethylamine with immobilisation via a urea linkage, where its enantiomeric ratio was determined. The method was found to be sensitive enough for use in human pharmacokinetic studies (Gimenez *et al.*, 1993).

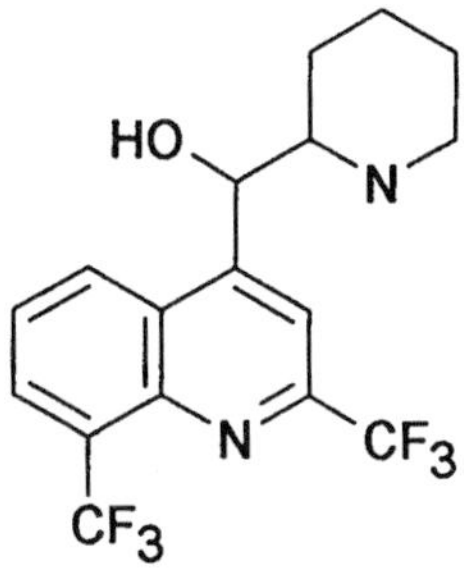

Fig. 5. The molecular structure of the
anti-malarial drug mefloquine.

The stereoisomers of leucovorin (LV, Fig. 6) and its metabolite 5-methyltetrahydrofolate (5-METHF), in plasma samples, were resolved by Wainer's group (Silan *et al.*, 1990). A CSP based on bovine serum albumin (BSA) was used to separate (6R)-LV and (6S)-LV, and (6S)-5-METHF from interfering plasma components, using a mobile phase composed solely of sodium phosphate (100 mM, pH 5.1).

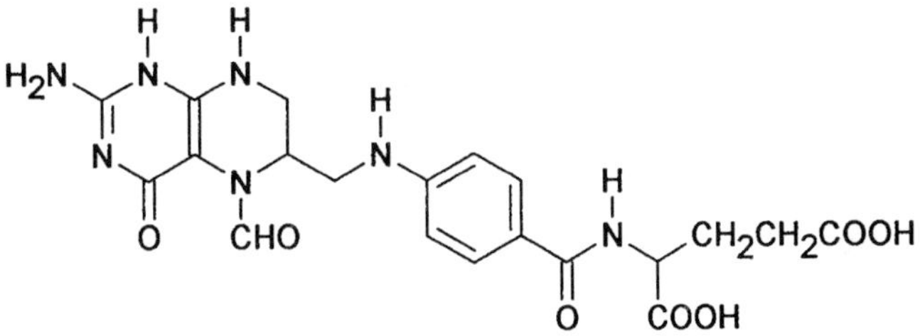

Fig. 6.   The molecular structure of leucovorin.

The eluent fraction from the CSP containing (6S)-LV was switched onto one C18 column, and the fraction containing (6R)-LV and (6S)-5-METHF (which co-eluted on the BSA-CSP) was switched to a second C18 column. The low eluting strength of the mobile phase used for the chiral separation meant that the target compounds were concentrated into very sharply defined bands at the head of each achiral column, and were not eluted any further. After the achiral columns were loaded, the solutes were sequentially eluted by rapidly increasing the eluting strength of the mobile phases in the achiral systems. This caused the solute bands to retain their sharpness and, at the same time, allowed the separation of (6R)-LV and (6S)-5-METHF.

This enhancement allowed detection of LV stereoisomers down to 5 ng/ml, an improvement of over 100-fold compared to a previous assay.

A further example of the column-switching approach is the work of McAleer *et al.* (1992) on the measurement of the (R)- and (S)- enantiomers of warfarin in patients undergoing anticoagulant therapy. Attempts to use direct injection of biological samples onto a Pinkerton internal-surface reversed phase column in the achiral stage of the method proved unsuccessful. Therefore, both solid phase sample preparation and achiral LC on a C8 column were required to separate warfarin from all endogeneous and metabolic interferences to allow switching of a "pure" warfarin peak onto the chiral column, an AGP-CSP. Total warfarin was determined on the achiral system using UV detection at 308 nm, while the ratio of enantiomers was determined on the column-switching system using fluorescence detection with excitation at 300 nm and emission at 390 nm (Fig. 7).

3. Use of On-line Reactor Columns

The use of automated multicolumn approaches in chiral bioanalysis is not restricted to the use of multiple separation columns. It is also possible to employ reactor columns to carry out on-line derivatisation. For bioanalysis where no enantiomer determination is required there are arguments for using either pre- or postcolumn reaction. However, for chiral bioanalysis, it is eminently more sensible to choose precolumn derivatisation since it is possible to introduce a moiety which will help facilitate the chiral separation as well as enhance detectability.

This approach has been demonstrated by Krull and coworkers (Gao and Krull, 1989). In early examples, they used immobilised FMOC-proline with the separation of the resultant diastereomers on an achiral reversed-phase system. Their automated system is shown in Fig. 8. Validation of this on-line derivatisation was carried out using d,l-amphetamine as an illustrative example (Fig. 9). It was found that linearity of the overall measurement was 3-4 orders of magnitude, 1.1% "enantiomeric contamination" could be detected, amounts as low as 50 ng.ml could be derivatised and detected, and good precision (RSD 1.8-6.4%) could be obtained for the direct injection of spiked urine samples. In a later example (Bourque and Krull, 1991), they used achiral derivatisation followed by chiral LC. A precolumn reactor column was used for dinitrobenzoylation and a π-basic Pirkle-concept column was used for separation. Although some problems were encountered in transferring aqueous samples derivatised in a predominantly aqueous system onto an essentially straight-phase

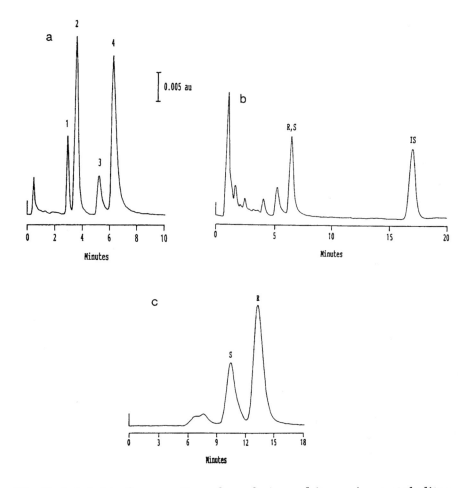

Fig. 7. (a) Achiral separation of warfarin and its major metabolites
on C8 column from spiked serum.   Mobile phase 0.015 M
phosphate buffer 15% v/v propan-2-ol (pH 7.0) at 0.65 ml/min.
(1) 6-Hydroxywarfarin; (2) 7-hydroxywarfarin and one warfarin alcohol; (3) second warfarin alcohol; (4) warfarin. (b)
Achiral separation of a patient sample. Conditions as for (a).
RS racemic warfarin and IS internal standard. (c) Chiral
separation of warfarin from patient sample shown in (b).
Mobile phase as for (a) at 0.9 m/min.

system, highly effective chiral separations of amines, amino alcohols
and amino acids were achieved.   The practical value of these
approaches was enhanced by the fact that it was very easy to
regenerate the reactor column once the reagent had been consumed.

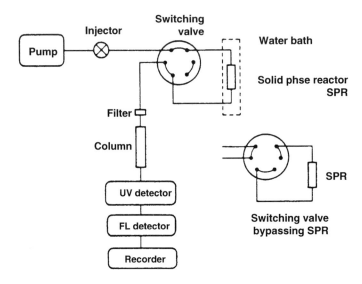

Fig. 8. Instrumentation for the on-line, precolumn solid phase derivatization LC-UV-FL detection.

It could be said that, to date, this approach has been restricted to acyl-transfer reactions. However, this is not necessarily a limitation. Acylations are very important with respect to derivatisation for chiral separations, especially in the case of amines in that they introduce a strong dipole in the form of an amide group, which is often critical in providing an additional analyte-CSP interaction to allow chiral recognition. Also, it has to be said that the feasibility of the technology has been established and, with modification, it may be extended to other reaction types. Furthermore, since the main advantage of the use of achiral derivatisation followed by LC on a Pirkle-concept column is that it allows for very rapid method development into a working, though not necessarily optimum, method with good enantioresolution and low limits of detection, this approach should be attractive for drug development in industry. Off-line derivatisation would allow a method to be set up quickly to cater to the early stages of development and then, once a compound had been established in development, the method could be converted to an on-line derivatisation system to cater to pharmacokinetic and other large studies.

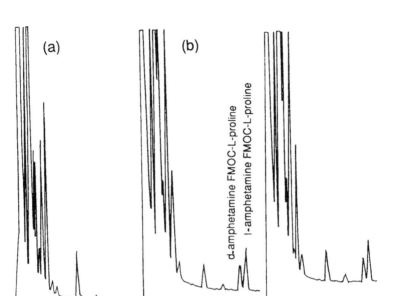

Fig. 9. (a) Urine blank; (b) repeated injections of spiked urine with *d,l*-amphetamine via on-line solid derivatizations.

4. Future Perspectives

The ultimate solution to most analytical problems is usually regarded as presenting the sample directly to the analytical instrumentation without any sample preparation steps and having the results produced from the analytical instrumentation without any further manual intervention. Taking this notion at face value, a fully-automated multicolumn approach to bioanalytical problem-solving has a lot to commend itself and, appropriately, is growing in popularity. Having said this, there are some potential problem areas that should not be overlooked. First and foremost, for each new bioanalytical problem, much needs to be done in the way of method development and optimisation. It is perhaps unrealistic to imagine that an expert system could ever take on this task on a routine basis. Secondly, since a fully-automated multicolumn LC approach involves a number of moving parts, care has to be taken to ensure physical robustness. Thirdly, it is also a composite of many individual computer-controlled steps and, even if each step seems to be reliable, the greater the number of steps, the greater the potential for error.

Because of these limitations, the current vogue for multicolumn approaches to chiral bioanalysis, while very proper, may not be enduring. The prospect of striving for simpler "black box" solutions may be compelling. For instance, capillary electrophoresis (CE) is already proving to be a very versatile vehicle for obtaining enantiomer separations in aqueous media (Noctor, 1993). Also in CE there is less need for the elimination of biological materials prior to analysis. Therefore, if detection limitations can be further overcome, CE approaches may eventually offer a more attractive practical solution to chiral bioanalysis than multicolumn LC. It is also worth bearing in mind that incorporation of CE steps may enhance the multicolumn approach to chiral bioanalysis.

As alluded to earlier, with the moves toward homochiral synthesis in drug development, there will eventually be less demand for chiral bioanalysis. However, in the meantime, there are sufficient studies that need to be carried out on existing drugs that have been marketed as racemates to ensure the need for multicolumn approaches to chiral bioanalysis for some time to come.

5. References

Brown, J. R. (1990). *Drug Info. J.*, 24, 117–120.
Bourque, A. J. and Krull, I. S. (1991). *J. Chromatogr.*, 537, 123–152.
Cooper, A. D. and Jefferies, T. M. (1990). *J. Pharm. Biomed. Anal.*, 8, 847–851.
de Camp, W. H. (1989). In: *Chiral Liquid Chromatography* (W. J. Lough, ed.), Chap. 2, pp. 14–22. Blackie, Glasgow.
Enquist, M. and Hermansson, J. (1989). *J. Chromatogr.*, 494, 338–345.
Gao, C.-X. and Krull, I. S. (1989). *J. Pharm. Biomed. Anal.*, 7, 1183–1198.
Gimenez, F., Dumartin, C., Wainer, I. W. and Farinotti, R. (1993). *J. Chromatogr.*, 619, 161–166.
Hermansson, J., Eriksson, M. and Nyquist, O. (1984). *J. Chromatogr.*, 336, 321–328.
Iredale, J. C. and Wainer, I. W. (1992) *J. Chromatogr.*, 573, 253–258.
Ishii, K., Banno, K.., Mitamato, T. and Kakimoto, T. (1991) *J. Chromatogr.*, 564, 338-345.
LeCorre, P., Gibassier, D., Sado, P. and LeVerge, R. (1988). *J. Chromatogr.*, 450, 211–216.
McAleer, S. D., Chrystyn, H. and Foondun, A. S. (1992). *Chirality, 4*, 488–493.

Noctor, T. A. G. (1993). *Chromatographic Society Bulletin*, *38*, 61–66.

Silan, L., Jadaud, P., Whitfield, L. R. and Wainer, I. W. (1990). *J. Chromatogr.*, *532*, 227-236.

Vigh, G. Y., Quintero, G. and Farkas, G. (1990) *J. Chromatogr.*, *506*, 481–494.

Wad, N. (1988). *J. Liq. Chromatogr.*, *11*, 1107–1116.

Part Three: Liquid Chromatography Methods for the Preparation of Drug Substances

CHAPTER 8

Applications of Preparative Liquid Chromatography in New Drug Discovery

ALFRED J. MICAL and MARK A. WUONOLA

DuPont Merck Pharmaceutical Company
Wilmington, DE 19880 U.S.A.

1. Introduction

A major goal in pharmaceutical research is the isolation of pure drug substances from complex mixtures. Many times, simple procedures can be used. When simple laboratory procedures fail, liquid chromatography (LC) is often effective. It is one of a very few techniques that can achieve the degree of purity necessary for valid bioactivity and toxicity measurements.

As the following table shows, increasing demands are placed on the chromatographer as the drug candidate proceeds through evaluation studies. The quantity and purity requirements also become more important since toxicology data will be the basis for the decision to continue or terminate a pharmaceutical program. The largest amounts will require the highest purity possible.

Study	Milligrams of Enantiomer
bioscreening	1–2
optical rotation	10–20
pharmacokinetics	1,000–2,000
toxicity	10,000–25,000

The isolation of pharmaceutical drugs represents a great challenge to preparative chromatographers. The compounds most frequently encountered are also the most troublesome in chromatography—

basic amines and their isomers. The difficulty lies in handling reactive compounds and separating them from undesirable compounds that show only minor chromatographic differences. Stability is a concern and can be enhanced by converting the compounds to salts. The salts are water soluble and an appropriate form for pharmacology. Forming salts provides an advantage in stability and pharmacology, but the chromatography may be more difficult because resolution between closely related compounds often becomes smaller.

2. Special Techniques: Resolving Isomers With Optimum Loading

Initial evaluation of solute for preparative chromatography uses a sequence of rules different from those in analytical work. An analytical method can be used effectively with marginal or even poor solvents because of spectroscopic sensitivity. The success of preparative work depends more on solvent selection, and so this is the first step. The specific column used will depend on the polarity of the solvent selected.

Figure 1 demonstrates the advantages of preparative scale-up derived from an analytical method. Solvent evaluation showed that chloroform was a better solvent than water for the regioisomers. These two isomers cannot be separated by adsorption mechanisms because of their molecular similarity. Hydrogen bonding and hydrophobicity mechanisms are the same for each. The only property that can be exploited for chromatographic resolution is the difference in the ionic dissociation of the carboxylic groups. The π-electrons that reside relative to the carboxylic group for each isomer act as a Lewis base and affect the dissociation of the acid. The closer the proximity to the acid group, the greater the dissociation effect.

The chromatographer has available a variety of columns that function using a single mechanism. One column that can provide more than one different resolution mechanism is the aminopropyl ($-NH_2$) column. It can function in normal and reversed-phase modes, and it can also provide anion exchange when needed. In the latter mode, the primary amine ($-NH_2$), covalently bonded to a silica backbone, is converted to the ammonium ion ($-NH_3^+$) under acidic conditions. This ion provides the exchange site. A variety of buffers or acids can produce the ion, but in preparative work volatile acids are preferred for this purpose. These modifiers can be removed by rotary evaporation, making recovery of the pharmaceutical compound easier.

It was stated earlier that the solvent of choice was chloroform. Under acidic conditions, this organic solvent can render high loading and produce the desired resolution by ion exchange. Adjustment of the acid concentration will control retention on the column and the degree of resolution. Substituting another solvent for chloroform can produce capacity changes that allow greater control of the method. This procedure can be tested or modified before column work using thin-layer chromatography plates coated with the NH_2 bonded phase.

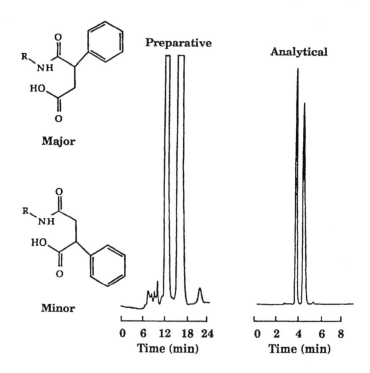

Fig. 1. Analytical conditions: column Zorbax™ NH_2, 4.6 i.d. × 250 mm; solvent 0.1% acetic acid in 50% chloroform:50% methanol; flow rate 2 ml min-1; detection, 254 nm. Preparative conditions: column Zorbax™ NH_2, 21.2 i.d. × 250 mm; solvent 0.1% acetic acid in 50% chloroform:50% methanol; flow rate 20 ml min-1; detection, 254 nm.

3. Preparative Separation of a Drug as a Free Base or a Salt

The preferred form for handling a basic compound is a stable salt. However, complex mixtures containing a large number of salts are not easily resolved chromatographically. The competitive mechanisms do not easily recognize subtle differences in large numbers of salt analytes, resulting in poor resolution. Converting a salt to a free base can present scale-up capabilities in a variety of chromatographic combinations using different bonded phases. In this condition, separation mechanisms can readily distinguish primary, secondary, and tertiary amines due to the large differences of dipole moment. There is, then, a direct correlation between column retention and drug polarity.

In normal phase chromatography, a number of bonded phases, such as CN, NH_2, polyethyleneimine (PEI) and even silica, have varying interaction potentials that allow the proper solvent combinations to be used for scale-up. These columns can be characterized as being acidic, basic, or neutral. Their distinct retention capabilities allow solvent selection while considering stability requirements. Drugs that are unstable in polar solvents can often be separated with less polar solvents on the CN column, which is less retentive. As shown in Fig. 2, bromolactone isomers are easily separated under the conditions indicated. The absence of polar solvents and the silanol sites of silica are necessary for the successful isolation of these compounds.

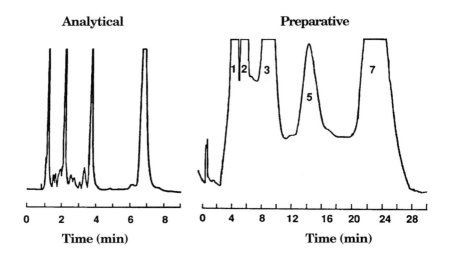

Fig. 2.   Analytical conditions: column Zorbax™ CN, 4.6 mm i.d. × 250 mm long; solvent 2% acetonitrile:98% n-butyl chloride; flow rate 3 ml min-1; detection, 230 nm. Preparative conditions: column Zorbax™ CN, 21.2 mm i.d. × 250 mm long; solvent 2% acetonitrile:98% n-butyl chloride; flow rate 20 ml min-1; detection, 230 nm; injection 185 mg.

The following classification will help in method development for all cases of drug separation that are dependent on the chromatographic environment for maintaining chemical integrity. Weak base groups, imidazole and pyridine, are common drug constituents. They interact differently with each bonded phase and have their strongest interaction with silica. Moderate and strong bases have such high interaction potentials with silica that contact should be avoided. These bases perform well on bonded phases in a normal mode. Figure 3 shows a separation of primary amines. Secondary amines are characterized by shorter retention times under similar conditions as shown in Fig. 4.

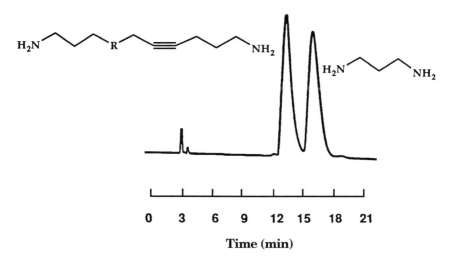

Fig. 3. Preparative conditions: column Zorbax™ NH₂, 21.2 mm i.d. × 250 mm long; mobile phase, linear gradient in 30 min; 0.2% triethylamine (TEA) in methyl-t-butyl ether to 0.2% TEA in methanol; flow rate 20 ml min⁻¹; detection, 220 nm; injection 150 mg.

The chemistry of water-soluble amino acids illustrates the behavior of a drug candidate as a salt. Natural and man-made amino acids are common constituents of many drugs. If the solution is made acidic, the acid group is the ammonium $-NH_3^+$, not the carboxylic group. The base group is $-COO^-$ and not $-NH_2$. This chemistry allows for easy separations on reverse-phase columns like C8 and C18 using acidic conditions. The reason for the acid is the inherent presence of unbonded silanol sites on all silica-based packings.

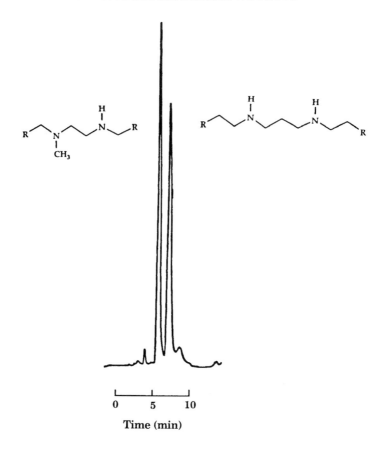

Fig. 4. Preparative conditions: column Zorbax™ NH₂, 21.2 mm
i.d. × 250 mm long; mobile phase, linear gradient in 15
min; 10% B to B, where A = *n*-butyl chloride and B =
methyl-*t*-butyl ether; detection 220 nm; injection 65 mg.

These sites are very acidic and have strong interactions with –NH₂
groups in neutral aqueous mobile phases. If this interaction is not
prevented, the analyte will complex with the silanol. This interaction
changes the performance of the column, and the analyte will itself be
subject to displacement. This chemistry can be exploited by the
chromatographer using acidic conditions to reduce potential interactions between silanol and amino groups. Other bonded phases such
as the aminopropyl (NH₂) column also exploit minor differences in
amino acid structures. A separation on an –NH₂ column that does
not work well on C8 or C18 columns is shown in Fig. 5.

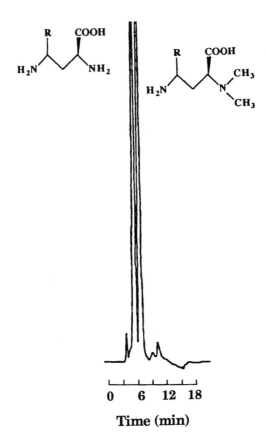

Fig. 5. Preparative conditions: column Dynamax™ NH_2, 41.4 mm i.d. 250 mm long; mobile phase, 0.2% trifluoroacetic acid in 50% methanol:50% water; flow rate 40 ml min[-1]; detection 215 nm; injection 100 mg.

4. Chiral Separations

Enantiomers of a pharmaceutical racemate often exhibit different biological activity and toxicity. This creates a need to study each enantiomer of a racemate as a separate drug candidate. Optically pure material can be obtained by forming diastereomers and separating them by physical methods such as distillation or fractional crystallization. A significant amount of research time can be spent in finding a diastereomeric derivative with colligative differences sufficient to provide separation. Asymmetric synthesis is another method used to obtain individual enantiomers. Both methods usually meet with only partial success.

A direct method of separation has been made possible with the availability of large-scale chiral columns. A wide range of enantiomers used in cardiovascular, central nervous system, inflammatory, cognition enhancement, and anti-cancer research have been purified in this manner and are being studied as separate pharmaceuticals.

A number of analytical columns on the market can achieve chiral recognition for most of the pharmaceutical racemates. This discussion focuses on those that are available in preparative sizes, such as the derivatized celluloses marketed by Daicel. A simple methodology has been developed for these columns that allows any chromatographer to obtain bulk isolations for pharmaceutical studies. This method extends the range of chiral recognition by manipulating the mobile phase and temperature. Effectively exploiting these parameters reduces the need for a large inventory of chiral columns.

An outline of considerations important for quick and successful separations is:

Procedure
   1. Achiral Analysis and Preparative Isolation of Racemic Pair
   2. Chiral Separation

Evaluation
   1. Confirmation of Isolated Target by Characterization
   2. Early Use of NMR, MS and Optical Rotation

Optimizing Preparative Loadings
   1. Effects of Increasing α
   2. Subambient Column Work
   3. Effects of Increasing Efficiency
   4. Contributions of Automation

Peak Symmetry in Preparative Work
   1. Observed Separations of Fused Peaks
      a. Neutral Analyte
      b. Basic Analyte

Column Selectivity
   1. Improving Resolution
   2. Order of Elution

Solvent Interactions
   1. Base Modifiers
   2. Acid Modifiers

Column Size
   1.   Selection of Appropriate Length
   2.   Influence of Analyte Stability
Column Care
   1.   Precautions
   2.   Storage

5. Chiral Separations on Derivatized Cellulose

The following simple methodology has been used successfully for development of rapid analytical chiral separations. The method is easily transferred to a larger preparative column using the same mobile phase but a faster flow rate. Before committing to the chiral development work, a prudent researcher should conduct an achiral analysis to determine the presence of impurities that may be present in addition to the racemic pair. This step may require the chemist to attempt additional purification by routine laboratory methods.

Figure 6 shows contamination of racemic starting material in the preparative chromatogram. Peaks 2 and 4 are the desired product enantiomers; peaks 1 and 3 represent the enantiomers of the starting material. The conditions used for the separation are adequate for large loadings because of the high alcohol content in the mobile phase and the well-resolved peaks of interest, 2 and 4. The presence of racemic precursors, peaks 1 and 3, limits additional loading and even influences the quality of the isolations for 2 and 4. Removal of the precursors would provide a significantly improved separation. Working with a pure racemic pair allows the chromatographer to confirm separation of target enantiomers. The presence of other compounds in the chromatogram will make this determination more difficult.

Most drugs possess functional groups that are characterized chemically as bases. The most successful Daicel column that has achieved chiral recognition for this class is the Daicel OD column. The chiral network is a cellulose carbamate (3,5-dimethylphenyl carbamate) coated on silica gel. All method development starts by trying the sample with the highest allowable alcohol and temperature for the column. For the Daicel OD column, this is 100% ethanol and 40°C. (For this column, the flow rate must not exceed 1 ml min^{-1} or produce a column pressure exceeding 50 bars.) This is a good starting point, and some chiral separations have been achieved using both a high alcohol concentration and elevated temperatures. If the racemates elute too quickly, the following steps should be followed:

Step 1   Reduce the alcohol content by half by diluting with hexane. After equilibration, try the separation again.

Step 2   Repeat step 1 until the separation is adequate or the alcohol concentration is reduced to less than 5%.

Step 3   If enantiomers are not resolved using ethanol: hexane, try 50% isopropanol:50% hexane.

Step 4   Repeat step 1.

Step 5   If enantiomers are not resolved, reduce temperature to ambient and begin these steps again with 100% ethanol.

Step 6   If enantiomers are not resolved, reduce temperature to 0°C and begin this procedure again with 100% ethanol.

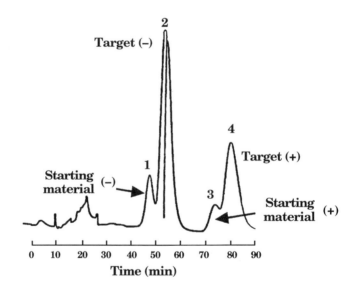

Fig. 6.   Preparative conditions: column, Chiracel OJ, 20 mm i.d. × 500 mm; solvent, 40% ethanol:60% hexane; column temperature, 24°C; flow rate, 5 ml min-1; detection, 275 nm; injection, 95 mg.

Some of the Daicel columns cannot tolerate ethanol; others cannot exceed an alcohol content of more than 50% isopropanol. Use the maximum allowable alcohol content as the method development starting point. A column temperature of 40°C will help elute highly retained enantiomers and should be used during the first steps of method development.

If the best result attained is a partial chiral resolution on the OD column, switch to a Daicel OG column and start with Step #3. This sequence has provided a large number of separations (see Fig. 7). Chiracel OJ, OB, and OF columns have provided chiral separations for other classes of racemates.

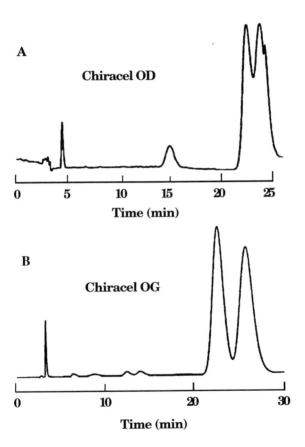

Fig. 7. **A**. Analytical conditions: Column, Chiracel OD, 4.6 mm i.d. × 250 mm; mobile phase, 20% isopropanol:80% hexane; column temperature, 20°C; flow rate, 1 ml min-1; detection, 220 nm. **B**. Analytical conditions: Column, Chiracel OG, 4.6 mm i.d. × 250 mm; mobile phase, 40% isopropanol:60% hexane; column temperature, 40°C; flow rate, 1 ml min-1; detection, 220 nm.

6. Solvent Modifiers

It is important to realize that column selection should not be based solely on the classification of functional groups in the analyte structure. This classification can be a starting point, but successful chiral recognition is shape-dependent. The cellulose carbamate is well-suited for separating basic and acidic racemates. Separations depend on acid or base modifiers, and examples will be shown for each case. Figure 8 shows the improvement in peak symmetry of the basic enantiomers achieved by adding 0.1% triethylamine (TEA) to the mobile phase. The compound hydroxychloroquine is available as

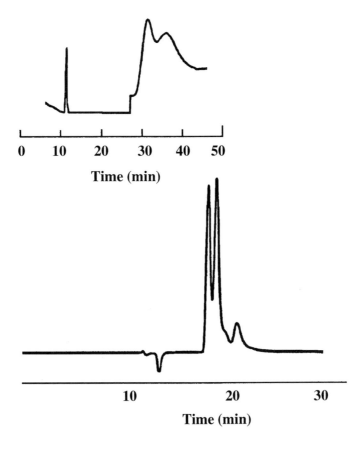

Fig. 8. (a) Preparative conditions: Chiracel OD, 20 × 250 mm; mobile phase, 20% ethanol:80% hexane; column temperature, 20°C; flow rate, 4 ml min-1; detection 200 nm. (b) LC conditions same as (a) with 0.1% triethylamine added.

a water-soluble salt. In this stabilized form, chiral recognition was not attained on any chiral column scouted. A separation was found by free-basing the salt and following the steps of the method previously described. The free base may allow an important chiral mechanism, hydrogen bonding, to take place. This bonding cannot take place with the amine salt.

Basic analytes are usually characterized by tailing in a chromatogram. To some extent bare silica sites exist even on highly chiral-coated columns. The acidity of silanol groups will provide strong interactions with the analyte base groups, and this mechanism will result in a degraded resolution. The triethylamine modifier neutralizes the silanol sites, eliminating their interaction with the analyte. The sharp peaks observed in Fig. 8 are free from silanol interactions. This solvent modification allows the high efficiency of the column to be used with the intended adsorption and chiral mechanism.

Another example of the importance of solvent modification is shown in Fig. 9. The enantiomers of carbobenzyoxyalanine (CBZ-alanine) do not elute from the Chiracel OD column without an acid modifier in the mobile phase. In this case trifluoroacetic acid was used, but formic or acetic acid may also be useful. Amino acids in

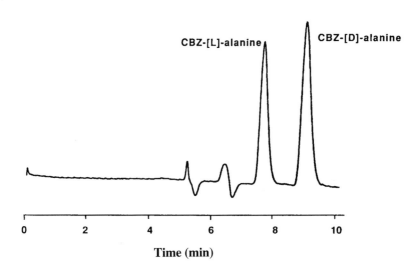

Fig. 9. Analytical conditions: column, Chiracel OD, 4.6 mm i.d. × 250 mm; mobile phase, 0.1% trifluoroacetic acid in 20% ethanol:80% hexane; column temperature, 40°C; flow rate, 0.5 ml min⁻¹; detection, 250 nm.

aqueous solution act as dipoles. In this normal phase environment, the acid modifier suppresses the dissociation of the carboxylic group in the amino acid. This reduces the dipole movement potential and creates an easily eluted analyte. The effect of the acid modifier in aqueous and nonaqueous environments on the amino acid analyte can be shown as

$$[\text{acid}]\ NH_3^+ \ + \ [\text{base}]\ COO^- \quad \text{in reverse phase} \qquad (1)$$

$$[\text{salt}]\ NH_3^+ \ + \ [\text{acid}]\ COOH \quad \text{in normal phase} \qquad (2)$$

6.1. Preparative Loadings

By necessity, analytical, semipreparative, and preparative LC differ in the size of the column and the amount of analyte used per run. In this section, preparative work will refer to the action taken that produces bulk enantiomers in one or in multiple injections on 1" i.d. preparative columns. The chromatographer has several choices for obtaining large amounts of enantiomers. The amount needed may dictate the most effective chromatographic procedure.

In most cases, preparative work proceeds from an analytical separation. However, in scaling up and overloading the column, more consideration should be given to the separation selectivity, α. α is calculated by the expression

$$\alpha \ = \ \frac{[\text{retention time peak 2} \ - \ \text{dead volume time}]}{[\text{retention time peak 1} \ - \ \text{dead volume time}]} \qquad (3)$$

On most chiral columns, reduction of alcohol content from initial analytical conditions increases retention time for each enantiomer. The retention ratio also increases and provides for larger volume and weight injections. This is shown in Fig. 10.

Because the lower alcohol content creates a weaker mobile phase with transport restrictions, the peaks are very broad. There is a definite limitation to improved loading with solvent selectivity adjustments. The alcohol content can be reduced to a level that does not support a dissolved analyte. This reduction results in precipitation of sample and severe tailing, which degrades the quality of isolations, and should be avoided. For this example a weight of 175 mg could be separated with an α of 1.95 and a run time of 110 minutes. After increasing the α by decreasing the alcohol content, larger weight injection should have been possible. However, these chromatograms showed a significant loss of resolution, presumably due to solubility effects.

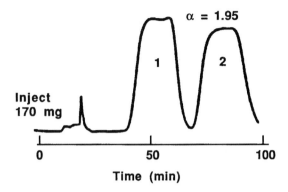

Fig. 10. Preparative conditions: column, Chiracel OJ, 20 × 500 mm; mobile phase, 20% ethanol:80% hexane, column temperature, 20°C; flow rate, 5 ml min^{-1}; detection, 290 nm.

A separation that demonstrates the use of high solvent strength, fast elution, and high column efficiency for loading optimization appears in Fig. 11. This set of conditions is characterized by shorter retention times and by sharper peaks. The loading per injection is not as high as that in the previous example, but study shows that injections of 30, 60, and 90 mg can be easily handled. If loadings are increased to the loss of the baseline separation, the amount injected is 105 mg. The running time is 55 minutes. A simple calculation shows that automating this method for repetitive injections will yield higher productivity than the other condition using a higher α.

Using the same chromatographic conditions, sample loadings of 30, 60, and 90 mg produce α values of 1.66, 1.60 and 1.55, respectively. This shows that increased loadings can influence analyte retention. The sample mass will require a specific portion of the column surface area, and increasing this mass will require a larger portion of the column surface area. This column effect increases with loading and at some loading level will cause loss of resolution. Solubility and alcohol content are involved in this phenomenon, and the loading limits must be found for each sample. Figure 12 shows a case characterized by high solubility loading. The 1000-Å pore silica has a low surface area, but high loadings are achieved because of high chiral coating and good solubility.

Another method for improving separations by increasing α is chiral chromatography at subambient conditions. Figure 13 shows the improvement of enantiomer resolution with a 20°C temperature decrease. This method is not restricted to analytical column develop-

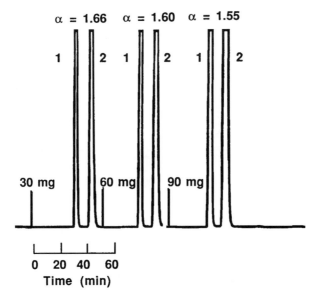

Fig. 11. Preparative conditions: column, Chiracel OJ, 20 × 500 mm; mobile phase, 50% ethanol:50% hexane; column temperature, 20°C; flow rate 4.5 ml min-1; detection, 245 nm.

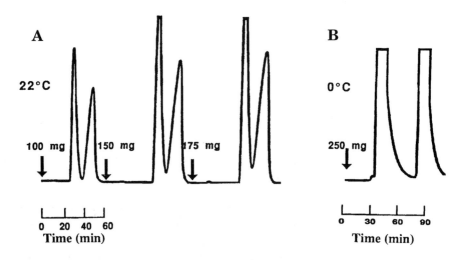

Fig. 12. A. Preparative conditions: column, Chiracel OD, 20 × 250 mm; mobile phase, 65% isopropanol:35% hexane; column temperature, 20°C; flow rate, 4 ml min-1; detection, 240 nm. B. Preparative conditions: column, Chiracel OD, 20 × 500 mm; mobile phase, 50% ethanol:50% hexane; column temperature, 0°C; flow rate, 3.5 ml min-1; detection, 225 nm.

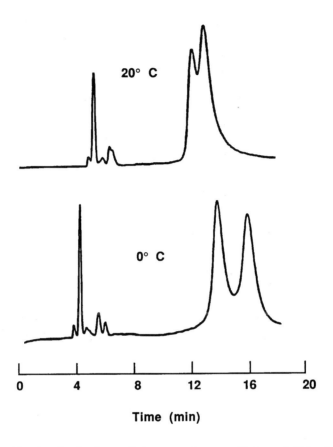

Fig. 13. Analytical conditions: column, Chiracel OD, 4.6 ×
250 mm; mobile phase, 0.1% triethylamine in 10%
isopropanol:90% hexane.

ment. This figure shows a preparative run that was conducted at two
temperatures. Partial resolution is obtained at a column temperature of 20°C. This resolution is enhanced when the column temperature is lowered to 0°C. Even at 0°C, recycling was required to obtain
high purity fractions for each enantiomer. Figure 14 shows recycled
peaks with improved resolution. Using peak shaving and recycling,
the chromatographer has at his disposal a variety of methods to
choose from. This choice will depend on the configuration of the
equipment.

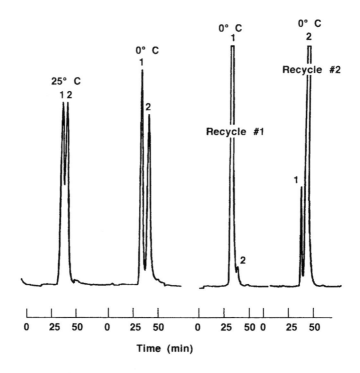

Fig. 14. Preparative conditions: column, Chiracel OD, 20
× 500 mm; mobile phase, 30% isopropanol:70%
hexane; detection, 275 nm.

6.2. Column Care

Chiral coatings on the Daicel Chiracel columns are susceptible to
shear when subjected to high flow rates or high column pressures.
The recommendations of the column manufacturers should be
adhered to for these limitations. Flow rates that are appropriate at
room temperature may not be used at reduced temperatures. Viscosity is dependent upon temperature and will increase dramatically with
subambient work.

CHAPTER 9

Application of Liquid Chromatography

to the Purification of Proteins and Peptides

SUNANDA R. NARAYANAN

Drug Metabolism, Oread Laboratories Inc.
1501 Wakarusa Drive, Lawrence, KS 66047, U.S.A.

1. Introduction

With the advent of recombinant DNA technology and the potential of recombinant-DNA-based protein pharmaceuticals, high resolution separation strategies are gaining widespread attention. Today chromatography is considered an essential step in the purification of insulin and other hormones, various growth factors, interferons, monoclonal antibodies and plasma proteins. Since stringent quality control procedures are imposed on therapeutic-grade protein products, separation procedures are also required to address the removal of contaminants that are commonly present in these preparations and are responsible for toxic side effects.

In 1959, Sephadex® was used for the first time as a gel filtration medium, and since then chromatography has been the backbone of purification protocols (Porath and Flodin, 1959). With the exception of highly specialized affinity supports, a large majority of chromatographic supports are commercially available (Janson and Kristiansen, 1990; Narayanan and Crane, 1990; Unger, 1990; Narayanan, 1993). Several excellent reviews on the application of LC to peptides and proteins have been published with a wealth of information (Horvath, 1983; Henschen *et al.*, 1985; Fallon *et al.*, 1987; Huang and Guichon, 1988; 1989; Hammond and Scawen, 1989; Kennedy *et al.*, 1989; Mant and Hodges, 1991a).

The purification of peptides and proteins is traditionally regarded as a difficult and laborious task. Since no two proteins are alike, each purification strategy has to be derived empirically and optimized individually. This chapter is intended to be a guide for researchers who are already familiar with the techniques of purification, but it is by no means intended to be comprehensive.

2. Modes of Separation

There are three main modes of chromatography in the purification of peptides and proteins (Huang and Guichon, 1989; Hancock, 1990). These are:

1) Frontal chromatography
2) Elution chromatography
3) Displacement chromatography

In frontal analysis, extensive regeneration of the column is required prior to the next cycle of purification. For this reason, frontal chromatography is rarely used for purification purposes. In elution chromatography, the retention of analytes depends on the rate of variation of the mobile phase composition. Elution chromatography is further classified into isocratic and gradient modes. Gradient elution can be linear or nonlinear (Huang and Guichon, 1989). In the elution mode, production rate increases with increasing volume and increasing concentration of the sample introduced into the column. Since the band width of the eluates also increases in both cases, there is a limit to the sample size which can be effectively separated. The effect of experimental conditions on the production rate and yield of several compounds has been reported (Huang and Guichon, 1989). Reports on optimization of experimental conditions of these model compounds are also available.

The mode of operation is selected based on the nature of the sample to be separated, the purity required and the recovery and rate of expected production. In the last few years Horvath and collaborators have introduced displacement chromatography as an alternative means of preparative purification (Horvath et al., 1983). Displacement chromatography is based on the same retention mechanism as elution but the sample is pushed out of the column by introducing a displacer. This results in the mixture being separated as a result of competition for the binding sites on the support material. This is a nonlinear chromatographic technique that allows higher sample loadings than are possible in the elution mode. The advantage of this mode of chromatography can be illustrated in the separation of about 20 mg of growth hormone on a C_{18} reversed-phase column (0.4 × 15 cm) in the displacement mode compared to 0.2 mg in the elution mode. Displacement chromatography has also been applied in the separation of β-lactoglobulin A and B on an anion exchange column using chondroitin sulfate as the displacer (Liao et al., 1987).

There are some disadvantages in using the displacement mode. The regeneration of the column is a lengthy process, which affects the production rate. Optimizing the displacer is often tedious and expensive. Since researchers are more familiar with the elution mode and feel comfortable about scaling up a separation strategy which works at the analytical level, displacement chromatography has not gained much attention. It has nevertheless been used in the purification of peptides, proteins and several other compounds (Horvath *et al.*, 1983; Torres *et al.*, 1984; 1987; Cramer *et al.*, 1987; Liao *et al.*, 1987; Katti and Guichon, 1988).

Several studies have addressed the issue of band spreading in chromatography and have suggested that band spreading is predominantly caused by intraparticle diffusion (Regnier, 1991). Longitudinal diffusion, solute diffusion in the mobile phase, and adsorption-desorption kinetics play a less significant role. To overcome this problem, a new chromatographic support material based on the concept of perfusion chromatography has been introduced which has particle transecting pores of 6,000–8,000 Å diameter to allow liquid flow through the sorbent. The reported advantage of this perfusion process is that intraparticle flow convectively transports solutes to the interior of the particle. Using this new support (POROS), proteins can be separated at velocities 10–30 times higher than those used in high performance liquid chromatography (Afeyan *et al.*, 1990). Fig. 1 illustrates an analytical reversed-phase separation of standard test proteins on POROS support. This technique has proven useful in routine analyses where large numbers of samples are processed and in a process environment where enhanced intraparticle mass transport will increase throughput and productivity.

3. Method Selection, Optimization and Scale Up

The requirements of a purification protocol are determined mainly by the nature and quality of the desired final product and its intended use. For example, proteins for therapeutic use must be extremely pure to minimize the risk of unwanted side effects or immunogenic response.

Four factors dictate the suitability of different separation procedures. These are:

1) the concentration of desired product in the starting material,
2) the composition of other components in the starting material along with their physical and chemical properties,
3) the desired product purity,
4) the volume of starting material to be processed.

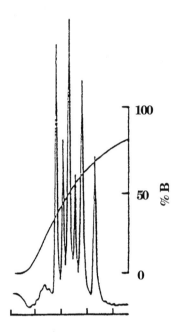

Fig. 1.   Analytical reversed-phase separation of standard test
            proteins on POROS R/H perfusion column (5 x 6 mm
            i.d.). Sample consists of ribonuclease A, lysozyme, β-
            lactoglobulins A and B and ovalbumin. Mobile phase is
            0.1% TFA in water. Gradient is 4–75% acetonitrile in
            19 column volumes. Flow rate is 4.0 ml min-1 (850
            cm/hr). Detection is at 280 nm. (Reproduced with
            permission from Afeyan et al., 1990.)

In order to design a purification strategy, the physicochemical and
biological properties of the sample are determined after the initial
extraction of the sample from growth media (in the case of recombinant proteins) (Fish and Lilly, 1984; Strathmann, 1985; Datar, 1986;
Low, 1990). In general, steps are chosen to give an effective concentration of the product with minimal loss of material.

In the initial stages of purification, the separation strategy is
focused on concentrating or enriching the desired product, which is
usually present in dilute amounts. This step is also targeted at
handling large volumes of the unpurified extract. Ion-exchange, affinity or hydrophobic interaction chromatography (HIC) are usually the
methods of choice as an initial step. If the sample is present in excess
salt after a precipitation/extraction step, HIC is well suited since the
sample can be directly loaded onto an HIC column.

In general, adsorption methods are favored during the initial stages of purification, since the volume to be processed usually remains large. These intermediate steps are important in removing major contaminants while maintaining the biological activity of the protein of interest. As few steps as necessary are introduced at this stage to keep the recovery high. Affinity chromatography is a popular method as an intermediate step in a purification process. Table 1 illustrates the guidelines that are followed in choosing chromatographic steps in a purification process.

In the final stages of purification, the product is separated from low-concentration and low molecular weight contaminants. Again, ion exchange, affinity or reversed-phase chromatography are used to remove the low concentration contaminants. Size-exclusion is generally used to remove the low molecular species or in some instances the polymeric form of the desired product.

Bonnerjea (1986) reports that ion-exchange chromatography is the most widely used method, followed by affinity chromatography. Gel filtration is used as a purification step (as distinct from a desalting step) in half of the cases evaluated. HIC is used in less than a third of all the analyses. In most cases, the trend is to use ion-exchange chromatography followed by affinity and finally size-exclusion. Bonnerjea also reports that affinity techniques are nearly an order of magnitude more effective than the other techniques; the average purification achieved is over 100-fold. In terms of recovery, ion-exchange methods recover about 66% while affinity methods recover about 61 % on average.

The biological activity of a protein molecule depends on the integrity of its three-dimensional structure. The maintenance of its structure is dependent on its environment, which includes the type of buffer, the pH and ionic strength of the buffer and other necessary additives which may be crucial for stability (*i.e.*, reducing agents such as mercaptoethanol, dithiothreitol and others). Ideally, the purification method should be fast and produce high purity product in good yield while maintaining biological activity.

Mazsaroff and Regnier (1986) have outlined an economic analysis of preparative LC of proteins and peptides, taking into account the properties of the starting material, the characteristics of the fractionation scheme, the required product purity, the throughput and the cost of the process. Several review articles deal with applications of liquid chromatography to the purification of biopolymers (Janson, 1984; Rahn *et al.*, 1986; Johnson, 1987; Guichon and Katti, 1987; Huang and Guichon, 1988). Berkowitz and coworkers (Berkowitz *et al.*, 1987) have suggested a three-step purification strategy using silica-based columns. These involve:

Table 1: Chromatographic methods used in purification of biological molecules

Molecular property	Method of separation	Characteristics	Best suited for
Size and shape	Size exclusion	Low resolution and speed in fractionation, but excellent for desalting. Capacity limited by sample volume	Removing known contaminants
Charge	Ion-exchange	Resolution, speed and capacity are good	Large sample volumes in the initial stages
Isoelectric point	Chromato-focusing	Resolution, speed and capacity are good	Later stages when sample is relatively pure
Hydrophobicity	Reversed-phase	High resolution, good speed and capacity no sample limitation	Small molecules but also used for proteins and peptides
	Hydrophobic	Resolution and speed good, capacity high	Toward the end of the process following ion-exchange
Biological affinity	Affinity	Excellent resolution, high capacity and good speed	Low concentration samples in a large volume

1) the use of a short column made of 3–5 μm particles to obtain rapid data prior to a preparative run

2) the use of a preparative column made of 40 μm particles to process large volumes of the starting material and to concentrate the molecule of interest

3) the use of high efficiency columns made of 3–15 μm particles to achieve the desired purity as a final step

The choice of separation media is the subject of several recent reviews (Narayanan and Crane, 1990; Regnier *et al.*, 1991; Unger, 1990; Hjertan, 1991) and so this chapter will not address that issue.

Optimization of each step in a purification protocol is important and contributes to the overall success of the operation (Janson and Hedman, 1987). Several key parameters such as resolution, capacity and speed dictate the success of the purification method. Resolution determines the quality of the final product and is an important aspect of therapeutic-grade protein products. For example, in the case of affinity chromatography, resolution depends on the specificity of the interaction between the immobilized ligand and the molecule to be separated. Capacity and speed together determine the productivity and cost effectiveness of the process. Speed by itself is crucial in product recovery since most biological molecules are unstable or are subjected to proteolytic degradation if such contaminants are present. Again, support materials designed for LC are operated under high flow rates giving rise to rapid equilibration times and lower process times. However, there are some common factors that determine the success of a chromatographic process, whether it is based on affinity, ion-exchange, hydrophobic or size-exclusion. Table 2 lists these pertinent parameters.

Table 2. Major factors in the success of a chromatographic process

Resolution (selectivity and efficiency)
Recovery
Throughput
Reproducibility
Stability
Maintenance
Economy
Convenience

Large scale purification of recombinant products is acclaimed to be a success following the introduction of chromatography. There are two main approaches to scaling up.

1) Optimize the separation in an analytical scale keeping in mind maximum product purity. Scaling up is achieved by increasing the column volume and the sample volume in proportion.

2) Optimize the separation keeping in mind maximum purity and maximum sample loading. Scaling up is achieved with the use of larger columns.

In the first approach, the available capacity of the larger column is not fully exploited, while in the second approach maximum productivity is achieved. Several groups have developed theoretical approaches to aid in calculations of factors important in preparative adsorption chromatography (Chase, 1985; Kato, 1987; Chase, 1988; Liapis, 1989; Chase and Draeger, 1992). In affinity chromatography, attention has been focused on the optimization of the adsorption, washing and elution phases (Chase, 1988).

4. Separation Procedures

Classical liquid chromatography can be classified according to the selectivity of the support materials and their retention mechanism. The five principal types or modes of chromatography are:

1) Size-exclusion chromatography
2) Ion-exchange chromatography
3) Reversed-phase chromatography
4) Hydrophobic interaction chromatography
5) Affinity chromatography

The key properties of proteins exploited in their purification are size, charge, hydrophobicity/solubility and biological affinity. These properties form the basis of the five major types of chromatography.

4.1. Size-Exclusion Chromatography

Size-exclusion or gel filtration chromatography (SEC) is based on the differences in the molecular size and shape of biomolecules and their selective permeation in the intraparticle column volume of the chromatographic support. The traditional gel filtration media are based on cross-linked dextran (for example, Sephadex, Pharmacia, Piscataway, NJ) or polyacrylamide (Biogel P, Biorad, Richmond, CA). These are available in a wide range of porosities. Numerous other gel filtration media are commercially available today (Preneta, 1989; Regnier et al., 1991; Unger et al., 1991).

The basic principle of size-exclusion is that molecules are partitioned between solvent and a stationary phase of defined porosity. If a protein is to be separated from relatively low molecular weight contaminants (less than 5000 daltons), then a size-exclusion support with small pores is needed. If different proteins of more closely related molecular size are to be separated, a support of the correct fractionation range is necessary. Smaller beads give better resolution, but are compressed under high flow rates. For large scale applications, rigid and larger beads are used.

Size-exclusion chromatography is best carried out in the buffer in which the protein is most stable. Cofactors, metal ions and other additives are often added in the elution buffer to stabilize the molecule of interest. Resolution is maximum when the sample is applied in a small volume, typically around 1–5% of the total bed volume. For desalting of a sample, however, much larger sample volumes (25–30% of the total column volume) can be loaded without losing resolution. Maximum resolution is also attained using long columns and slow flow rates (a flow rate of 2 ml cm^{-2} h^{-1} is usually used). In the case of Sephacryl, which is rigid, flow rates up to 30 ml cm^{-2} h^{-1} are reported (Preneta, 1989).

In one study, the effect of mobile phase on the recovery of proteins from a size-exclusion column was reported to be significant (Kelner et al., 1989). Silica-based supports reportedly provide higher recoveries of protein mass and activity than polymer-based material in SEC. The hydrophobicity of the mobile phase is reduced by the addition of glycerol or increased by the addition of salts, which also affects the recovery of protein samples.

SEC is often used as the final step to assess the purity of the preparation. In general, SEC has been reported to be useful in desalting a sample, in determining molecular weights, in calculating equilibrium constants for protein binding or complex formation, or in assessing formation of aggregates, breakdown or denatured products.

The desired characteristics of a high performance SEC support have been reviewed (Gooding and Freiser, 1991). General guidelines for successful SEC are also outlined. The influence of flow rate on the separation of a standard protein mixture by SEC has been reported by Engelhardt and Schön (1986). An increase in resolution is observed with decreasing interstitial velocity. An interstitial velocity around 0.1 mm s^{-1} is optimum for molecules with molecular weights up to 200,000 and particle diameters of 3–5 μm. For molecules with molecular weights up to 1,000,000, optimum interstitial velocity has to be reduced to 0.01 mm/s. Interstitial velocity of 2 mm s^{-1} corresponds to the usual standard velocity of 1 ml min^{-1} in the case of a 4 mm i.d. column.

In the preparative purification of monoclonal antibodies, SEC is employed to achieve highly purified product or to have the product in a defined environment (Ostund, 1986; Kenney, 1989; Schmidt, 1989). SEC is also used in the purification of recombinant human interferon (Zhang *et al.*, 1992). Chromatography on Sephacryl S-100 is used as the final step in the purification protocol. The last step in the purification of interferon α–2a is based on size-exclusion chromatography on Sephadex G-50 from Pharmacia as described elsewhere (Hochuli, 1988). A key step in the purification of insulin is size-exclusion chromatography to remove dimers (Sofer, 1986). Ninety-six liters of support material packed in sectional columns are used to process about two liters of sample. The column is run at 14 l h-1 for 7 hr to produce 56 g of insulin per cycle. For insulin-like growth factor (IGF–1), fractionation by size-exclusion is followed by ion-exchange chromatography, which is then followed by buffer exchange and fractionation by a second size-exclusion step (Sofer, 1986).

Purification of plasminogen activator includes a size-exclusion step in the range of 5,000–300,000 to fractionate the sample components (Binder *et al.*, 1981; Dingeman and Collen, 1981). In the purification of recombinant tumor necrosis factor (rTNF), the size-exclusion chromatography step serves the purpose of desalting, where the phosphate buffer containing ammonium sulfate is exchanged for buffers suitable for further biochemical characterization of the protein (Lin and Yamamoto, 1987).

Recombinant proteins for therapeutic use adhere to strict standards of purity. Since purification of proteins requires the addition of various chemicals to maintain the stability of the protein and to perform various separation steps, there is always concern about trace contaminants in the final product. Size-exclusion chromatography is most used in the final stages of purification to remove some of these additives.

Recently SEC has been applied to a limited degree in the separation of peptides in the 200–5,000 Dalton or 2–50 residue range. However, SEC is also used as the first step in a multistep process in the resolution of a complex peptide mixture (Mant and Hodges, 1985; 1989). Standard run conditions for SEC of peptides are outlined in Hodges and Mant (1991). The addition of urea (*e.g.*, 8 M urea) to maintain or reform a particular conformation as opposed to the random coil configuration is recommended to predict retention behavior of peptides in SEC.

Ideal separation of peptides by SEC occurs only when there is no interaction between the peptide molecules and the support material. Since most SEC columns are weakly anionic and slightly hydrophobic, deviations from ideal size-exclusion behavior are observed during chromatography of peptides. In certain cases, these interactions are exploited and superior separation strategies are established. Recently the need for peptide standards to address the resolution of peptides by SEC has been addressed (Mant and Hodges, 1991b; Mant *et al.*, 1987). The use of peptide standards in the same manner as protein standards in SEC greatly facilitates an accurate comparison of data from different laboratories. They also help in identifying non-specific interactions between peptides and the support material. These standards help in monitoring the efficiency and resolution of the column, the effect of mobile phase composition and other chromatographic conditions such as flow rate and column dimensions on retention.

4.2. Ion-Exchange Chromatography

Ion-exchange chromatography (IEC) is one of the earliest forms of chromatography and it is widely used on both an analytical and preparative scale. The low cost of ion-exchange media compared to others, particularly affinity, and the ease of use and scale up makes it the most commonly practiced chromatographic method of protein purification. The concept of IEC is based on the interaction between charged molecules on the support material and the charged molecules and ions in the mobile phase. In anion-exchange chromatography, negatively charged molecules compete for the positive sites on the support material. These are usually tertiary ammonium moieties such as diethyl amino ethyl (DEAE). In the cation-exchange mode, positively charged molecules compete for the negative sites on the support material. Cationic groups are often sulfonic acid moieties such as sulfo-propyl or carboxymethyl. The functional groups that determine the variation of surface charge with pH are used to classify the various categories of ion exchangers (Table 3). Mixed forms of ion exchangers have combinations of two functional groups such as carboxysulfone. Separation is achieved on the basis of retention; weakly retained molecules elute before the strongly retained ones.

Table 3.   Classification of ion exchange media based on functional groups

Interaction	Anion exchange media	Cation exchange media
Weak	Polyamino	Carboxymethyl
Moderate	Diethylamino ethyl (DEAE)	Phospho
Strong	Quaternary ammonium	Sulfo

In IEC, the degree of retention of molecules can be finely controlled by the pH of the mobile phase. The pH of the system regulates the degree of ionization of the support material and the ionogenic groups on the protein. Raising the pH will cause basic molecules to elute rapidly from a cation-exchange column and a drop in pH will cause acidic molecules to elute rapidly from an anion-exchange column. Under extreme conditions of pH, when the molecules are fully charged or uncharged, pH has no effect. A pH value between the two pK_a values of two molecules brings about the greatest selectivity. Since strong cation and anion exchangers can be used over a wide pH range, they are used more often than their weak counterparts. On the other hand, weak ion-exchangers are used for highly charged molecules, which are otherwise too strongly retained.

In selecting the most suitable functional group for a IEC-based purification, the pH stability of the protein molecule should be taken into account. If the protein is more stable above its isoelectric point (pI), then an anion exchanger is used. On the other hand, if the protein is more stable below its pI, then the obvious choice is a cation-exchange column. This concept is applied to the purification of α-amylase, which has a pI of 5.2 and is unstable below pH 5.0. The effective use of cation-exchange chromatography below pH 5.0 is therefore not possible; thus, the method of choice is anion-exchange chromatography. Retention vs. pH maps of a variety of proteins on ion exchangers are available (Kopaciewicz et al., 1983).

Although there are some criteria based on pI by which protein retention on ion exchangers can be predicted, it is reported that retention is also governed by the charge density of the support material and nonspecific interactions between proteins and the support material (Kopaciewicz et al., 1985). Cooperative hydrophobic-ionic interactions are suggested in the retention mechanism of a weak cation exchanger (Kopaciewicz and Regnier, 1986). Studies have been carried out to correlate retention and mass recoveries with the nature of the cross-linker used to stabilize coated polymeric ion exchangers. Several studies have documented extensively the influence of mobile phase on the chromatographic behavior of proteins separated on several different ion-exchange resins. The effect of pH and ionic strength of mobile phase and displacer salt on protein retention is discussed (Rounds and Regnier, 1984; Hearn et al., 1988; Hodder et al., 1989; Henry, 1990; Aguilar and Hearn, 1991).

Ion-exchange chromatography is practised in the isocratic elution mode, stepwise elution or gradient elution mode. An increasing pH gradient is used in the cation-exchange mode, while a decreasing pH gradient is used for anion exchangers. However, a continuous pH gradient is seldom recommended due to the titration of both protein and ion exchanger ionogenic groups as the pH is altered. At constant ionic strength, a changing pH gradient is difficult to maintain. Therefore, pH variation during elution is usually used as a stepwise method. Both pH and ionic strength of the elution buffer can be combined to advantage. A variety of ion-exchange supports are commercially available for use in low-to-medium pressure chromatography, as well as in the LC mode (Roe, 1989; Henry, 1990; Nowlan and Gooding, 1991). Selected examples of proteins and peptides purified by IEC are listed elsewhere (Aguilar and Hearn, 1991).

Anion-exchange chromatography in conjunction with size-exclusion and hydrophobic interaction (HIC) is used in the preparative purification of recombinant tumor necrosis factor (Lin and Yamamoto, 1987). Elution of rTNF is achieved by applying an increasing linear gradient of NaCl. Ion-exchange chromatography on DEAE-Sepharose is the first step in the separation of human interleukin 1β (IL–1β) (Casagli et al., 1987). In the case of isolation of recombinant human malaria vaccine candidates from E. coli, cation-exchange chromatography is used to remove degradation products, followed by reversed-phase LC (Wasserman et al. 1987).

Trace amounts of cytokines (TNFα and TNFβ) are isolated from large volumes of cell culture media by a multistep process. Both bind to DEAE-cellulose ion-exchange column and elution is affected by 50 to 100 mM NaCl in the equilibration buffer (Aggarwal, 1987). Further purification is achieved using a Mono-Q ion-exchange column and Mono-P chromatofocusing column. In the case of connective tissue metalloproteinases, chromatography on DEAE sepharose is one of the initial steps of purification. Elution is carried out with a gradient of NaCl (from 0 to 0.5). Collagenase and stromelysin do not bind to this column, whereas gelatinase does.

Anion-exchange chromatography has been traditionally used in the purification of immunoglobulins. However, phenol red from the cell culture media and albumin, the major contaminant in both ascites and tissue culture media, bind strongly to anion-exchangers, reducing the capacity of the column and rendering it more difficult for re-use. Cation-exchangers such as S-Sepharose have been used to isolate a monoclonal antibody from tissue culture supernatant. Albumin and transferrin elute earlier than the antibody in the equilibration buffer. Antibody is collected in a linear gradient (10 column volumes) from 0 to 0.5 M NaCl in the same buffer. Any aggregation of the antibody is screened by SEC. The advantages of cation-exchange chromatography for the purification of antibodies are the following:

1) Better recovery compared to Protein A columns
2) Less inactivation of the antibody compared to Protein A columns

3) Cation-exchange chromatography is applicable to all Ig classes and IgG isotypes

4) In the case of ascites, cation-exchange chromatography can separate the monoclonal antibody from host immunoglobulins.

5) In the case of tissue culture supernatant, cation-exchange chromatography is capable of concentrating the antibody fraction to a high degree.

The disadvantage of cation-exchange chromatography is that each individual antibody purification must be optimized; in addition, dilution of the sample prior to chromatography is required to lower the ionic strength. This may be a problem for process-scale operations.

A novel form of ion-exchanger which has gained attention in the last few years is a mixed-mode support containing both weakly anionic and weakly cationic functional groups (Nau, 1990). ABx (J.T. Baker) has high binding capacity (150 mg IgG/g ABx), increased mechanical strength (since it is based on polymer-coated silica) and increased column lifetime. ABx does not bind the common contaminants such as albumin, transferrin, proteases and indicator dyes or other protein contaminants. It has been used both in the analytical and preparative-scale purification of antibodies (Nau, 1990). It has also been demonstrated that ABx is useful for purifying pyrogen-free antibodies of any class or subclass (IgG_1, IgG_2, IgG_3, IgG_4, IgA_1, IgA_2, IgM, IgD and IgE) from serum/plasma, ascites fluid, cell culture supernatant and chicken egg yolk. ABx is also useful in the purification of antibodies from several species (human, mouse, rat, rabbit, horse, cow, pig, sheep, goat, guinea pig etc.).

Viral proteins have been subjected to IEC in the presence of denaturants, strong detergents or organic solvents to prevent aggregation. A list of viral proteins purified by IEC is available (Welling and Wester, 1991). Anion-exchange LC of viral proteins such as the hemagglutinin-neuraminidase protein (HN) and the fusion protein (F) present in Sendai virus has been described (Welling and Wester, 1991). IEC on Mono Q (Pharmacia LKB) is the method of choice in this case.

High performance IEC is gaining attention in the field of peptide purification as well (Alpert, 1991; Mant and Hodges, 1991a). The use of peptide standards for monitoring ideal and non-ideal behavior in cation-exchange chromatography has been addressed (Mant and Hodges, 1991c). Strong cation exchangers are preferred for peptide separation in the IEC mode since anion-exchange chromatography would require very basic pH to be of general utility. Weak cation exchangers have functional carboxyl groups of the carboxymethyl type; these are predominantly uncharged at pH 3 and, hence, unsuitable for peptide purification. The effect of polypeptide chain length and charge density on peptide retention in cation-exchange chromatography has been reported (Mant and Hodges, 1991c). Addition of a low concentration of acetonitrile (10–20% v/v) to the mobile phase buffers prior to sample application is recommended to suppress any hydrophobic interactions with the support. A strong cation-exchange column, polySULFOETHYL Aspartamide (A), is available from Poly-LC Inc. A mixture of 54 peptides has been separated on polySULFO-ETHYL A (Crimmins *et al.*, 1988). Fig. 2 illustrates the use of acetonitrile to separate the acid form of substance P from its native form.

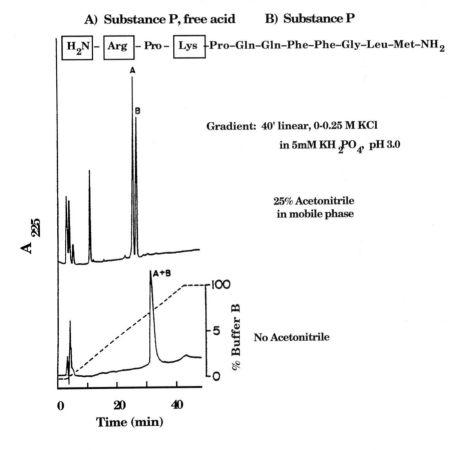

Fig. 2. Effect of organic solvent on selectivity: resolution by strong cation-exchange chromatography of substance P and its free acid form. Reproduced with permission from Alpert and Andrews (1988).

A very elegant application of chromatofocusing is the detection of transferrin variants in the serum of alcoholic subjects (Storey *et al.*, 1985). A Mono P column (Pharmacia LKB) is used at a pH gradient of 6.5 to 4.5, and the presence of an abnormal transferrin (desilylated transferrin) is determined. This technique has several advantages over isoelectric focusing, being faster and more accurate.

4.3. *Reversed-phase Chromatography*

In reversed-phase chromatography (RPC), separation is generally achieved on an inert column packing which is covalently bonded with a high density of hydrophobic functional groups such as linear hydrocarbons 4, 8 or 12 residues in length. For hydrophobic proteins, the use of relatively more polar phenyl packings with lower retention has been introduced. Resolution on RPC is achieved on the basis of the total hydrophobicity of the sample, which is determined mainly by solvent effects since the stationary phase is generally an inert hydrocarbon (Fallon *et al.*, 1987). These solvophobic interactions gave rise to the solvophobic theory, and suggest that the energetics of the behavior of a solute consists of an interaction in the gas phase and a solvent effect (Horvath *et al.*, 1976). The solvent effect consists of the positive free energy required to form a cavity for the solute within the solvent and is derived from the surface tension of the solvent and the surface area of the solute molecules (Fallon *et al.*, 1987). Although the most commonly used mobile phases for RPC are water alone or water plus an organic modifier, much interest has recently been expressed in non-aqueous mobile phases. The popularity of RPC is due primarily to the fact that a wide variety of mobile phases can be used to alter the effects of pH, temperature and solvent on separation.

Several factors affect the resolution of reversed-phase columns, such as the characteristics of the support material, the carbon loading on the material, carbon chain length, porosity and particle size of the material. The characteristics of the gradient elution process such as the mobile phase, organic modifier and pH also affect resolution. Traditionally, RPC columns are made of silica, which offers high resolution and superior mass transfer characteristics. However, the support itself is not very stable at extreme pH, and residual silanol groups which are sometimes encountered on silica-based RPC columns are responsible for peak tailing phenomena (Glajch *et al.*, 1987). These drawbacks have been overcome with the introduction of polymer-based chromatographic columns which show superior chemical stability. However, polymer-based RPC materials are reported to adsorb certain proteins, although they are displaced with minimal conformational and structural changes (Imai *et al.*, 1986). The retention relationships of proteins in solvophobic chromatography have recently been addressed (Aguilar and Hearn, 1991). The dependence of protein retention on organic solvent mixtures and salt concentration was also studied by Aguilar and Hearn.

The properties of RPC support materials in terms of pore diameter and specific surface area and how they can affect resolution are discussed by Esser and Unger (1991). A minimum pore diameter of 150 Å is required to separate peptides of 10 KDa, while for proteins, pore diameters of 500–1000 Å are required. The use of non-porous silica for RPC columns is suggested to circumvent the limited access, slow mass transfer and exclusion phenomena (Maa and Horvath, 1988). The particle diameter of the support material also affects resolution and speed of separation. For non-porous material, 2–3 µm is used with columns of 50 mm in length for fast separations while with porous materials, 5 µm is the optimum particle size for analytical-scale separations (Jilge et al., 1987; Maa and Horvath, 1988). The length of n-alkyl chains of RPC material is important in the separation mechanism, since with longer chains the strength of hydrophobic interactions increases, which sometimes results in loss of activity for certain proteins. A general strategy is to use C_4 and C_8 bonded columns for protein separations and C_8 and C_{18} columns for peptide separations. The effect of particle diameter, flow rate and column length on resolution has been discussed elsewhere in detail (Snyder and Stadalius, 1980).

RPC has made significant progress in the field of biomolecular isolation and purification. Some of the early applications of RPC in the separation of biomolecules and guidelines for the selection of stationary phases were addressed by Unger and Lork (Unger et al., 1991). Using a Vydac 214 TP column, recombinant interleukin-2-muteins have been purified by RPC (Kunitani et al., 1986). RPC is also used in the separation of platelet activation factor in plasma and in the isolation of histones from calf thymus chromatin among others (Salari, 1986; Lindner et al 1986).

The advent of wide-pore high performance reversed-phase supports for macromolecules (> 30 KDa) has increased the use of RPC for proteins of all sizes (Lewis et al., 1980; Pearson et al., 1982). Several proteins have been purified by RPC, despite being present in sub-nanomole amounts in vivo. Some of these are structural proteins such as collagen, blood proteins, hormones, glycoproteins and therapeutically important proteins such as human growth hormone and epidermal growth factor (Goeddel et al., 1979; Nice and O'Hare, 1979; Kato et al., 1982; Strickler et al., 1982; Newman and Kahn, 1983). RPC has also been used to examine conformational changes in proteins and to identify peptides which contain cysteine in protein hydrolysates (Fullmer, 1984; Luiken et al., 1984).

Since shorter columns do not lose resolution of proteins in the reversed phase mode, it has been suggested that the process of retention is not controlled by simple partition chromatography alone (O'Hare et al., 1982). Selected examples of proteins purified by RPC are documented (Aguilar and Hearn, 1991). Several growth factors such as insulin-like growth factor, fibroblast growth factor, heparin-binding growth factors and bone growth factor have been purified using C_8–C_{18} RPC columns. RPC of insulin and growth hormone is the subject of a review in which the effect of mobile phase such as changes in buffer components, pH, organic modifier, ion-pairing agents and temperature is addressed (Welinder et al., 1991). The most commonly used mobile phase additives for RPC of insulin in the high performance mode are acid phosphates, ammonium salts, alkyl-ammonium salts and TFA. Acetonitrile is used as the organic modifier in most cases. RPC analyses of human growth hormone are less often reported than those of insulin, reflecting the increasing difficulties in handling a 22–KDa hydrophobic protein under reversed-phase conditions (Christensen et al., 1991). These authors also address the phenomena of protein adsorption to RPC and HIC support materials.

In spite of the extensive use of RPC for the separation of macromolecules with biological activity, rarely is the loss of bioactivity ascribed to the chromatographic conditions used. The bioactivity of insulin and iodinated insulin after RPC has been studied (Welinder et al., 1990). Table 4 summarizes some of the common causes of denaturation of bioactive polypeptides during RPC. A detailed account of the experimental conditions for several RPC purified proteins and

Table 4.   Factors responsible for the denaturation of polypeptides during RPC

---

1. Organic solvents or low pH of the mobile phase (*e.g.*, thyroid-stimulating hormone) (Bristow *et al.*, 1983)

2. Binding of mobile phase components to the sample (such as organic modifiers) (*e.g.*, growth inhibitory glycopeptide) (Sharifi *et al.*, 1985)

3. Contamination of leachates from the bonded phase (*e.g.*, mono-iodoinsulins) (Welinder *et al.*, 1990)

4. Strong or irreversible binding to the bonded phase (*e.g.*, interleukin-2) (Kniep *et al.*, 1984)

5. The process of isolation of the molecule from the mobile phase (Interferon-β) (Johannsen and Tan, 1983)

---

attempts to increase biological activity has been given by Welinder (1991).

Although RPC is not the ultimate method of choice in the purification of proteins, it is undoubtedly a very useful technique that can be optimized and used to advantage. However, in the analysis and structure elucidation exercise, RPC is a mature, well-established technique (Esser and Unger, 1991). RPC has been successfully used in peptide mapping (Regnier, 1987; Simpson et al., 1989). In the case of therapeutic proteins, retention on a RPC column is one of the several techniques used for routine confirmation of protein product purity (Riggin and Farid, 1990). Micropellicular support materials for rapid RPC analysis of proteins and peptides are reported to be very useful in the range of 1.5–7 μm particle size (Kalghatgi and Horvath, 1990).

A wide variety of support materials are commercially available for RPC separation of peptides (Unger et al., 1991; Nugent, 1991). The requirement of peptide standards to monitor column performance and the effect of column dimensions, mobile phase, organic modifiers and temperature on RPC is emphasized by Mant and Hodges (1991d). These authors have outlined the advantages of using peptide standards routinely to monitor peptide separations by RPC. These standards are also useful for monitoring non-ideal behavior due to column chemistry (for example, the presence of underivatized silanols on the bonded phase) (Mant and Hodges, 1991e).

The effect of flow rate and gradient on resolution of peptides by RPC has been evaluated (Burke et al., 1991). The effect of peptide retention times, peak height, peak width and overall resolution is discussed in terms of optimizing a separation strategy. Detection sensitivity is yet another parameter that can be manipulated to improve resolution. Bennett (1991) has addressed the manipulation of mobile phase pH and ion-pairing reagents to maximize the performance of RPC. The effect of three commonly used ion-pairing agents, namely, TFA (trifluoroacetic acid), PFPA (pentafluoropropionic acid) and HFBA (heptafluorobutyric acid), on the separation of seven natural and synthetic peptides has been reported (Bennett, 1991). The basic peptides are more retained in the presence of HFBA and least retained in the presence of TFA since HFBA is a strong ion-pairing agent compared to TFA. All three ion-pairing agents react with the basic amino acids such as lysine, arginine and the terminal amino group which in turn mask the basic charge and cause the peptide to bind more strongly to the column.

In order to exploit the acidic nature of peptides, the pH is raised above the pKa for the carboxylic acids (which is at least pH 5.0), usually by addition of 0.01 M triethylamine acetate (TEA), which is a weak hydrophobic ion-pairing agent. TEA increases the pH by ionizing the carboxyl side chains of glutamic acid and aspartic acid and the terminal carboxyl group. Peptides have an increased hydrophilicity in the presence of TEA, and a decrease in retention follows. For the maximum exploitation of the ionic character of a given peptide or protein, the use of tetrabutylammonium phosphate at pH 7 is suggested. At high pH, tetrabutyl ammonium phosphate reacts with carboxylic acids which are fully charged and available for ion-pairing; this results in greater retention of the peptide (Bennett, 1983). The effect of various concentrations of anionic and cationic ion-pairing reagents on the resolution of synthetic peptide standards on a C_8 or C_{18} RPC column has been reported (Mant and Hodges, 1991f). The degree of shift in retention time with different ion-pairing agents is dependent on the hydrophobicity and concentration of the counterion and the type of bonded phase (Gaertner and Puigserver 1985).

RPC is also used for the separation of enantiomers after derivatization with 1-(9)fluorenyl ethyl chloroformate and peptide diastereoisomers are separated (Einarsson and Hansson, 1991; Hruby et al., 1991). RPC is a very useful tool in the sequence and amino acid analysis of peptides and proteins (Kossmann et al., 1989). Discontinuous RPC has recently been reported to be very effective in the separation of complex protein mixtures, and can be used both in the analytical and preparative scale (Grün and Reinhardt, 1991). Discontinuous RPC consists of at least two different columns which are combined in optimum sequence. For example, proteins are eluted from the first column at lower concentrations of the B eluent than from the next, thereby facilitating an enrichment process which minimizes band broadening. Detailed accounts of this particular mode of RPC are described elsewhere (Grün and Reinhardt, 1987; Grün et al., 1988). RPC is thus a well-established technique with a wide variety of applications in the separation of lipids, carbohydrates, prostaglandins, leukotrienes, steroids, vitamins and antibiotics.

4.4. Hydrophobic Interaction Chromatography

Hydrophobic interaction chromatography (HIC) is very similar to RPC except that the separation is carried out in aqueous solvents and the proteins generally retain their native structure. The basis of optimization of HIC is also very much the same as for RPC. HIC differs from RPC in the mobile phase, ligand density and hydrophobicity of the support material (Ingraham, 1991).

HIC is carried out in medium to high salt concentration using weakly hydrophobic columns. Non-polar moieties of proteins interact with the hydrophobic ligands of the column and elute in the order of increasing hydrophobicity since a descending salt gradient weakens hydrophobic interactions. The support materials for HIC are usually alkyl or phenyl derivatives of either agarose or microparticulate material. Protein retention is reported to increase in the following order: Hydroxypropyl < methyl < benzyl = propyl < isopropyl < phenyl < pentyl. In the case of hemoglobin, binding increased in the order of methyl < ethyl < propyl (Alpert, 1986). Ligand density has a profound effect in HIC. When the effect of binding is examined as a function of increasing phenyl ligand density, it is reported that for small proteins a decrease in retention occurs with decreasing number of phenyl groups. The larger proteins also show a decrease in retention, while the very large molecules are not eluted from the high phenyl density column. This phenomenon is attributed to the surface area of proteins accessible to the phenyl groups for interaction. It is suggested that for proteins with little hydrophobicity, strongly hydrophobic columns are more useful (Ueda et al., 1987).

As in other modes of chromatography, separation on a HIC support can be manipulated by adjusting the initial buffer components such as salt (type and concentration), pH, temperature and mobile phase modifiers (Fausnaugh et al., 1984). The effect of gradient rate on resolution has been examined by Kato et al. (1984). The potential effects of solvents and hydrophobic supports on the secondary, tertiary and quaternary structures of proteins have been compiled (Mant and Hodges, 1991g).

The purification of polypeptides by HIC was reviewed by Kato and Kitamura (1990). Several proteins such as lipoxidase, phosphoglucose isomerase and α-amylase are separated on a preparative scale on a HIC column. About 200 mg of lipoxidase is purified on a phenyl bonded phase with a 120 min linear gradient of ammonium sulfate from 1.5 to 0 M (Kato et al., 1985). The use of HIC in the separation of steroid receptors has been addressed (Hyder et al., 1991). Two different supports, namely a propyl-bonded and a polyether-bonded, are used with recoveries in the range of 75–100%. The usefulness of HIC in the analysis, separation and identification of different receptors has been discussed. The wide range of proteins and peptides separated by HIC are listed (Aguilar and Hearn, 1991). Under mild binding conditions (0.7 M ammonium sulfate instead of 1.5 or 2 M) on a propyl-bonded phase, human or mouse IgG monoclonal antibodies from fetal bovine serum-supplemented cell culture fluids can be selectively separated with near quantitative recovery if the initial step is based on an ion-exchange mode (Nau, 1990).

4.5. Affinity Chromatography

Biospecific affinity chromatography (AC), where the immobilized ligand is a biomolecule such as Concanavalin A and pseudoaffinity chromatography, where the affinity ligand is a nonbiological molecule such as a dye or metal, are both varieties of liquid chromatography that are now widely used in the separation of proteins, peptides and polynucleotides (Narayanan and Crane, 1990). Affinity chromatography is based on the unique specificity inherent in a ligand-biomacromolecule interaction. The last decade has seen an increase in the use of affinity-based techniques, not only in the separation field but also in the analysis of biological macromolecules (Angal and Dean, 1989; Chaiken, 1990; Josic et al., 1991).

The properties of support materials suitable for affinity chromatography has been the subject of several reviews (Janson and Kristiansen, 1990). The commercial availability of affinity chromatographic supports is also documented (Janson and Kristiansen, 1990; Narayanan, 1993). Characterization of support material in terms of pore and particle size and accessible surface area is emphasized since it plays a key role in the success of an affinity-based chromatographic purification (Narayanan et al., 1990).

In order for an affinity purification protocol to be cost effective, it is important that the column be re-usable. The stability of the matrix-ligand linkage is of primary concern in devising the coupling chemistry. Several schemes for activating supports with functional groups have been extensively reviewed (Janson and Kristiansen, 1990; Narayanan, 1993). Some of these activated ready-to-use supports are commercially available (Janson and Kristiansen, 1990). Ligand leakage is a serious limitation of affinity chromatography since in high affinity systems such as hormone-receptor interactions, the isolation of small amounts of protein is compromised by the release of ligand from the matrix (Angal and Dean, 1989). In therapeutic-grade protein products, the presence of ligand can result in serious consequences.

Ligand density is another parameter that can play an important role in the performance of affinity chromatography. The degree of substitution or ligand density influences the loading capacity or the amount of protein that can be specifically adsorbed by the affinity ligand. The degree of substitution needs to be carefully controlled, since it can lead to nonspecific interactions. In affinity chromatography, non-specific interactions are generally characterized as hydrophobic or ionic and also include any other interactions that depend on gross physicochemical properties. Hydrophobic interactions result from non-polar side chains of proteins and the support material, the spacer arms between support and ligand, or the ligand itself. The use of long spacer arms between the ligand and the support or incomplete attachment of ligands, leaving unreacted functional groups, can also result in hydrophobic interactions. Ionic interactions can arise due to the polyelectrolytic nature of proteins or from the support material, spacer arm, ligand or the coupling chemistry such as the CNBr activation. Moderate concentrations of salt (0.25–0.5 M NaCl/ KCl) can suppress ionic effects, while high ionic strength can enhance hydrophobic interactions.

Affinity ligands can be classified as general or specific depending on whether they are tailored for one particular protein purification (e.g., antibodies) or can be exploited for the purification of a class of closely related proteins (e.g., Concanavalin A). Specific ligands can be further classified into biological (e.g., antibodies) or synthetic (e.g., antisense peptides). Group-specific ligands can also be biological (e.g., Concanavalin A, Protein A or G) or synthetic (e.g., dyes, metals and amino acids). The use of group-specific ligands is advocated for several reasons. While natural ligands have an inherent biological affinity for a molecule, synthetic ligands are often made specific by varying the binding and elution conditions. Natural ligands are often fragile and unstable and so their use is limited at ambient or elevated temperature or under harsh conditions. Stringent cleaning-in-place procedures are also not compatible with these biological ligands. The synthetic ligands, however, are robust and not susceptible to denaturation. They are also more stable under strong elution conditions and more cost effective than their biological counterparts (Vijayalakshmi, 1989; Narayanan, 1993).

Several chapters have dealt with the applications of affinity chromatography and the guidelines to the use of the method in the purification of proteins (Angal and Dean, 1989; Clonis, 1989; Chaiken, 1990; Jones, 1990; Josic et al., 1991; Van Eyk et al., 1991; Narayanan, 1993). Group-specific affinity ligands have gained popularity in the last few years, and a recent review deals with the use of Protein A and G, lectins and heparin in the purification of immunoglobulins, horseradish peroxidase and antithrombin III, respectively. Detailed schemes using these ligands are discussed with special emphasis on the high performance mode (Josic et al., 1991).

Affinity chromatography is also used in the analysis of biological macromolecules to study structure-function relationships of interaction processes and to design recognition molecules (Chaiken, 1990; Chaiken et al., 1992; Van Eyk et al., 1991). Affinity interactions on immobilized ligands are useful for discovering, characterizing and designing recognition molecules (Chaiken et al., 1992). AC is reported to be very useful in measuring low or weak affinity interactions. A column of immobilized anti-glycoside monoclonal IgG at a high density is able to differentiate a mixture of glucose-containing tetrasaccharides with K_a values of 5×10^3, 2×10^3, 1×10^3 and 8×10^2 M^{-1}. At present, AC is considered to be the only tool that can provide information on interactions with such low affinity.

Recently, Ohlson *et al.* introduced the concept of weak-affinity chromatography (Ohlson *et al.*, 1989; Zopf and Ohlson, 1990). In weak affinity chromatography, the dissociation step is more rapid than that observed in conventional affinity interactions and dissociation can be brought about using low concentrations of displacing buffer. AC can thus be accomplished under mild, isocratic conditions where similar molecules can be separated in the same manner as in ion-exchange or RPC (Stevenson, 1990). This concept is extended in the design of short peptides called paralogs that mimic the binding characteristics of moderate affinity antibodies. These ligands offer a spectrum of specificities between traditional immunoaffinity chromatography and ion-exchange chromatography (Kauvar *et al.*, 1990). Paralogs are designed to simulate the overall surface of the paratope of an antibody by complementing properties such as charge, bulk and hydrophobicity in the linear structure of a peptide. The authors note that enzymes and antibodies have some tolerance to the specific amino acid composition, but are very dependent on the overall characteristics, especially those mentioned above, in order to retain binding. With recent advances in the preparation of synthetic peptides, the concept of paralogs can have a promising future.

An elegant extension of weak affinity-interaction is found in the design of antisense peptides as affinity agents for separation as well as for other biotechnology applications (Chaiken, 1992). Antisense peptides (AS peptides) are sequences of amino acids encoded in the antisense strand of DNA, and are synthesized only in a few organisms such as phage and viruses. Synthetic AS peptides have been shown to interact selectively with the corresponding sense peptide, such as in the case of ACTH, which reacts selectively with the antisense peptide encoded in the strand of DNA antisense to the sense strand coding ACTH. Although this interaction has not been found in all cases studied, AS peptides have been used as affinity ligands in the purification of several peptides and proteins, notably recombinant human interferon-β (Scapol *et al.*, 1992). Immobilized AS peptides are reported to perform similar to a monoclonal antibody column in the purification of human interferon-β. The possibility of deducing the AS peptide sequences on the basis of sequences of proteins to be purified permits wide application of this technology.

A new synthetic affinity ligand, Avid Al, has been introduced for the purification of IgG. The advantage of totally synthetic low molecular weight affinity ligands compared to Protein A or G is their ability to withstand stringent treatments with acid, base, organic solvent, autoclaving and proteolytic enzymes (Ngo and Khatter, 1992). Avid Al is synthesized by reacting sepharose with 3,5-dichloro-2,4,6-trifluoropyridine followed by 2-mercaptoethanol. Elution of IgG on Avid Al is carried out with a mild neutral buffer, and preparative runs of about 12 g IgG from serum have been demonstrated. One of the principal drawbacks of affinity chromatography, the use of expensive and unstable biological ligands, can be surmounted by the use of synthetic or pseudospecific ligands as mentioned above. The use of small peptides offers yet another solution to this problem.

Validation of affinity chromatography in therapeutic-grade protein purification has recently gained attention due to the introduction of these new synthetic ligands. Affinity chromatography is also used for removing toxic substances from the plasma of patients, in addition to separation of molecules intended for clinical or biological use. Antigenicity, transformation due to contaminating DNA, the presence of transmissible diseases, and pyrogenicity due to bacterial endotoxins are the four main concerns when considering approval of a therapeutic-grade protein product (Bristow, 1990). In line with these concerns, therapeutically useful proteins are required to conform to a range of regimens which include the types of contamination that may occur and the possible consequences of various types of contaminants.

In the case of affinity supports, particularly with antibody ligands, validation involves extra steps in quality control and assurance. The Food and Drug Administration (FDA) requires proof that affinity ligands, which often leach off the column, are fully characterized and removed from the final product. The FDA also demands extensive documentation of support materials used in the chromatographic process. Several manufacturers therefore have drug master files on their supports. The regulatory and legal requirements associated with chromatographic procedures must be carefully studied before a process is actually scaled up.

5. Future Challenges

As improvements in the technology of chromatographic support materials are developed, the art of separation science is gaining momentum, and today the literature abounds with choice in separation media and alternative methods of purification. The pharmaceutical and biotechnology industries cater to the need for small quantity, high value products and the commodity chemicals of exceptionally low cost but large volume. When dealing with high value protein products, the requirements for purity become more stringent. In this regard, purification is one of the main cost factors, accounting for up to 80% of the total manufacturing cost of the therapeutic product (Hacking, 1986). In many cases, while the commercial production of proteins consists of highly automated fermentor systems, the subsequent recovery and purification is primarily a manual operation with a limited level of automation and no control, resulting in a large labor requirement. This is a serious bottleneck in the large scale production of therapeutic-grade protein products.

One of the most significant contributions to the overall reduction of processing costs lies in the development of automated and controlled downstreaming operations. Although some progress has been made in automation, such as switching of valves in a chromatographic system at given times, the technology is far behind in actually controlling the process in terms of actively monitoring the events and responding to what is happening. Control and automation can also speed up the development of chromatographic methods for use in process-scale operations.

Micro-isolation methods such as LC are increasingly being used because of their speed, good sample recovery and high resolution. The original silica-based support materials have been improved to give higher recovery of proteins and improved stability above pH 7.0. The polymer-based supports offer chemical stability over a wide pH range, but exhibit some nonspecific adsorption of proteins. Superior solvents with high UV transmission in the 206–210 nm range and good ion-pairing characteristics at high pH are needed for RPC of proteins and peptides (Shively, 1989). The issues of particle size, pore size and surface area in chromatography are being debated in terms of non-porous and perfusive supports on one hand and wide pore, 1000-Å pore diameter supports on the other.

About 25–50% of the total processing cost can be attributed to the chromatographic steps of a purification protocol (Knight, 1990). The need to reduce the number of processing steps is becoming very important since this increases efficiency and also means fewer steps to validate. The use of affinity chromatography capable of purifying a solution 3,000-fold in a single run is a step toward reducing the number of different separation strategies.

As more biopharmaceuticals have entered the clinical testing stage, regulatory issues have gained significant attention. Validation of a product by the FDA without undue delay can mean the difference between success and failure in the marketplace. Improvements in assay sensitivity to aid in this process hence are also very important.

6. References

Afeyan, N. B, S. P. Fulton, N. F. Gordon, I. Mazsaroff, L. Varady, and F. E. Regnier (1990). *Biotechnology, 8*, 203–206.

Aggarwal, B. B (1987). In: *Protein Purification: Micro to Macro*, (R. Burgess, ed.) pp. 17–26, Alan R. Liss, New York.

Aguilar, M. I. and M. T. W. Hearn (1991). In: *High Performance Liquid Chromatography of Proteins, Peptides and Polynucleotides*, (M. T. W. Hearn, ed.) pp. 247–277. VCH Publishers, Deerfield Beach, FL.

Alpert, A. J. (1986). *J. Chromatogr., 359*, 85–94.

Alpert, A. J. (1991). In: *High Performance Liquid Chromatography of Peptides and Proteins: Separation, Analysis and Conformation* (C. T. Mant and R. S. Hodges, eds.), pp. 187–195. CRC Press, Boca Raton, FL.

Alpert, A. J. and P. C. Andrews (1988). *J. Chromatogr., 445*, 85.

Angal, S. and P. D. G. Dean (1989). In: *Protein Purification Methods: A Practical Approach* (E. L. V. Harris and S. Angal, eds.), pp. 245–290. IRL Press, Oxford.

Bennett, H. P. J. (1983). *J. Chromatogr., 226*, 501–512.

Bennett, H. P. J. (1991). In: *High Performance Liquid Chromatography of Peptides and Proteins: Separation, Analysis and Conformation* (C. T. Mant and R. S. Hodges, eds.), pp. 319–327. CRC Press, Boca Raton, FL.

Berkowitz, S. A., M. P. Henry, D. R. Nau and L. J. Crane (1987). *Am. Lab., 19* 33–42.

Binder, B. R., G. Reissert and R. Beckmann (1981). *Thromb. Haemost., 46*, 11–20.

Bonnerjea, J. (1986). *Biotechnology, 4*, 954–958.

Bristow, A. F. (1990). In: *Protein Purification Applications: A Practical Approach* (E. L. V. Harris and S. Angal, eds.), pp. 29–44. IRL Press, Oxford.

Bristow, A. F, C. Wilson and N. Sutaliffe (1983). *J. Chromatogr., 270*, 285–292.

Burke, T. W. L, C. T. Mant and R. S. Hodges (1991). In: *High Performance Liquid Chromatography of Peptides and Proteins: Separation, Analysis and Conformation* (C.T. Mant and R.S. Hodges, eds.), pp. 307–319. CRC Press, Boca Raton, FL.

Casagli, M. C., M. G. Borri, C. D'Ettore, C. Baldari, C. Galeotti, P. Bossù, P. Ghiara and G. Antoni (1987). In: *Protein Purification: Micro to Macro* (R. Burgess, ed.), pp. 421–427. Alan R. Liss, New York.

Chaiken I. W. (1990). In: *High Performance Liquid Chromatography in Biotechnology* (W. J. Hancock, ed.), pp. 289–301, John Wiley & Sons, New York.

Chaiken, I. W. (1992). *J. Chromatogr.*, *597*, 29–36.

Chaiken, I., S. Rose and R. Karlsson (1992). *Anal. Biochem.*, *201*, 197–210.

Chase, H. A. (1985). In: *Discovery and Isolation of Microbial Products* (M. S. Verall, ed.), pp. 129–147. Ellis Harwood, Ltd, Chichester.

Chase, H. A. (1988). *Makromol. Chem., Macromol. Symp. 17*, 467–482.

Chase, H. A. and N. M. Draeger (1992). *J. Chromatogr.*, *597*, 129–145.

Christensen, T., J. J. Hansen, H. H. Sorensen and J. Thomsen (1991). In: *High Performance Liquid Chromatography of Proteins, Peptides and Polynucleotides* (M. T. W. Hearn, ed.), pp. 191–205. VCH Publishers, Deerfield Beach, FL.

Clonis, Y. D. (1989). In: *HPLC of Macromolecules: A Practical Approach* (R.W.A. Oliver, ed.), pp. 157–182. IRL Press, Oxford, New York.

Cramer, S. M. and C. Horvath (1988). *J. Chromatogr.*, *1*, 29–49.

Cramer, S. M., Z. E. Rassi and C. Horvath (1987). *J. Chromatogr.*, *394*, 305–314.

Crimmins, D. L., J. Gorka, R. S. Thoma and B. D. Schwartz (1988). *J. Chromatogr.*, *443*, 63–74.

Datar, R. (1986). *Process Biochemistry*, *21*, 19–26.

Dingeman, R. C. and D. Collen (1981). *J. Biol. Chem.*, *256*, 7035–7041.

Einarsson, S. and G. Hansson (1991). In: *High Performance Liquid Chromatography of Peptides and Proteins: Separation, Analysis and Conformation* (C. T. Mant and R. S. Hodges, eds.), pp. 369–379. CRC Press, Boca Raton, FL.

Engelhardt, H. and U. M. Schön (1986). *Chromatographia, 22*, 388–399.

Esser, U. and K. K. Unger (1991). In: *High Performance Liquid Chromatography of Peptides and Proteins: Separation, Analysis and Conformation* (C. T. Mant and R. S. Hodges, eds.), pp. 273–278. CRC Press, Boca Raton, FL.

Fallon, A., R. F. G. Booth and L. D. Bell (eds.) (1987). *Applications of HPLC in Biochemistry*, Elsevier, Amsterdam.

Fausnaugh, J. L., E. Pfannkoch, S. Gupta, and F. E. Regnier (1984). *Anal. Biochem.*, *137*, 464–472.

Fish, N. M. and M. D. Lilly (1984). *Biotechnology, 2*, 623–627.

Fullmer, C. S. (1984). *Anal. Biochem. 142*, 336–339.

Gaertner, H. and A. Puigserver (1985). *J. Chromatogr.*, *350*, 279–291.

Goeddel, D. V., D. G. Kleid, F. Bolivar, H. L. Heyneker, D. G. Yansura, D. G. R. Crea, T. Hirose, A. Kraszewski, K. Itakura and A. D. Riggs (1979). *Proc. Natl. Acad. Sci. 76*, 106–110.

Glajch, J. L, J. J. Kirkland and J. Kohler (1987). *J. Chromatogr.*, *384*, 81–90.

Gooding, K. M. and H. H. Freiser (1991). In: *High Performance Liquid Chromatography of Peptides and Proteins: Separation, Analysis and Conformation* (C. T. Mant and R. S. Hodges, eds.), pp. 135–144. CRC Press, Boca Raton, FL.

Grün, J. R. and R. Reinhardt (1987). *J. Chromatogr.*, *397*, 327–338.

Grün, J. R. and R. Reinhardt (1991). In: *High Performance Liquid Chromatography of Peptides and Proteins: Separation, Analysis and Conformation* (C. T. Mant and R. S. Hodges, eds.), pp. 409–421. CRC Press, Boca Raton, FL.

Grün, J. R., B. Kossmann and R. Reinhardt (1988). *Chromatographia*, *25*, 189–198.

Guichon, G. and A. Katti (1987). *Chromatographia*, *24*, 165–189.

Hacking, A. J. (1986). *Economic Aspects of Biotechnology*, Cambridge University Press, New York.

Hammond, P. M. and M. D. Scawen (1989). *J. Biotechnol.*, *11*, 119–134.

Hancock, W. S. (1990). In: *High Performance Liquid Chromatography in Biotechnology*, W. S. Hancock (ed.), John Wiley & Sons, New York.

Hearn, M. T. W., A. N. Hodder, P. G. Stanton and M. I. Aguilar, (1988). *J. Chromatogr.*, *443*, 97–118.

Henry, M. P. (1990). In: *High Performance Liquid Chromatography in Biotechnology* (W. S. Hancock, ed.), pp. 21–63. John Wiley & Sons, New York.

Henschen, A., K-P. Hupe, F. Lottspeich and W. Voelter (eds.) (1985). *High Performance Liquid Chromatography in Biochemistry*, Weinheim: VCH, Deerfield Beach, FL.

Hjerten, S. (1991). In: *High Performance Liquid Chromatography of Proteins, Peptides and Polynucleotides* (M. T. W. Hearn, ed.), pp. 119–149. VCH Publishers, Deerfield Beach, FL.

Hodder, A. N., M. I. Aguilar and M. T. W. Hearn (1989). *J. Chromatogr.*, *476*, 391–411.

Hodges, R. S. and C. T. Mant (1991). In: *High Performance Liquid Chromatography of Peptides and Proteins: Separation, Analysis and Conformation* (C. T. Mant and R. S. Hodges, eds.), pp. 11–22. CRC Press, Boca Raton, FL.

Horvath, C. (ed.) (1983). *High Performance Liquid Chromatography: Advances and Perspectives*, Vol. 4. Academic Press, New York.

Horvath, C. and Z. E. Rassi (1986). *Chromatography Forum*, *1*, 49–56.

Horvath, C., W. Melander and Z. Molnar (1976). *J. Chromatogr.*, *125*, 129–156.

Horvath, C., J. Frenz and Z. E. Rassi (1983). *J. Chromatogr.*, *255*, 273–293.

Hochuli, E. (1988). *J. Chromatogr.*, *444*, 293–302.

Hruby, V. J., A. Kawasaki and G. Toth (1991). In: *High Performance Liquid Chromatography of Peptides and Proteins: Separation, Analysis and Conformation* (C. T. Mant and R. S. Hodges, eds.), pp. 379–389. CRC Press, Boca Raton, FL.

Huang, J-X. and G. Guichon (1988). *BioChromatography*, *3*, 140–148.

Huang, J-X. and G. Guichon (1989). *J. Chromatogr.*, *492*, 431–469.

Hyder, S. M., J. Dong, P. Folk and J. L. Wittliff (1991). In: *High Performance Liquid Chromatography of Peptides and Proteins: Separation, Analysis and Conformation* (C. T. Mant and R. S. Hodges, eds.), pp. 451–475. CRC Press, Boca Raton, FL.

Imai, H., G. Imai and S. Sakura (1986). *J. Chromatogr.*, *371*, 29–35.

Ingraham, R. H. (1991). In: *High Performance Liquid Chromatography of Peptides and Proteins: Separation, Analysis and Conformation* (C. T. Mant and R. S. Hodges, eds.), pp. 425–435. CRC Press, Boca Raton, FL.

Janson, J. C. (1984). *Trends in Biotechnology, 2*, 31–38.

Janson, J. C. and P. Hedman (1987). *Biotechnology Progress, 3*, 9–13.

Janson, J. C. and T. Kristiansen (1990). In: *Packings and Stationary Phases* (K. K. Unger, ed.), pp. 747–781. Marcel Dekker, Inc. New York and Basel.

Jilge, G., R. Janzen, H. Giesche, K. Unger, J. N. Kinkel and M. T. W. Hearn (1987). *J. Chromatogr.*, *397*, 71–80.

Johannsen, S. and Y. H. Tan (1983). J. *Interferon Res.*, *3*, 473–477.

Johnson, R. D. (1987). *Dev. Ind. Microbiol. 27*, 77–83.

Jones, K. (1990). *Amer. Biotechnol. Lab.*, October 26-30.

Josic, D., A. Becker and W. Reutter (1991). In: *High Performance Liquid Chromatography of Proteins, Peptides and Polynucleotides* (M. T. W. Hearn ed.), pp. 469–495. VCH Publishers, Deerfield Beach, FL.

Kalghatgi, K. and C. Horvath (1990). In: *ACS Symposium Series, 434*, (C. Horvath and J. G. Nikelly, eds.), pp 162-180. American Chemical Society, Washington, DC.

Kato, S. (1987). *Trends in Biotechnol., 5*, 328–331.

Kato, Y. and T. Kitamura (1990). In: *HPLC in Biotechnology* (W.S. Hancock, ed.), pp. 279–289. John Wiley & Sons, New York.

Kato, Y., K. Komiya and T. Hashimoto (1982). *J. Chromatogr.*, *246*, 13–22.

Kato, Y., T. Kitamura and T. Hashimoto (1984). *J. Chromatogr.*, *298*, 407–416.

Kato, Y., T. Kitamura and T. Hashimoto (1985). *J. Chromatogr.*, *333*, 202–210.

Katti, A. M. and G. A. Guichon (1988). *J. Chromatogr.*, *449*, 25–40.

Kauvar, L. M., P. Y. K. Cheung, R. H. Gomer, and A. A. Fleischer, (1990). *Paralog Chromatography*, *8*, 204–209.

Kelner, D. N., J. W. Mayhew and J. S. Hobbs (1989). *American Biotechnol. Lab.*, Feb. 40–43.

Kennedy, J. F., Z. S. Rivera and C. A. White (1989). *J. Chromatogr.*, *9*, 83–106.

Kenney, A. C. (1989). *Adv. Biotechnol. Process*, *11*, 143–160.

Kniep, E. M., B. Kniep, W. Grote, H. S. Conradt, D. S. Monner and P. F. Muhlradt, (1984). *Eur. J. Biochem.*, *143*, 199–293.

Knight, P. (1990). *Biotechnology*, *8*, 200–201.

Kopaciewicz, W. and F. E. Regnier (1983). *Anal. Biochem.*, 133, 251–259.

Kopaciewicz, W. and F. E. Regnier (1986). *J. Chromatogr.*, *358*, 107–117.

Kopaciewicz, W., M. A. Rounds, J. Fausnaugh and F. E. Regnier, (1983). *J. Chromatogr.*, *266*, 3–11.

Kopaciewicz, W., M. A. Rounds and F. E. Regnier (1985). *J. Chromatogr.*, *318*, 157–172.

Kossmann, B., R. Richter, J. R. Grün and R. Reinhardt (1989). *3 Symp*, *1–3*, Pharmacia-LKB, pp. 67–75, Gmbh, Freiburg.

Kunitani, M., D. Johnson and L. R. Snyder (1986). *J. Chromatogr.*, *371*, 313–322.

Lewis, R. V., A. Fallon, S. Stein, K. D. Gibson and S. Udenfriend *Anal. Biochem.*, *104*, 153–159.

Liao, A. W., Z. E. Rassi, D. M. Le Master and C. Horvath (1987). *Chromatographia*, *24*, 881–885.

Liapis, A. I. (1989). *J. Biotechnol.*, *11*, 143–160.

Lindner, H., W. Hellinger and B. Pushendorf (1986). *J. Chromatogr.*, *357*, 301–310.

Lin, L. S and Yamamoto (1987). In: *Protein Purification: Micro to Macro* (R. Burgess, ed.), pp. 409–419. Alan R. Liss, New York.,

Low, D. (1990). In: *High Performance Liquid Chromatography in Biotechnology* (W. S. Hancock, ed.), pp. 117–171. John Wiley & Sons, New York.

Luiken, J., R. Vander Zee and G. W. Welling (1984). *J. Chromatogr.*, *284*, 482–486.

Maa, Y.-F. and C. Horvath (1988). *J. Chromatogr.*, *445*, 71–83.

Mant, C. T. and R. S. Hodges (1985). *J. Chromatogr.*, *326*, 349-362.

Mant, C. T. and R. S. Hodges (1989). *J. Liq. Chromatogr.*, *12*, 139–151.

Mant, C. T. and R. S. Hodges (eds.) (1991a). *High Performance Liquid Chromatography of Peptides and Proteins: Separation, Analysis and Conformation*, CRC Press, Boca Raton, FL.

Mant, C. T. and R. S. Hodges (1991b). In: *High Performance Liquid Chromatography of Peptides and Proteins: Separation, Analysis and Conformation* (C. T. Mant and R. S. Hodges, eds.), pp. 125–134. CRC Press, Boca Raton, FL.

Mant, C. T. and R. S. Hodges (1991c). In: *High Performance Liquid Chromatography of Peptides and Proteins: Separation, Analysis and Conformation* (C. T. Mant and R. S. Hodges, eds.), pp. 171–187. CRC Press, Boca Raton, FL.

Mant, C. T. and R. S. Hodges (1991d). In: *High Performance Liquid Chromatography of Peptides and Proteins: Separation, Analysis and Conformation* (C. T. Mant and R. S. Hodges, eds.), pp. 289–297. CRC Press, Boca Raton, FL.

Mant, C. T. and R. S. Hodges (1991e). In: *High Performance Liquid Chromatography of Peptides and Proteins: Separation, Analysis and Conformation* (C. T. Mant and R. S. Hodges, eds.), pp. 297–307. CRC Press, Boca Raton, FL.

Mant, C. T. and R. S. Hodges (1991f). In: *High Performance Liquid Chromatography of Peptides and Proteins: Separation, Analysis and Conformation* (C. T. Mant and R. S. Hodges, eds.), pp. 327–341. CRC Press, Boca Raton, FL.

Mant, C. T. and R. S. Hodges (1991g). In: *High Performance Liquid Chromatography of Peptides and Proteins: Separation, Analysis and Conformation* (C. T. Mant and R. S. Hodges, eds.), pp. 437–450. CRC Press, Boca Raton, FL.

Mant, C. T., J. M. R. Parker and R. S. Hodges (1987). *J. Chromatogr.*, *397*, 99–112.

Mazsaroff, I. and F. E. Regnier (1986). *J. Liq. Chromatogr.*, *9*, 2563–2583.

Narayanan, S. R. (1994). *J. Chromatogr. A 658*, 237-258.

Narayanan, S. R. and L. J. Crane (1990). *Trends in Biotechnology*, *8*, 12–16.

Narayanan, S. R., S. Knochs, Jr. and L. J. Crane (1990). *J. Chromatogr.*, *503*, 93–102.

Nau, D. (1990). In: *High Performance Liquid Chromatography in Biotechnology* (W. S. Hancock, ed.), pp. 399–531, John Wiley & Sons, New York.

Newman, P. J. and R. A. Kahn (1983). *Anal. Biochem.*, *132*, 215–218.

Ngo, T. T. and N. Khatter (1992). *J. Chromatogr.*, *597*, 101–109.

Nice, E. C. and M. J. O'Hare (1979). *J. Chromatogr.*, *162*, 401–407.

Nowlan, M. P. and K. M. Gooding (1991). In: (M. T. W. Hearn, ed.). *High Performance Liquid Chromatography of Proteins, Peptides and Polynucleotides*, pp. 203–215, VCH Publishers, Deerfield Beach, FL.

Nugent, K. D. (1991). In: (C. T. Mant and R. S. Hodges, eds.) *High Performance Liquid Chromatography of Peptides and Proteins: Separation, Analysis and Conformation*, pp. 279–289, CRC Press, Boca Raton, FL.

O'Hare, M. J., M. W. Capp, E. C. Nice, N. H. C. Cooke and B. G. Archer, (1982). *Anal. Biochem.*, *126*, 17–28.

Ohlson, S., L. Hansson, M. Glad, K. Mosbach and P-L. Larsson (1989). *Trends in Biotechnology*, 7, 179–186.

Ostlund, C. (1986). *Trends in Biotechnology*, November, 288–293.

Pearson, J. D., N. T. Lin and F. E. Regnier (1982). *Anal. Biochem.*, *124*, 217–230.

Porath, J. and P. Flodin (1959). *Nature, 183*, 1657–1659.

Preneta, A. Z. (1989). In: (E. L. V. Harris and S. Angal, eds.) *Protein Purification Methods: A Practical Approach*, pp. 293–305, IRL Press, Oxford.

Rahn, P., W. Joyce and P. Schratter (1986). *Am. Biotechnol. Lab. 4*, 34–43.

Regnier, F. E. (1991). *LC-GC, 5*, 392.

Regnier, F. E. (1991). *Nature, 350*, 634–635.

Regnier, F. E., K. K. Unger and R. E. Majors (1991). *J. Chromatogr.*, *544*, 3–413.

Riggin, R. M. and N. A. Farid (1990). In: *ACS Symposium Series, Analytical Biotechnology, Capillary Electrophoresis and Chromatography*, (C. Horvath and J. G. Nikelly, eds.), pp 113-127. American Chemical Society, Washington, DC.

Roe, S. (1989). In: (E. L. V. Harris and S. Angal, eds.) *Protein Purification Methods: A Practical Approach*, pp. 175–242, IRL Press, Oxford.

Rounds, M. A. and F. E. Regnier (1984). *J. Chromatogr., 283*, 37–45.

Salari, H. (1986). *J. Chromatogr., 382*, 89–98.

Scapol, L., P. Rapuoli and G. C. Viscomi (1992). *J. Chromatogr., 600*, 235–242.

Schmidt, C. (1989). *J. Biotechnol., 11*, 235–252.

Sharifi, B. G., C. C. Bascom, V. K. Khurana T. C. and Johnson (1985). *J. Chromatogr., 324*, 173–180.

Shively, J. E. (1989). *Pharmaceutical Technology*, February, 32–42.

Simpson, R. J., L. D. Ward, M. P. Batterham and R. L. Moritz (1989). *J. Chromatogr., 467*, 345-357.

Snyder, L. R. and M. A. Stadalius (1980). In: (Cs. Horvath, ed.) *High Performance Liquid Chromatography: Advances and Perspectives*, *Vol. 4*, pp. 195–221, Academic Press, New York.

Sofer, G. K. (1986). *Bio / Technology, 4*, 712–715,

Stevenson, R. (1990). *Biotechnol. Lab.*, March, 6–7.

Storey, E. L., U. Mack, L. W. Powell and J. W. Halliday (1985). Clin. Chem., *31*, 1543–1545.

Strathmann, H. (1985). *Trends in Biotechnology, 3*, 112–118.

Strickler, M. P., J. Kintzios and M. J. Gemski (1982). *J. Liq. Chromatogr., 5*, 1921–1933.

Torres, A. R., B. E. Dunn, S. C. Edberg and E. A. Peterson (1984). *J. Chromatogr., 316*, 125–132.

Torres, A. R., S. C. Edberg and E. A. Peterson (1987). *J. Chromatogr.*, *389*, 177–182.

Ueda, T., Y. Yasui and Y. Ishida (1987). *J. Chromatogr.*, *476*, 363–371.

Unger, K. K. (1990). *Packings and Stationary Phases in Chromatographic Techniques*, Marcel Dekker, Inc., New York and Basel.

Unger, K. K., K. D. Lork and H. J. Wirth (1991). In: (M. T. W. Hearn, ed.) *High Performance Liquid Chromatography of Proteins, Peptides and Polynucleotides*, pp. 59–119, VCH Publishers, Deerfield Beach, FL.

Van Eyk, J. E., C. T. Mant and R. S. Hodges (1991). In: (C. T. Mant and R. S. Hodges, eds.) *High Performance Liquid Chromatography of Peptides and Proteins: Separation, Analysis and Conformation*, pp. 479–499, CRC Press, Boca Raton, FL.

Vijayalakshmi, M. A. (1989). *Trends in Biotechnology*, *7*, 71–76.

Wasserman, G. F., R. Inacker, C. C. Silverman and M. Rosenberg (1987). In: (R. Burgess, ed.) *Protein Purification: Micro to Macro*, pp. 337–354, Alan R. Liss, New York.

Welinder, B. S. (1991). In: (C. T. Mant and R. S. Hodges, eds.) *High Performance Liquid Chromatography of Peptides and Proteins: Separation, Analysis and Conformation*, pp. 343–350, CRC Press, Boca Raton, FL.

Welinder, B. S., K. R. Hejnaes and B. Hansen (1990). In: (W. S. Hancock, ed.) *High Performance Liquid Chromatography in Biotechnology*, pp. 79–91, John Wiley & Sons, New York.

Welinder, B. S., H. H. Sorensen, K. R. Hejnaes, S. Linde and B. Hansen (1991). In: (M. T. W. Hearn ed.) *High Performance Liquid Chromatography of Proteins, Peptides and Polynucleotides*, pp. 495–555, VCH Publishers, Deerfield Beach, FL.

Welling, G. W and S. W. Wester (1991). In: (C. T. Mant and R. S. Hodges, eds.) *High Performance Liquid Chromatography of Peptides and Proteins: Separation, Analysis and Conformation*, pp. 223–229, CRC Press, Boca Raton, FL.

Zhang, Z., K. T. Tong, M. Belew, T. Petterson and J.-C. Janson (1992). *J. Chromatogr.*, *604*, 143–155.

Zopf, D. and O. Ohlson (1990) *Nature*, *346*, 87–88.

Part Four: Development and Validation of Analytical Methods in Pharmaceutical and Biomedical Research

CHAPTER 10

Development and Validation of Liquid Chromatographic Assays for the Regulatory Control of Pharmaceuticals

RONALD J. BOPP, TIMOTHY J. WOZNIAK
and SALLY L. ANLIKER

Eli Lilly and Company
Indianapolis, Indiana 46285 U.S.A.

and

JOHN PALMER

MAC-MOD Analytical, Inc.
Chadds Ford, Pennsylvania 19317 U.S.A.

1. Introduction

Drugs that are being tested in clinical trials or that are approved for marketing require analytical methods for the control of identity, quality, purity, and strength as well as for the determination of drug stability characteristics. The requirements for analytical methodology in the United States are outlined in the Current Good Manufacturing Practices (cGMPs), which are legally enforceable federal articles that set minimum standards for the pharmaceutical industry. The cGMPs also require that analytical test methods used to assess compliance of pharmaceutical products with established specifications must meet proper standards of accuracy and reliability (Current Good Manufacturing Practice Regulations, 1990). Over the past twenty-plus years, liquid chromatography has emerged as the technique of choice for assigning the strength (potency) and controlling the purity of new drug candidates. However, challenges such as quality-speed and chiral purity have presented opportunities for analytical scientists to utilize new technologies for methods development and validation in the pharmaceutical industry.

315

1.1. Quality-Speed in the Pharmaceutical Industry

The current political and economic climate is forcing scientists in the pharmaceutical industry to re-examine old standards and procedures for developing and validating analytical methods. The majority of new drug candidates that are tested in clinical trials do not survive the rigorous requirements for safety and efficacy necessary for marketed products. Thus, research-based pharmaceutical companies must be able to pay for expensive clinical trials and pre-clinical research with profit from the few successful compounds they produce. The increased regulatory control and cost of drug development are forcing pharmaceutical companies to streamline their businesses in order to remain competitive.   "Quality-speed" strategy is being adopted industry-wide to meet these new challenges. The concept of quality-speed involves adopting new technologies and new standards of efficiency to get high-quality products approved in the shortest possible time.

In the analytical laboratory, quality-speed translates to faster, more efficient development of methodology for new drug entities. In addition, since most compounds tested in clinical trials will not evolve into marketable products, analytical methods developed to support early clinical evaluation may not need to meet the rigorous standards of ruggedness required for methods used to support marketed products. The reason for this is that early clinical evaluation is typically supported within a single laboratory for a relatively short period of time. In contrast, marketed products are generally supported in several quality control laboratories in different countries for a period of many years. Development of extremely rugged methods for use in a quality control laboratory setting may be delayed until clinical testing has shown strong evidence of safety and efficacy. Methodology to support early phases of clinical testing may use techniques that are not well-suited for use in the quality control setting because they are more expensive and/or less transferable (*e.g.*, LC-MS). These techniques provide faster solutions to problems even though they may not be practical for long-term use.

1.2. Leveraging Technology

One way to implement quality-speed in the analytical laboratory is to use advanced technologies to streamline methods development, *i.e.*, leveraging technology. A variety of tools are available to aid analytical chemists in development of liquid chromatographic methods that are rapidly replacing the "use your favorite column" approach.

Some of these new techniques include:

- software programs that use the results from a relatively small se, of experiments to suggest the optimum liquid chromatographic conditions (*e.g.*, DryLab®, Diamond®),
- improved commercial columns that offer greater consistency in the silica support as well as smaller particle sizes (3–5 micron),
- column switching,
- statistical approaches to method optimization (*e.g.*, Plackett-Burman experimental design), and
- ancillary and complementary techniques to simplify the determination of peak homogeneity including capillary electrophoresis (CE), diode array detection, and LC-MS.

Diode array detectors and LC-MS have been available for a number of years, but are being used more frequently due to improved user friendliness and other enhancements in the hardware and software.

1.3. Chiral Purity

An additional complicating factor in analytical methods development in the pharmaceutical industry is the current emphasis on development of single-enantiomer drug candidates versus the previous practice of developing racemates. For the analytical chemist, the development of single-enantiomer drugs requires the ability to control and monitor both the chiral and the achiral purity. This usually involves developing additional assays beyond those that are needed for racemic compounds. Liquid chromatographic assays using chiral columns to resolve the chiral drug from its enantiomeric impurity are becoming routine for compounds containing a single stereogenic center. Chiral HPLC methods can usually be validated using the same criteria as for achiral chromatographic assays (Wozniak *et al.*, 1991).

1.4. Focus of This Chapter

This chapter describes development and validation of liquid chromatographic methods for the control of small molecule (non-protein) pharmaceuticals. The chapter emphasizes the current regulatory environment and the new technologies that are available. It should also be noted that this chapter encompasses only control of the active drug and does not discuss methodology for control of excipients, starting materials, etc. The discussion does, however, include drug substance (bulk chemical) and drug product (tablets, capsules, etc.) methodology. Most of the terminology is consistent

)rug Administration (FDA) guidelines and with the
aarmacopoeia (USP). Similar requirements exist in
tries; however, terminology and exact regulations

2. Methods Development

Method development refers to the experiments that are
performed to obtain a procedure capable of qualifying or quantifying a
given substance. Before investing laboratory time to develop and
validate an HPLC method, the analyst must determine the objectives
of the method (*e.g.*, potency, stability, purity, rate of release). If the
customer is developing a chemical synthesis for manufacturing a drug
substance, the analyst will have to provide a method that monitors
the impurity profile of the drug substance. The method may need to
be modified to measure additional impurities as the chemical process
evolves. An assay to support this endeavor will be different from that
supporting the validated process, that is, the final process will have a
limited number of analytes to measure based on knowledge of a
controlled process. To determine the proper approach to method
development, the analytical chemist must devise a quality control
system for a particular problem. This system will evaluate what
analytes must be measured and the optimum analysis time for the
requisite methods. Once this control system is defined, then the
analyst will choose appropriate control parameters and techniques to
determine the basic requirements for the methods to be developed.

2.1. Systems to Control Quality

As noted in the regulatory guidelines, appropriate quality control
procedures must include identity, purity, quality, and strength for the
drug substance and drug product. The development of analytical
methodology to support a quality control system will depend on
whether the manufacturing process is still under development or
whether it is well-defined. An analyst must define which analytical
properties are critical to accurately assess the quality of a product
(*e.g.*, identity, potency, purity, volatiles, salt form, polymorphism,
particle size). Next the analyst must determine which analytical tests
are appropriate to measure the requisite analytical properties (*e.g.*,
identity by infrared spectroscopy, potency by titrimetry, purity by
liquid chromatography, volatiles by loss on drying). Some properties
can often be measured by a single assay (*e.g.*, identity), whereas
others such as purity may require several assays.

If an analyst is supporting the development of a chemical process, then the analytical method must be able to monitor a wide range of potential process impurities and degradation products. In this case, the requirement for a rapid assay is not justified and the analyst may need to incorporate a gradient HPLC procedure to discern all the potential analytes. Conversely, if the manufacturing process has already been validated, then the impurity profile is usually well-established. In this instance, the analytical chemist may modify the existing gradient method and provide a rapid isocratic assay to monitor the desired elements. Some compounds, *e.g.*, those produced by a fermentation process, contain literally dozens of small impurities. These compounds are more likely to require a more complex gradient assay throughout the lifetime of the process. Thus, by determining the end use of the analytical data (*i.e.*, development or routine quality control), the appropriate control system can be devised.

Once it has been determined that an HPLC method is needed and what the end use of the method will be, method development can begin. Prior to selecting a separation system, the analyst must assess the chemical and physical properties of the analyte. An early assessment of this data should include the relevant physicochemical properties, such as pK_a, solubility, UV absorptivity, chirality, temperature, and light stability of the sample solutions. Additionally, the analyst must determine if existing methodology can be adapted for other tasks. For example, the methodology developed for the drug substance may be adapted to control the drug product.

Solubility properties will often determine the selection of sample preparation and separation criteria. The presence of a salt or functional group will be a primary consideration in determining what separation mode is desired (*i.e.*, normal phase, reversed-phase, ion exchange, ion-pairing, etc.). Selection of the appropriate mobile phase pH could allow the analyst to perform either a reversed-phase, ion-pair, or ion-exchange separation.

The spectroscopic characteristics of a compound must also be considered. If a compound has a UV chromophore, then UV detection is usually employed. In other cases, non-UV detectors such as polarimeters, fluorometers, refractive index, evaporative light scattering, mass spectrometry, etc., may be required.

Many drug candidates contain a single stereogenic center. For this type of compound, additional analytical tests are required to determine enantiomeric purity. Usually, an HPLC assay utilizing a chiral column is required to control the trace enantiomer. However, in cases where drug candidates contain multiple stereogenic centers, the need to develop a chiral assay decreases as the number of stereogenic centers increases. For example, consider a compound that contains three stereogenic centers where each starting material contains one stereogenic center (Fig. 1). The introduction of additional stereogenic centers has the potential to produce diastereomeric impurities rather than enantiomers. Since diastereomers have different chemical and physical properties, they may be readily separated and quantitated by achiral HPLC methodology.

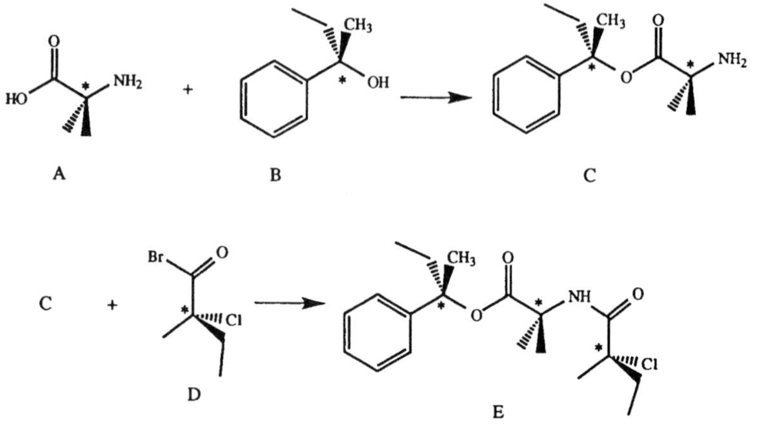

Fig. 1. Example of synthesis of a compound with three chiral centers.

Sample solution stability may define whether the separation needs to be carried out in a non-aqueous, buffered, temperature-controlled, or reduced light conditions. Attempts to initiate method development without knowledge of these physiochemical properties can often lead to ineffective or non-rugged HPLC methods. Troubleshooting of such methods can be very difficult in many cases.

Once the analyst has defined suitable separation conditions to meet the criteria defined by the customer, the ruggedness of the methodology must be investigated. Ruggedness can be defined as the ability of the method to perform within a range of method parameter deviations. The final stages of development, prior to validation, require the analyst to determine the critical method parameters. These parameters can significantly impact method performance and require strict control during the method operation. Critical method parameters can be determined by a series of mapping experiments, in which the analyst systematically varies each potential parameter to determine the impact of minor changes. For example, for a reversed-phase method using an ion-pair reagent, the following conditions can be mapped for their effect on capacity factor(s) or resolution of a critical pair of analytes as a function of deviation from the original method conditions:

- Mobile phase pH,
- Buffer concentration,
- Ion pairing reagent concentration,
- Percent organic composition,
- Column temperature,
- Injection volume,
- Gradient dwell volume, and
- Column lots or column manufacturers.

The effect of many of these parameters can be investigated through laboratory experiments as well as by using many of the method optimization software packages. By investigating the sensitivity of the method to the variation of these parameters, the analyst can determine which properties are the critical method parameters. The method can be further refined or redefined to eliminate the severity of these changes, or the control of critical parameters can be carefully defined in the laboratory procedure. For example, although the defined column gives adequate retention and peak shape, the addition of an amine modifier might allow the analyst to use several different manufacturers' columns without introducing problematic tailing or differences in retention times. The definition of the critical parameters may allow the method user to determine, for example, that mobile phase pH is critical and must be controlled to within ±0.5 units. Inadequate understanding of the factors affecting ruggedness will often prevent adequate validation of the methodology.

2.2. Typical Method Development Strategy for a New Drug Candidate

Typically, the first step in characterizing a new drug candidate is to develop a chromatographic assay to determine the purity of the new drug substance so that the material can be used in toxicological studies and clinical trials. The goal of this initial chromatographic assay is to separate any potential impurities from the main component of interest. A reversed-phase separation with a C–18 (for less polar compounds) or C–8 column is often the first choice because these phases are traditionally the best characterized and most rugged. If good chromatography is obtained for the main compound, the analyst checks for interference from potential impurities. Relatively high concentrations of the compound are chromatographed with various gradient conditions to maximize the likelihood of seeing all impurities that may be present. Next, samples of the compound that have been subjected to conditions that cause degradation are assayed using a variety of gradient conditions. The goal of these experiments is to assess whether or not the assay is stability-indicating; that is, when the drug substance shows a loss in potency, the impurities assay should reflect an analogous increase in the number of and/or percent of impurities. Conditions of heat, light, acid, and/or base that are more severe than normal storage conditions are typically used to force degradation of the compound.

Complementary techniques such as capillary electrophoresis, photodiode array detection, and LC-MS are used to help ensure peak homogeneity. In addition, these techniques increase the likelihood that all impurities are being detected. Once it has been determined that the main peak is homogeneous, the gradient conditions can be optimized to reduce run time. Typically, gradient elution is used for assay of impurities in the new drug substance even if no late eluting compounds are observed. This practice minimizes the possibility that unexpected impurities or degradation products will be missed when the drug is placed on stability programs or as the manufacturing process changes. A separate isocratic assay usually assesses potency of the drug substance. In contrast to the assay for impurities, which is usually done as a "peak versus total" analysis, the potency assay utilizes a reference standard sample that has a well-established purity. Synthetic drug substance lots are assayed for potency by comparison to the reference standard. The potency is expressed as a weight percent.

For chiral compounds, a separate chiral assay may be required to determine the enantiomeric purity. Specification limits for the trace enantiomer depend on the pharmacological and toxicological properties of this trace isomer. Thus, the type of chromatography used must be able to control the impurity within meaningful specifications. Since chiral columns often do not allow loading large amounts of compound, the specification for enantiomeric purity may be driven by the limit of quantitation of the chiral assay. Usually, chiral assays are done as a "peak versus total" assessment in which a single injection of the compound is made and the area of the trace enantiomer is expressed as a percentage of the total area of the two peaks. Obviously, it is highly desirable to have the unwanted enantiomer elute first because a much larger resolution is required if the trace-level compound elutes after the main peak. In some cases, chiral columns are needed to separate diastereomers if the analyst is unable to find an achiral system to separate these isomers.

After development of an assay to measure potency and impurities in the drug substance, an assay is usually needed to measure the concentration of the drug in a variety of matrices such as acacia suspensions, feeds, or solutions. These formulations are used to dose animals in toxicology studies. In this case, the analyst needs to ensure that the matrix components do not interfere with the chromatographic peak of interest. In addition, the method must be capable of determining the stability of the drug in the matrix. Often, the impurities assay for the drug substance can be modified for use with toxicology formulations.

Similarly, an assay for the drug in the drug product (*e.g.*, capsules, tablets) is needed prior to the start of clinical trials. Again, matrix affects must be considered and the assay must be able to measure stability of the drug product for the duration of the clinical trial. Unlike biological samples (*e.g.*, plasma or urine), it is not common practice to use an internal standard in chromatographic assay methodology for drug products. Thus, recoveries must be essentially quantitative. Liquid extraction of ground tablets or capsule contents followed by rotary mixing or sonication is a typical sample preparation technique. Often, the volume of insoluble excipients is relatively large so that a known volume of extraction solvent must be added to a ground tablet or capsule contents rather than using the dilute-to-volume approach. In this way, the actual concentration can be accurately determined. It is necessary to know the exact concentration of sample preparations so that comparison of the sample to a reference standard (usually drug substance) is meaningful.

2.3. *Understanding HPLC Column Chemistry*

Most synthetic pharmaceutical samples are polar organic compounds that may be neutral, acidic, basic, or zwitterions. Reversed-phase HPLC is by far the most common chromatographic technique in routine use for pharmaceutical compounds. The reversed-phase chromatographic separation of pharmaceutical samples is controlled by partitioning sample molecules between the mobile phase and the column packing material as the components move through the column bed. This is complicated by the tendency of the molecule to interact with both the bonded silanes and the silica surface. This is especially true of basic compounds in organic/aqueous mobile phases. The process of sample-column packing interaction is further complicated by the differences in silica particle surface chemistries produced by the manufacturing methods used by various column producers. As illustrated in Fig. 2, it is unlikely that any two like bonded phase columns from different suppliers will perform the same under identical mobile phase conditions. In fact, related columns from the same manufacturer may exhibit different selectivities.

1. k' Anthracene–Nonpolar

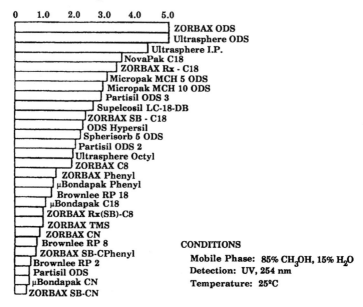

CONDITIONS

Mobile Phase: 85% CH_3OH, 15% H_2O
Detection: UV, 254 nm
Temperature: 25°C

2. k' Diethyl Phthalate–Polar

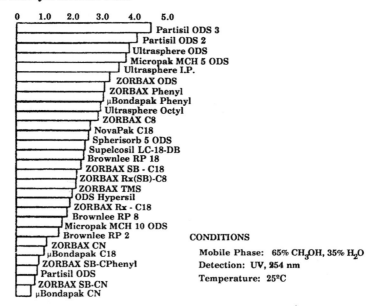

CONDITIONS

Mobile Phase: 65% CH_3OH, 35% H_2O
Detection: UV, 254 nm
Temperature: 25°C

Fig. 2. Comparison of capacity factors (k') for 1) anthracene
and 2) diethyl phthalate on various reversed-phase
HPLC columns.

2.4. Column Selection in Pharmaceutical Analysis

Manufacturers utilize organosilanes that have single (monofunctional, Fig. 3) or multiple (multifunctional, Fig. 4) reactive sites to facilitate bonding organosilane molecules to the silica surface of the particles. These silanes may in turn have one or more organic chains that will affect retention of sample components. The multifunctional silanes tend to produce phases with various degrees of polymerization. While the polymerized phases are less open to hydrolytic cleavage, it is more difficult to control the degree of polymerization that can lead to poor control of column selectivity. The monofunctional silanes react to form a single molecule layer (monolayer) of bonded phase on the particle surface. For this reason, monolayers are better defined and can be synthesized on a more reproducible basis. Recent development of sterically protected monolayer bonded phases has extended the longevity of HPLC columns. As shown in the schematic diagram in Fig. 5, sterically protected phases have bulky groups that surround the site of attachment. Chromatographers can operate with these stabilized columns at low pH values (pH = 2) without pre-

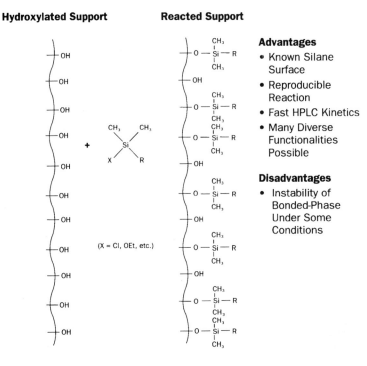

Fig. 3. Schematic diagram of a typical monolayer bonded phase.

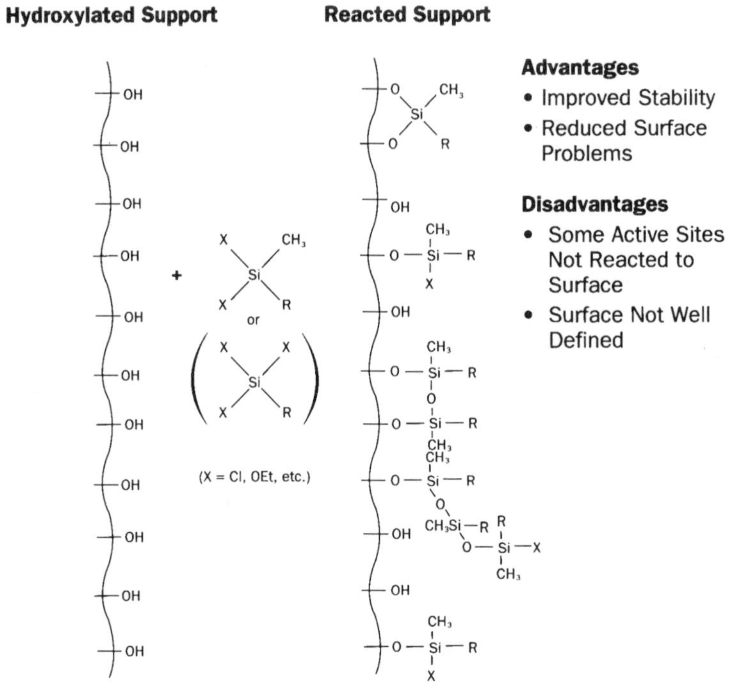

Hydroxylated Support   **Reacted Support**

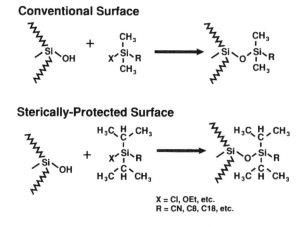

Advantages
- Improved Stability
- Reduced Surface Problems

Disadvantages
- Some Active Sites Not Reacted to Surface
- Surface Not Well Defined

Fig. 4. Schematic diagram of a typical polymeric bonded phase.

Conventional Surface

Sterically-Protected Surface

X = Cl, OEt, etc.
R = CN, C8, C18, etc.

Fig. 5.   Schematic representation of conventional and sterically protected bonded monolayer silica surfaces.

mature failure caused by hydrolytic cleavage of the siloxane bond (Glajch and Kirkland, 1990). These improvements have made the use of phenyl and CN columns, traditionally less stable phases, far more attractive for routine use. A plot that illustrates the enhanced column stability of sterically protected phases is shown in Fig. 6. The HPLC columns utilized in Fig. 6 have been stressed by pumping thousands of column volumes of an acetonitrile/trifluoroacetic acid solution (pH 2) gradient through the columns. The capacity factor (k') of 1-phenylheptane using an isocratic acetonitrile/trifluoroacetic acid system has been used as a marker of column performance as a function of column volumes of gradient mobile phase. It is apparent from the plot that the sterically protected bonded phase is relatively unaffected even after 4000 column volumes.

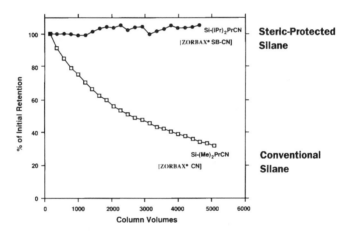

Fig. 6.   Comparison of column retentitiveness: conventional and sterically protected silane columns stressed with thousands of column volumes of an acetonitrile/ trifluoroacetic acid solution (pH 2) gradient. The capacity factor (k') of 1-phenylheptane (expressed as a percentage of the initial value) has been used as a marker of column performance. (Reproduced with permission from Kirkland, Glajch and Farlee, 1988.)

In utilizing different bonded phase functionalities or selectivities, it may be useful to think in terms of polarity and degree of hydrophilicity. As stated earlier, pharmaceutical compounds are often polar and somewhat hydrophilic in nature. The organic functionalities bonded to the silica particle surface vary in the amounts of hydrophilic character. In general, decreased carbon chain length bonded to the silica particle increases the hydrophilic nature of the phase. This generally produces more retention for the more water soluble compounds.

In light of this hydrophilic character, it is prudent to utilize C–8, phenyl, and CN bonded phases to effect adequate separation of most pharmaceutical compounds. C–18 phases are better suited to the more nonpolar, hydrophobic molecules that are occasionally encountered in drug analysis.

Most modern silica-based HPLC columns utilize spherical particles. The particle sizes generally range from 3 to 5 μm in diameter. The surface of an HPLC-grade silica particle (spherical or otherwise) is populated with a variety of silanol (–Si–OH) groups (Fig. 7). The free silanols are the primary cause of sample interactions that produce unwanted tailing for some sample types. This is especially true of amine groups that are routinely encountered in pharmaceutical compounds. Trace metals entrained in the silica matrix can also interact with samples as well as increase the activity of the free silanols. Silica particles synthesized from high purity raw materials will exhibit more predictable chromatography.

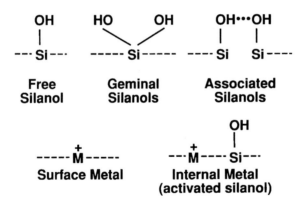

Fig. 7.   Types of silanol groups found on HPLC silica particle surfaces.

Free silanols are strongly acidic with pK_as between 3 and 4. This is more acidic than most organic acids. Since the silanols are easily ionized, a rather strong ion exchange mechanism can take place between these active surface sites and positively charged sample components (Fig. 8). Amines are likely candidates to suffer from this interaction process due to their higher pK_a values (>9). This problem can be addressed in a variety of ways:

- by reacting the residual free silanols with end-capping reagents after the initial bonding step,
- by applying a strongly adsorbed basic compound (*e.g.*, $(CH_3)_2NR$ where R = octyl or longer) after bonding, or
- by creating a high population of associated silanols on the surface prior to the bonding process.

It is often necessary to add a "silanol blocker" such as triethylamine (TEA) to the mobile phase.

$$-Si\text{-}O^- + R_3NH^+ \longleftrightarrow -Si\text{-}O^- \ ^+HNR_3$$

$$-Si\text{-}OH + R_3N \longleftrightarrow -Si\text{-}O^- \ ^+HNR_3$$

$$-Si\text{-}OH + AcO^- \longleftrightarrow -Si\text{-}O^- ...H^+... \ ^-O\,Ac$$

Fig. 8. Examples of ion-exchange interactions of analytes with free silanol groups.

Another method for controlling this surface ionization is through manipulation of mobile phase pH and ionic strength. By increasing the ionic strength and/or lowering the pH of the mobile phase, the chromatographer can decrease the degree of ion exchange on the silica surface of the column packing. These changes not only affect peak shape, but they also affect the degree of retention. The data in Table 1 illustrate the effects of various additives in the mobile phase on peak plate count. Obviously, pH and buffer concentration have a marked effect on the chromatography.

2.5. *Optimization of Peak Shape*

Many analysts wait until late in the development process to optimize peak shape; however, the better approach is to perform this exercise before optimizing peak separation. Most of the tools used to sharpen chromatographic peaks affect the amount of interaction the analyte has with the column packing. Consequently, the peaks may be sharper, but in all probability the selectivity will change as well. This is especially true if the sample components have amine functional groups. In addition, other problems such as injected solvent effects may contribute to peak asymmetry and may require additional optimization.

Table 1.   Number of theoretical plates (N) as a function of chromatographic conditions[1]

Conditions	Column Plate Number (N)		
	Morphine	Codeine	Oxymorphone
pH 6, no TEA, 2 mM acetate	2860	730	624
pH 6, no TEA,, 25 mM phosphate	2620	1160	2950
pH 6, 0.1% TEA, 2 mM acetate	3150	1470	2590
pH 3.5, 25 mM phosphate	7280	3950	5100
pH 3.5, 0.1% TEA, 2 mM phosphate	7000	5370	6630
computer-simulated N values	6180	11,200	12,500

[1]Column is a Zorbax Rx C–8. (Reproduced with permission from
Stadalius, Berus and Snyder, 1988.)

Figure 9 is an example of the effect that silicas of differing acidity
have on the separation of a polar (phenol), a non-polar (toluene), and
a basic compound (dimethylaniline, DMA). Since the purpose of the
experiment was to illustrate the contribution of sample-to-surface
silica interaction, no attempt was made to improve peak shape
through manipulation of pH, ionic strength, or addition of a "silanol
blocker." As expected, the basic compound, DMA, has a stronger
affinity for the more acidic silica. Not only does the asymmetry of the
peak increase as the acidity of the silica increases, but the retention
time increases as well.

Figure 10 shows a good example of how an isomer that is present
in significant amounts can be missed in the tail of a related isomer.
Without optimizing the peak shape, the small peak that is equal to
10% of the mass of the parent compound would have been lost in the
tail of the main peak. It is advisable to select conditions (column and
aqueous portion of the mobile phase) that have been optimized for
good peak shape before the organic portion is modified to obtain best
peak selectivity and resolution.

The following parameters have been shown to provide good peak
symmetry in many cases. Alternatively, some analysts utilize 0.1%
trifluoroacetic acid (TFA) adjusted to pH 2.5 with triethylamine (TEA)
as a starting point to provide initial peak symmetry.

Column:	less acidic silica support
Organic:	acetonitrile
Buffer:	25–50 mM phosphate
TEA:	50 mM
pH:	2–3
Sample Mass:	< 1 mg injected

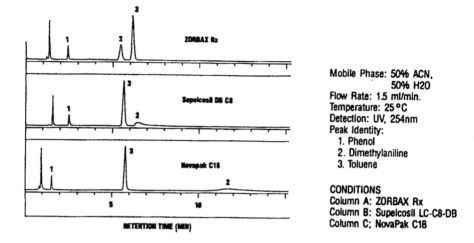

Mobile Phase: 50% ACN,
                     50% H2O
Flow Rate: 1.5 ml/min.
Temperature: 25 °C
Detection: UV, 254nm
Peak Identity:
  1. Phenol
  2. Dimethylaniline
  3. Toluene

CONDITIONS
Column A: ZORBAX Rx
Column B: Supelcosil LC-C8-DB
Column C; NovaPak C18

Fig. 9.   Chromatograms depicting the effect of differences in silica acidities on the peak symmetry and retention of an acidic (phenol), basic (dimethylaniline) and neutral (toluene) compounds.

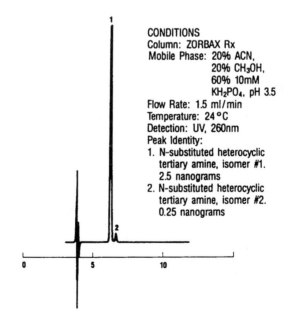

CONDITIONS
Column: ZORBAX Rx
Mobile Phase: 20% ACN,
                     20% CH₃OH,
                     60% 10mM
                     KH₂PO₄, pH 3.5
Flow Rate: 1.5 ml/min
Temperature: 24 °C
Detection: UV, 260nm
Peak Identity:
1. N-substituted heterocyclic
   tertiary amine, isomer #1.
   2.5 nanograms
2. N-substituted heterocyclic
   tertiary amine, isomer #2.
   0.25 nanograms

Fig. 10. Chromatogram showing the separation of two isomeric substituted heterocyclic amines that may not be baseline resolved without good peak shape optimization.

Figure 11 demonstrates how the conditions listed above can be used to optimize peak shape. The peak shape in the chromatograms in Fig. 11 shows a marked improvement when higher ionic strength and lower pH are used.

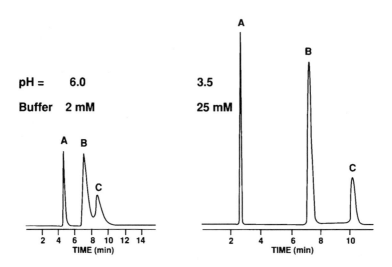

Fig. 11.   Examples of the effect of mobile phase pH and buffer concentration on the peak shape and resolution of A) morphine, B) codeine and C) oxymorphone.

2.6. Chromatographic Simulation Computer Programs

The exploration of selectivity changes based on varying the column or the mobile phase has been cumbersome and time-consuming in the past. These drawbacks have largely been removed with the advent of computer-aided chromatographic simulations. These simulation programs are available from several sources and can be used independently or integrated with data collection systems. While all of these programs offer time-saving advantages by reducing the number of actual chromatographic runs required, versatility varies among the various software packages. The most useful software programs are those that allow exploration of isocratic and/or gradient mobile phase variation coupled with column dimension optimization.

Regardless of the source of the simulation software program, the method of operation is largely the same. Two or three retention data points for each sample component are generated by a change in a mobile phase parameter such as organic modifier concentration. In some systems, the identity of individual peaks can be tracked using an integrated photodiode array detector. The simulation algorithm then calculates a straight line function that predicts peak movement over the range defined by the initial data points. This allows the user to view the separation of all peaks of interest simultaneously. Once the optimal separation has been chosen, a verifying chromatographic instrument run can be made. If the experimental chromatogram is satisfactory, then validation can be completed. The state-of-the-art simulation programs are capable of predictions that are not only accurate in the relative peak placement, but are able to predict retention times within 90% of the actual values. In addition to the ability to simulate the effects of mobile phase changes, computer programs such as DryLab® allow the ruggedness of a method to be evaluated. Parameters can be displayed visually in a resolution map that will allow the analyst to evaluate which chromatographic parameters are most critical for the given separation.

2.7. HPLC Method Development Utilizing Chromatographic Simulation Software

In the past, analysts were often overwhelmed by the sheer number of options available for chromatographic separations. This bewildering set of choices coupled with the need to perform innumerable HPLC runs often led researchers to choose a column and force it to fit the sample. Many times the chromatographer accepted the first method that "worked" with little knowledge of long-term ruggedness or whether it was the best analytical tool for complete resolution of the sample.

Using such a method for the regulatory control of pharmaceuticals will probably require time-consuming troubleshooting, redevelopment, and revalidation as the drug proceeds through different stages of development and use. Problems may occur in laboratory areas least equipped to solve difficult chromatography problems, such as the production quality control setting. Often the expedient method causes greater expenditure of time and resources than one resulting from thorough evaluation of alternatives.

Fortunately, methods that avoid most of these problems can be developed through efficient, logical experimentation. With the advent of new technology in HPLC bonded phases, chromatographic simulation software, and an understanding of chromatographic separation mechanisms, analysts can explore more options in less time than previously possible.

An efficient approach to developing rugged methodology involves evaluating HPLC columns of different polarity as well as individual mobile phase parameters to determine their effects on the separation of the sample. Use of a chromatographic simulation software package can usually greatly facilitate this approach. Typically, method development can be accomplished in time periods from two to five days depending on the appropriateness of the initial choices.

2.8. *A Template for Method Development Using Simulation Software*

By following the template for method development described below, an analyst can utilize simulation software to efficiently develop chromatographic conditions.

1. Choose columns of different bonded phase polarity.
2. Choose mobile phase parameters to give optimum peak shape.
3. Choose acetonitrile (lower back pressure and lower UV cutoff) or methanol (better solubility for higher buffer concentrations) as the organic portion of the mobile phase.
4. Run two gradient separations on each column.
5. Enter retention times and column/instrument parameters into the simulation software. Compare peak retention times and areas to identify movement of peaks due to selectivity changes between the two runs.
6. Choose between isocratic or gradient separation mode and proceed to that portion of the software.
7. Plot the resolution maps for each column.
8. Choose the best conditions by evaluating the information in the resolution maps.
9. If necessary, optimize the column (*e.g.*, particle size, column length) to minimize run time, etc.

Figures 12–15 give examples of using the above method development approach. The chromatograms in Fig. 12 show a separation of a sample on a phenyl column using a gradient of 5–80% acetonitrile over a period of a) 20 minutes and b) 60 minutes. Figure 13 shows the same experiment repeated using a CN column. The resolution maps for the chromatographic data from Figs. 12 and 13 are shown in Fig. 14. The resolution maps depict the separation of critical pairs of analytes as a function of time or rate of gradient change. The software determines which pair of analytes is the critical pair at any given gradient time (the x-axis is the time required to complete a 5–80% gradient). In Fig. 14a, the resolution map for the phenyl column contains several sharp "peaks." This indicates that the separation is extremely sensitive to small changes in the percentage of organic in the mobile phase in these regions. In contrast, Fig. 14b shows that

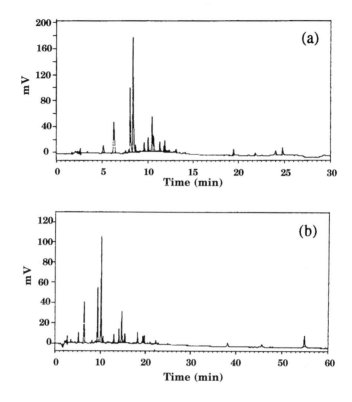

Fig. 12. Chromatogram for a sample on a phenyl
column using a gradient of 5–80% acetonitrile
over a) 30 min and b) 60 min.

the separation on the CN column is more rugged with respect to
changes in mobile phase composition. Figure 15 shows the "best"
conditions for each of the two columns derived from the resolution
maps in Fig. 14. The separation on the phenyl column (Fig. 15a) does
not contain as many peaks as the chromatogram in Fig. 15b (CN
column). Thus, the CN column is the most appropriate choice for this
separation from the standpoint of ruggedness as well as selectivity.

RONALD J. BOPP *et al.*

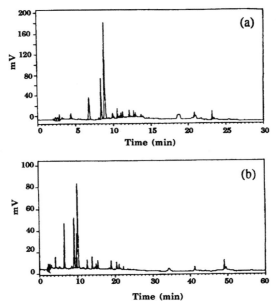

Fig. 13. Chromatogram for a sample on a CN column using a gradient of 5–80% acetonitrile over a) 30 min and b) 60 min.

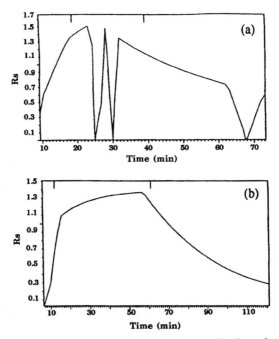

Fig. 14. Resolution maps generated by DryLab software using the chromatographic data of a) Fig. 12 and b) Fig. 13.

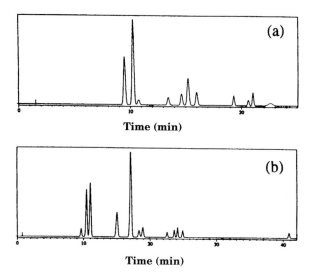

Fig. 15. a) Chromatogram for a sample on a phenyl column using the "best" mobile phase conditions from the resolution map in Fig. 14a. b) Chromatogram for a sample on a CN column using the "best" mobile phase conditions from the resolution map in Fig. 14b.

3. Method Validation

Once a method has been developed that is suitable for its intended use, a validation package must be generated before the method can be included in a regulatory document. The amount of data required to validate a method varies depending on the intended use of the method. For example, a method intended to support an initial regulatory filing in which permission to test a compound in humans is requested may need only a relatively brief validation package. In contrast, a method intended to support a marketed product requires an extensive validation package that demonstrates the degree of ruggedness that can be expected over long-term use.

In the USP XXII, validation of an analytical method is defined as the process by which it is established, by laboratory studies, that the performance characteristics of the method meet the requirements for the intended analytical applications. The USP lists eight typical parameters used in assay validation and describes several different types of assays as well as the parameters that should be considered

for each assay type. The eight parameters that are listed in the USP
are precision, linearity and range, accuracy, selectivity, ruggedness,
limit of quantitation, and limit of detection. Some practical definitions
of these terms follow.

Precision: An assessment of how repeatable the assay is, *i.e.*, if the
assay is performed multiple times, how likely is it that the same
result will be obtained?

Linearity and Range: An assessment of how well the changes in
response of the analyte with concentration can be described by an
equation for a line, *i.e.*, $y = mx + b$. How good is the line? Does it go
through zero? Over what concentration range is the line a good fit?

Accuracy: This term has two meanings. Sometimes it means how
well will two or more different techniques compare, *e.g.*, comparison
of an HPLC result with a titration result. "Accuracy" used this way
means "closeness to the true answer." The more common meaning
is recovery, *i.e.*, if an analyte is assayed with and without the matrix,
how do the results compare? An assessment of recovery is important
for drug products (*e.g.*, capsules) and toxicology preparations such as
feeds or suspensions.

Selectivity: This term refers to assurance that the response for the
assay is only due to analyte and not other related substances or
components.

Ruggedness: This term refers to how well the assay will perform
over time. It includes an assessment of the stability of standard and
sample solutions, *e.g.*, will the assay still work if the sample and/or
standard solutions are two days old? It also includes the use of
different equipment and different analysts, *e.g.*, do the results
compare well if two different laboratories assay the same sample on
different equipment?

Limit of quantitation (LOQ): This term refers to the lowest analyte
concentration (or amount) for which the assay gives reasonable
accuracy and precision information.

Limit of detection: This term is usually used to mean the same thing
as limit of quantitation for qualitative tests. For example, how low
can the analyte concentration be such that a TLC spot is still visible?

3.1. *Validation Parameters for Two Types of HPLC Assays*

For the purpose of this discussion, two types of HPLC assays that
are commonly performed in the pharmaceutical industry will be
considered. These are:
A. Main component assays
B. Impurities assays

The following method validation parameters should be considered for these two types of assays (United States Pharmacopoeia, 1990b).

Type A—Main Component Assays

Validation Parameters
- Precision
- Accuracy (not applicable to drug substance)
- Linearity (includes range)
- Selectivity
- Ruggedness (includes stability)
- LOQ - ONLY if range of interest is near sensitivity limit!

Examples
- Potency assays for drug substance or drug product
- Toxicology assays (*e.g.*, solutions, animal feeds or acacia suspensions)
- Preservative assays

Type B—Impurities Assays

Validation Parameters
- Precision
- Accuracy (not applicable to drug substance)
- Linearity (includes range)
- Limit of Quantitation
- Selectivity
- Ruggedness (includes stability)

Examples
- Related substances assays for drug substance and drug product
- Chiral purity assays

3.1.1. Main component assays

The determination of strength (potency) of a drug substance or drug product is of primary importance in the assessment of quality of pharmaceuticals. HPLC assays of this type are generally performed by comparison to a reference standard. The reference standard is a highly purified sample of the drug substance for which a careful determination of potency has been done. The potency assay assesses the strength of the subsequent lots of drug substance and drug product for comparison to established regulatory specifications. Commonly, it is necessary for a drug substance to assay between

97% and 102% to be acceptable for use in the manufacture of drug products. This rigorous requirement for absolute purity places a stringent need for a high degree of precision for the potency assay. Obviously, assay variability of greater than 2–3% will cause material that is of acceptable quality to fail specification testing. Additional replicates may help to reveal whether or not the material is truly unacceptable or whether assay variability is an issue. However, in a recent court case, FDA versus Barr Laboratories (Federal Supplement, 1993), Judge Wolin admonished a pharmaceutical company for performing additional replicates on material that failed to pass specifications, even though all results were averaged. Thus, it is important that analysts be able to place a high degree of confidence in the original result obtained by a given analytical method.

The following is a typical validation protocol for an HPLC drug substance potency assay for an initial IND.

Precision: Prepare 10 solutions at the nominal concentration from 10 separate weighings. Inject these solutions and determine the standard deviation and relative standard deviation of the peak areas (corrected for sample weight).

Linearity: Prepare a stock solution that is sufficiently concentrated to make serial dilutions that cover the range of interest. Use this stock solution to prepare 5 to 10 solutions in the concentration range of 10–200% of nominal. For example, if the nominal concentration for the assay is 0.1 mg ml^{-1}, prepare a 0.2 mg ml^{-1} stock and dilute to give 5 to 10 solutions in the concentration range 0.01–0.2 mg ml^{-1}. The free drug concentration should be used in the calculations rather than the salt or hydrate concentration because this is the component that is measured by the detector.

Selectivity: Ensure that the method separates the drug substance peak from any related substances. Demonstrate the homogeneity of the main peak (*e.g.*, using diode array detection). Also, assay some degraded material. Ensure that the potency assay shows a loss in potency for the main peak if the sample shows an increase in impurities through degradation. Note that degraded samples do not usually show good mass balance between the potency assay and the related substances assay (for impurities) because of response factor differences between the main compound and its related substances. This is particularly true when UV detection is employed. Also, some degradation products may not be observed at all.

Ruggedness: Test the stability of the sample and standard solutions from the precision study using appropriate time points and conditions that will show good stability (use information from development experiments). Additionally, other experiments may be done to demonstrate method ruggedness. For example, run the precision solutions on two different instruments using two different columns.

The type of validation data represented by the above protocol provides a "snapshot" of the method performance under the relatively ideal conditions of method validation. These data are a useful predictor of how well the assay will perform long-term. For example, if the precision data show an RSD of greater than 3%, it is likely that the long-term precision of the assay will be fairly high and that meeting a 97–102% purity specification on a routine basis may be problematic.

For a method that is being submitted in an NDA and that will be used in the quality control laboratory setting, a validation protocol might look very much the same as that for an initial IND, except that more extensive ruggedness testing is generally required. This may involve testing the method on the same samples in many different laboratories using different brands of instrumentation and columns from different manufacturers. In this way, there is a much better prediction of the long-term method performance.

3.1.2. Impurities assays.

HPLC assays for impurities, usually called "related substances assays," are key indicators of the impurity profile and stability of pharmaceutical compounds. These assays are usually performed at a higher concentration of drug substance than potency assays and are therefore more sensitive to the presence of small impurity peaks. They are often not done relative to an external standard and are thus not an indicator of absolute purity on a weight percent basis. Usually, a related substance assay is done as a single injection of a relatively high concentration of the drug substance and impurities are calculated as "peak versus total." That is, all peaks in the chromatogram are integrated and the percentage of related substances is calculated as the sum of the areas of the impurity peaks divided by the total peak areas in the chromatogram. Greater sensitivity is possible if a "high-low" approach is used in which the peak area of the main compound is estimated by injection of a diluted sample (Inman and Tenbarge, 1988). This technique has the advantage of eliminating the problematic integration of the main peak at or near concentrations that show effects of column overload. The validation protocol for an HPLC drug substance related substances assay is very similar to that for a potency assay. The difference is that the assessment of linearity is usually required to cover a much broader concentration range and a limit of quantitation must be determined (usually 0.1% or below). Also, the linearity of related substances may

d by separate studies if synthetic samples are available so
ıg experiments can be performed. Precision studies include
peak areas of impurity peaks as well as the precision of the main
peak area.

3.2. *Statistical Approaches to Method Optimization*

Experimental design studies, such as the Plackett-Burman
approach, can be used in conjunction with validation studies to
determine critical method parameters. These parameters can be
attributed to either the sample preparation or the separation
processes. Such studies provide a systematic evaluation of the critical
parameters that must be controlled to ensure the ruggedness of the
assay. A thorough understanding of key operational parameters will
facilitate the transfer of the method to other laboratories.

For example, consider an assay for drug potency in a tablet
formulation. The variables that might be considered for the tablet
extraction procedure include extraction solvent composition,
temperature, pH, the tablet grinding procedure, the filtration
procedure, and the concentration of the analyte. Using a modified
Plackett-Burman experimental design, as many as 18 parameters
can be investigated in 24 experiments.

These tools can be easily integrated into the method development
and validation process to provide a more consistent, systematic
approach to optimization of method operating parameters. In turn,
the ruggedness of the analytical method will facilitate transfer to
other laboratories using the assay.

3.3. *System Suitability*

Following successful validation of the method, an appropriate
system suitability test must be designed. As noted previously, the
validation of an analytical method consists of a series of experiments
to confirm that the method produces results that have the desired
accuracy and precision for the method's intended use. Typically, the
validation exercise is performed over a short period of time and in
itself is unable to confirm the method's continuing suitability for use
with different systems and analysts over a long period of time.
Therefore, the analyst must define system suitability tests to be run
each time the method is performed. These tests will assure the user
that the system is able to perform as intended.

System suitability tests were proposed by several FDA analytical
chemists in the mid-1970s (King *et al.*, 1974). Subsequent to that

time, the USP has also described several parameters for determining the suitability of a method setup (United States Pharmacopoeia, 1990a). These criteria can include reproducibility of replicate injections, peak tailing factor, resolution, number of theoretical plates, and peak retention time. Wallich and Carr (1990) have recently included additional parameters, including those determined through experimental design to significantly affect the method's suitability. These criteria might include the use of control sample to evaluate accuracy, comparison of standards throughout the run to monitor system stability, and standards at different concentrations to evaluate linearity.

Regardless of the criteria selected, the system suitability test must accurately assess the method's ability to perform properly. For example, column aging will affect the retention and resolution of critical analytes (Wilson and Fogarty, 1988). Selection of a system suitability test mixture that monitors the resolution of a critical pair may be appropriate to ensure that the column is still able to perform adequately for the method. The test would consist of replicate injections of this test mixture. Injection precision, retention times, resolution, and peak tailing could be monitored and limits could be established. Determination of the rejection criteria would depend on how the separation of the test analytes reflects the system's ability to separate and retain all analytes observed during the validation studies (*i.e.*, degradation products, process impurities). It is crucial that the system suitability test and acceptance criteria are defined to reflect the variances that affect the quality of the data generated. In addition, the method must indicate what an analyst should do in the event of a system suitability test failure.

4. Conclusion

Method development and validation have advanced beyond the simple "trial and error" processes of the past. With the evolution of analytical technologies, the analyst has a series of tools that facilitate the method development process. By the interactive use of these tools, liquid chromatographic methods can be developed in a faster, more systematic fashion. The coupling of experimental design, expert systems, photodiode array detection, LC-MS detection, and improved column technologies allow for a better understanding of method performance and ruggedness. Unfortunately, many of these advances are still underutilized or misunderstood, resulting in reliance on the "tried and true" methods of the past.

To facilitate the timely development of new pharmaceutical entities, the analytical chemist must learn to rely on these new technologies. Technology itself does not offer final solutions. Therefore, the analyst must also understand the objectives of the method and have a thorough knowledge of the chemistries of the molecule and column. Full use of these tools will lead to faster, more systematic development of liquid chromatographic methods. In addition, improved method performance is likely to occur as a result of better understanding of the critical parameters that affect the suitability of a method for its intended use.

5. References

Current Good Manufacturing Practice Regulations (1990) 21 CFR 211.194(a)

Federal Suppl. 812, United States of American, Plaintiff, v. Barr Laboratories, Inc. *et al.*, Defendants (Civ. A. No. 92-1744) pp. 458–492, 1993.

Glajch, J. L. and J. J. Kirkland (1990) *LC-GC, 8,* 140–144.

Inman, E. L. and H. J. Tenbarge, (1988) *J. Chromatogr. Sci. 26,* 89–94.

King, R. H., L. T. Grady and J. J. Reamer (1974) *J. Pharm. Sci., 63,* 1591–1596.

Kirkland, J. J., J. L. Glajch and R. D. Farlee (1988) *Anal. Chem., 61,* 2–11.

Stadalius, M. A., J. S. Berus and L. R. Snyder (1988) *LC/GC, 6,* 494–500.

United States Pharmacopoeia (1990a) *<621> Chromatography,* USP XXII, 1566–1567.

United States Pharmacopoeia (1990b) *<1225> Validation of Compendial Methods,* USP XXII, 1710–1712.

Wallich J. C. and G. P. Carr, (1990) *J. Pharm. Biomed. Anal., 8,* 619–623.

Wilson, T. D. and D. F. Fogarty, (1988) *J. Chromatogr. Sci. 26,* 60–66.

Wozniak, T. J., R. J. Bopp and E. C. Jensen (1991) *J. Pharm. Biomed. Anal. 9,* 363–382.

CHAPTER 11

A Comprehensive Method Validation Strategy

for Bioanalytical Applications

in the Pharmaceutical Industry

J. RONALD LANG

Glaxo, Inc., Five Moore Drive,
Research Triangle Park, NC 27709 U.S.A.

SANFORD M. BOLTON

College of Pharmacy and Allied Health Professions
St. John's University, Jamaica, NY 11439 U.S.A.

1. Introduction

Method validation is an important component in determining the reliability and reproducibility of a bioanalytical method and is a requirement of any regulatory submission (FDA, 1985). FDA policy states that for each analytical method used to quantitate drug concentrations from biological fluids, specific analytical parameters must be determined with respect to accuracy, linearity, precision, sensitivity, specificity, and recovery (Purich, 1980).

Since the bioavailability-bioequivalence regulations were introduced in 1977, the quality of FDA submissions has significantly improved. However, inadequate analytical documentation describing validation data has remained a major cause of deficient biopharmaceutic submissions (Shah, 1987). Therefore, in designing a validation strategy, it is important that it be defensible and satisfy regulatory requirements.

Method-validation publications have generally defined analytical terms identified as important to a regulatory submission (Vanderwielen and Hardwidge, 1982; Cardone, 1983; Fisher, 1984; Darioush and Smyth, 1986; Williams, 1987; NCCLS, 1989). In designing the present validation strategy, the approach taken was for it to satisfy the following criteria: The method should be defensible with respect to regulatory requirements and reliable by incorporating statistical analyses to evaluate its performance. It should also incorporate continuous validation to monitor the reproducibility of the method over time and be a timely procedure to generate adequate validation data.

2. Validation Strategy

The overall validation strategy consists of four components which are the prevalidation, validation proper, study proper, and statistical analysis. These components constitute the platform upon which to evaluate the reliability and ruggedness of a bioanalytical method. The validation flowchart summarizes the strategy and sequence of events (Fig. 1).

3. Prevalidation

Prior to initiating the validation proper, a prevalidation is performed by the primary analyst. This component provides the analyst with an opportunity to obtain some practical experience with the method and helps identify the optimum chromatographic conditions. It is recommended that the following studies be conducted prior to initiating the validation proper. The appropriate peak response to use in quantitating drug concentrations should be selected. This involves comparing internal and external methods to evaluate the reproducibility and ruggedness. The optimum standard curve range and the number of calibrators should be established. The appropriate regression model which best fits the data is then selected.

The extraction scheme and its recovery should be optimized to give insight into the limit of quantitation and to help determine if the extraction procedure is reproducible. If possible, potential compounds that could interfere with the chromatography should be evaluated. The remaining system suitability parameters are then investigated by selecting an appropriate chromatographic column, optimizing the mobile phase composition, and determining the temperature effects on the chromatographic separation.

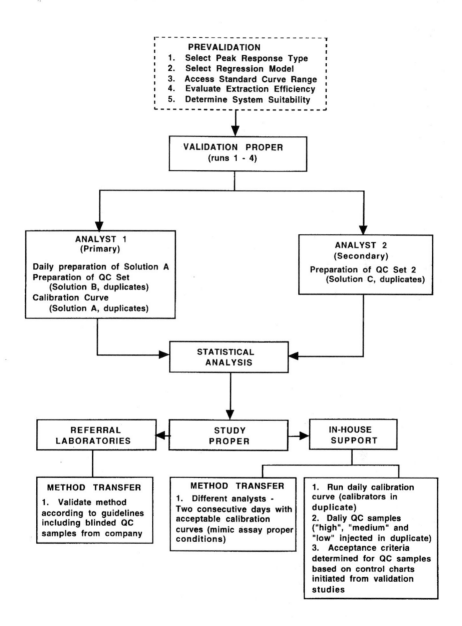

Fig. 1 Method validation strategy.

4. Validation Proper

The validation proper consists of four analytical runs generated on separate days involving two analysts. Each analyst has specific responsibilities that help determine the performance and reproducibility of the method. Each validation run emulates the analysis conditions and expected run time of the study proper. This strategy helps the primary analyst develop a daily routine to use during the study proper and anticipate potential assay problems.

Suggestions for additional samples to be analyzed during each validation run include samples fortified with metabolites and potentially interfering compounds to identify chromatographic behavior and ensure assay specificity; spiked biological samples to determine extraction characteristics and recovery; stability samples to begin generating information pertinent to sample processing and storage; additional biological matrices to identify and reserve for use in future clinical studies; and pre-dose study samples to determine if there are co-eluting contaminants in the samples to help prevent analytical difficulties during the sample analysis.

From the data generated, specific analytical parameters, including linearity, accuracy, precision, recovery, and limit of quantitation, are reported. From each standard curve, the slope, intercept, correlation coefficient, and variance are monitored. Statistical analyses are used to determine within- and between-run variances and to demonstrate how the method can be expected to perform on a daily basis. Based on the initial quality control data concentration results, acceptance criteria are established. Subsequent analytical runs are monitored during the study proper using these acceptance criteria to determine if the data generated are valid.

4.1. Preparation of stock solutions

During the method validation, two analysts are involved in the preparation of stock solutions. Analyst A, the primary analyst, is responsible for preparing stock solutions A and B from different weighings. During the validation, solution A will be made daily to use in preparing the calibration standards. This will serve to incorporate realistic variability into the validation runs and to demonstrate the ruggedness and reproducibility of the assay. The second analyst, Analyst B, will prepare stock solution C from a separate weighing. From solutions B and C, separate quality control sample sets are prepared by the two analysts. From these stock solutions, each analyst will prepare either calibration standards or quality control samples.

4.2. Preparation of calibration standards

The primary analyst is responsible for preparing daily calibration standards during the method validation. Calibrators comprising the standard curve are prepared by making serial dilutions from solution A and spiking them into the appropriate biological matrix. It is recommended that at least 5 calibrators, evenly spaced, be in the standard curve. If the standard curve range is broad, additional calibrators can be included.

4.3. Preparation of quality control samples

Quality control samples are prepared by Analysts A and B. The primary analyst prepares quality control samples from Solution B and the secondary analyst prepares them from Solution C. The quality control samples prepared by the second analyst will verify the controls prepared by the primary analyst. Solutions B and C may be used repeatedly during the validation if additional quality controls are required. Three different quality control concentrations (low, medium and high) are required and are prepared from these two stock solutions. Before preparing the quality control samples, the biological matrix should be screened for endogenous components that might interfere with the chromatography. When acceptable lots of the biological matrix have been identified, adequate volumes should be reserved and used during the method validation and study proper.

Sufficient quantities of each quality control concentration are prepared by Analysts A and B for approximately twenty-five analytical runs. These control batches are separated into aliquots, frozen in appropriate containers, and used in the method validation and subsequent study proper analyses. When additional quality control samples are required for a study proper, they are prepared before the original quality control sample sets are depleted. Both the original and freshly prepared quality control samples are analyzed concurrently to determine if they are statistically equivalent.

4.4. Method specificity

The method must be specific, with no endogenous components interfering with the separation and quantitation of the principal analyte, and should be capable of resolving co-administered drugs and metabolites from the parent drug. Retention times will be identified for all compounds and included in the validation report.

4.5. Recovery

Recovery of the analyte from the biological matrix must be determined to ensure adequate and consistent recoveries. The recovery will be documented throughout the standard curve range. The recovery will be calculated by comparing the interpolated (extracted) from the theoretical (non-extracted) concentration.

4.6. Stability

For new chemical entities, stability data are generated in biological fluids, sample containers, freeze-thaw studies, and under the appropriate chromatographic conditions. Quality control samples (low, medium and high) prepared by the primary analyst can be used to initiate the stability study. Quality control charts may be used to monitor the stability characteristics of the drug under the above-mentioned conditions.

4.7. Method transfer

If the method has been validated and is to be performed by another analyst using the same chromatographic conditions, the analyst must generate at least two consecutive analytical runs with acceptable quality control results. Both runs must emulate the validation run conditions.

If a contract research organization performs the sample analysis and has not used the procedure previously, a complete method validation is performed using the described validation procedure. During the validation exercise, blinded quality control samples are analyzed to verify the assay results. Blinded quality control results are analyzed in duplicate in two analytical runs and are subject to the validation acceptance criteria.

4.8. Method cross validation

If a referral laboratory intends to use an alternative analytical procedure that has not been validated, the following cross validation guideline outlines the necessary procedure to verify that the method is reliable and reproducible.

Prior to beginning the method cross validation, the referral laboratory will provide a copy of the methodology and their validation results. After approving the procedure and validation results, the blinded quality control samples (low, medium, and high) are analyzed in two different runs in at least duplicate. The acceptance criteria for the quality control samples are based on the established control chart limits determined for that method. The acceptance of the standard curve and study proper results is subject to the criteria determined by the statistical analyses.

From each standard curve, the following cross validation data are reported:

> *Standard curve statistics*
> > Correlation coefficients
> > Slopes
> > Intercepts
> > Interpolated standard concentrations
> > Summary standard deviation for each calibrator
> > > concentration
> > Summary relative standard deviation for each calibrator
> > > concentration
>
> *Quality control results*
> > Individual quality control data
> > Blinded quality
> > Summary standard deviation for each calibrator
> > > concentration
> > Summary relative standard deviation for each calibrator
> > > concentration
>
> *Chromatograms*
> > Complete standard curve
> > Quality control samples
> > Blank

5. Study Proper

Daily standard curves are generated to determine clinical sample concentrations. All calibrators and quality control samples are analyzed in duplicate. Clinical sample concentrations are based on a single determination. To ensure that the assay and chromatographic system are working properly, calibrators are placed at the beginning and end of the analytical run. Quality control samples are analyzed in duplicate and evenly interspersed among the clinical samples. The quality control sample sequence is carefully monitored for systematic errors. For each standard curve, the slope, intercept, variance, correlation coefficient and interpolated calibrator concentrations are reported.

Acceptance of the assay results are determined by monitoring the quality control results. If the interpolated concentrations are within the confidence limits of the control charts, established during the method validation, the data are considered valid. Upon completing a study proper and accepting the analytical runs, the quality control results are incorporated into their respective databases to update their confidence limits.

When subject samples are re-analyzed to verify the drug concentration, final data are reported using the described flowchart strategy. When possible, samples are re-analyzed in duplicate and, based on these results, a final concentration is reported (Fig. 2).

6. Statistical Analyses

Many potential problems can be encountered in the validation process that relate to the statistical evaluation of bioanalytical data. The statistical analyses are meant to be flexible and allow for modifications and additions that may be required as each situation demands. The following statistical analyses and data interpretations are discussed: standard curve raw data analysis, limit of quantitation, regression analysis for the standard curve, quality control sample data analysis, control charts for QC samples and outliers. A flowchart describing these processes is shown in Fig. 3.

In these analyses, a significance level of 5% indicates that an analytical problem may exist. Significance at the 1% level requires the analyst to carefully examine the data relating to the significant effect and to make an appropriate decision regarding further action, such as deleting outliers or repeating analyses.

6.1. Standard Curve Raw Data Analyses

The daily standard curve consists of standard prepared and analyzed in duplicate. The raw data may be presented in a format similar to that shown in Table 1.

6.1.1. Analysis of variance (ANOVA)

The purpose of this analysis is to assess the within- and between-day consistency of the calibration data. The ANOVA is performed on a logarithmic (log) transformation of all calibration data and includes "days," "replicates" and "concentration" as factors. The log transformation equalizes the variance of the observations that have relatively constant standard deviations. The terms of interest in the ANOVA are "replicate" and "replicate × concentration" interaction (Table 2).

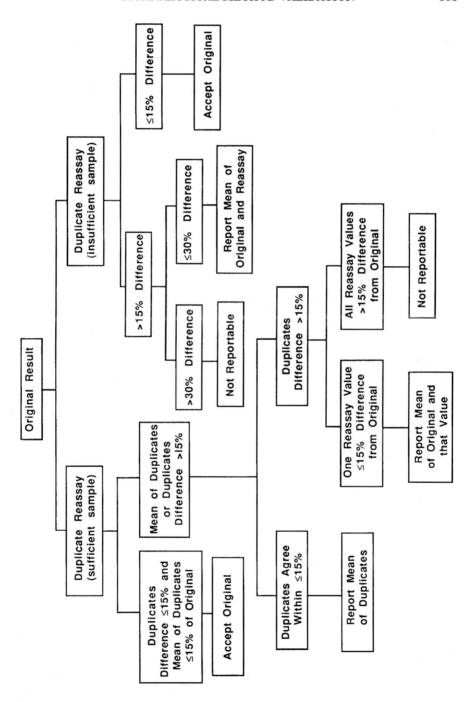

Fig. 2. Criteria for re-analysis decisions.

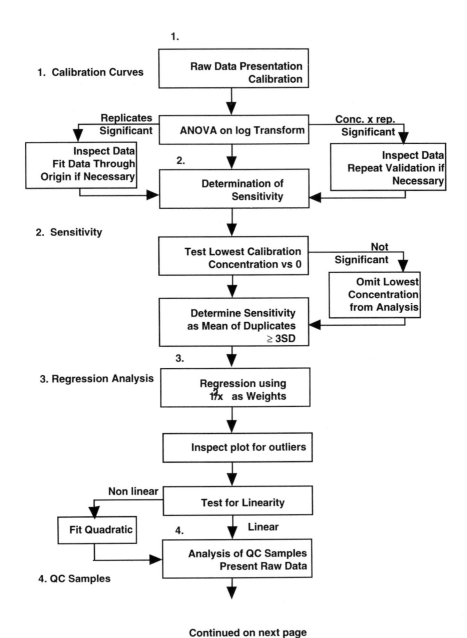

Fig. 3. Flowchart for statistical analysis

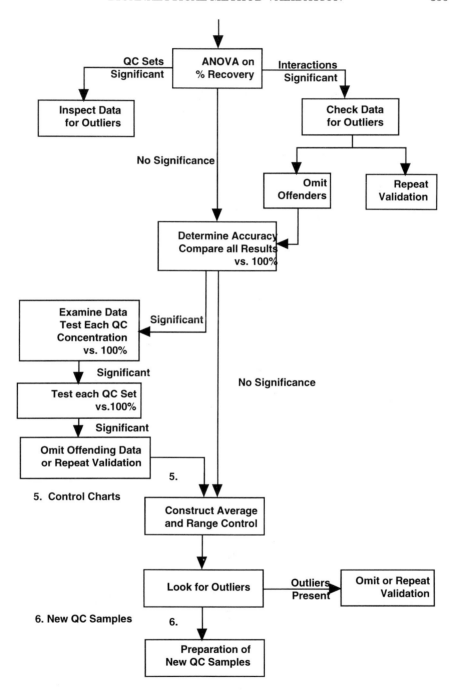

Table 1. Calibration results for hypothetical drug

Theoretical Concentration	Peak Area		Average	RSD (%)
	Rep. 1	Rep. 2		
Day 1				
1	770316	800684	785500	2.73
2	1613058	1599858	1606458	0.58
5	5024084	3962054	4493069	16.71
10	10003846	8664332	9334089	10.15
20	18182916	16979228	17581072	4.84
40	36696072	35977024	36336548	1.40
60	54620832	54699344	54660088	0.10

Table 2. ANOVA table for cefuroxime response variables: LOG (area ratio)

Source	DF	Sum of Squares	Mean Square	F-Ratio	Prob<F
A (Day)	3	3.5531590	1.1843860		
B (Replicate)	1	0.0004380	0.0004380	1.38	0.3255
C (Conc)	8	298.5447000	37.3180800		
AB	3	0.0009553	0.0003184		
AC	24	0.0922100	0.0038420		
BC	8	0.0119132	0.0014890	0.24	0.9782
ABC	24	0.1476226	0.0061500		
Total	71	302.3510000	0.0061500		

A significant "replicate × concentration" effect in the ANOVA can discredit the validation. The data must be examined for errors or outliers before initiating a re-validation. Outliers should be eliminated from the data set and the statistical analysis repeated.

One way of monitoring the variability of points around the standard regression line is to construct a control chart for the standard deviation computed from the residual sum of squares from the least-squares plot. If a standard curve shows a standard deviation out of limits, the plot should be inspected and the outlier or outliers eliminated. The remaining data can than be re-analyzed.

6.1.2. Test of linearity

A statistical test of linearity is performed for each curve separately using a weighted analysis of variance. An F test comparing the deviation mean square of the means from the line compared with the within mean square from the duplicates ($p < 0.05$) are tested. Because separate tests are performed from each analytical run, linearity is considered to exist if (a) all four calibration curves show linearity or (b) three show linearity and one curve shows non-linearity at the 0.05 level but is not significant at the 0.01 level.

If the fit shows non-linearity, not attributable to the lowest or highest concentration, a quadratic model may be considered. If either the lowest or highest concentration is causing the non-linearity, the data are re-analyzed, omitting outliers.

The following calculations are used to compute the deviation and within mean squares for the ANOVA. The weighted sum of squares will be computed from the duplicates at each concentration.

$$\frac{(Y_1 - Y_2)^2}{2X^2} \tag{1}$$

where Y_1 is the concentration of one duplicate and Y_2 is the concentration of the other duplicate.

Sum these values for each concentration and divide by the number of different calibrator concentrations to give the within mean square. Obtain the error sum of squares from the regression analysis and subtract the within sum of squares (SS) calculated. This is the "deviation SS." The deviation mean square (MS) is the deviation SS divided by the number of calibrator concentrations minus 2. Perform an F test of (deviation MS)/(within MS).

Within and deviation sums of squares are calculated as follows:

$$SS = \Sigma wy^2 - \frac{(\Sigma wy)^2}{\Sigma y} \tag{2}$$

In the present case, $$w = \frac{1}{X^2} \tag{3}$$

$$\text{weighted within SS at each concentration} = \frac{(Y_1 - Y_2)^2}{2X^2} \tag{4}$$

where X = concentration and Y_1 = observed response.

Using the data from Table 1,

at $X = 1$, $Y_1 = 770316$ and $Y_2 = 800684$ (5)

weighted SS $= \dfrac{(770316 - 800684)^2}{2 \times 1^2} = 4.611E + 08$ (6)

Repeat at each concentration, X, and sum:

Concentration	Weighted SS
1	4.6111E+08
2	2.1780E+07
5	2.2558E+10
10	8.9715E+09
20	1.8111E+09
40	1.6157E+08
60	8.5613E+05
Total	3.3986E+10 (within SS)

Dev SS + within SS = error SS (7)

Dev SS + (3.3986E+10) = (4.1640E+10) (8)

Dev SS = 7.6544E+09 (9)

Dev MS $= \dfrac{7.6544E + 09}{5} = 1.5309E + 09$ (10)

Within MS $\dfrac{3.3986E + 10}{7} = 4.8551E + 09$ (11)

$F_{5,7} = \dfrac{\text{Dev MS}}{\text{Within MS}} = 0.315$ ($F_{5,7} = 3.97$ at 0.05 level) (12)

6.2. *Limit of Quantitation*

The limit of quantitation is defined as the smallest concentration included in the standard curve and used to interpolate unknown sample concentrations. The criteria for determining this concentration is based on two factors involving background interferences (signal-to-noise) and the reproducibility of the response.

The response difference between the lowest concentration and the background sample is determined for each analytical run. The mean of these response differences is tested statistically against the mean background response. If this difference is not significant, this calibrator will not be included in the standard curve. Additionally, the variability of the response is evaluated by comparing the mean response of the lowest concentration to the standard deviation calculated from the responses at that concentration. If the mean response is not equal to or greater than 3 S.D., this concentration is not accepted as the limit of quantitation.

From Table 3, the mean response at the low concentration is 0.00425 with a standard deviation of 0.00155. A t-test with 3 degrees of freedom shows this is significantly different from 0 (t = 2.353 at p = 0.05). Three standard deviations are: 3 x 0.00155 = 0.00465. Therefore, this concentration was not accepted as the limit of quantitation, and cannot be included in the standard curve.

Table 3. Calculation of limit of quantification for cefuroxime ($0.05\ \mu g\ ml^{-1}$)

Day	Average Response
1	0.0035
2	0.0040
3	0.0065
4	0.0030
---	---
Mean	0.00425
S.D.	0.00155

$$T = (0.00425) / 0.00155 / \sqrt{4} = 5.47$$

If the value equal to 3 S.D. is considerably smaller than the mean average response at the lowest calibrator concentration, the analyst can test the limit of quantitation at a lower concentration and repeat the analysis (Table 4).

Table 4. Calculation of limit of quantification for GR43175C
         (1 ng ml)

Day	Average Response
1	785500
2	1183261
3	1119620
4	1230865
Mean	1079812
S.D.	201431

$T = (1079812)/(201431/\sqrt{4}) = 10.72$

$3 \times S.D. = 604293$

6.3. *Regression Analysis for the Standard Curve*

Regression analysis is performed using a weighted least-squares, with weights equal to $1/X^2$, where X is the theoretical concentration. The slope, intercept, correlation coefficient, variance, and interpolated concentration are recorded for each standard curve. A representative standard curve is included in the validation report. A visual inspection of these daily plots is used to identify trends or outliers that could cause problems in the data analysis.

6.4. *Quality Control Data Analysis*

During the validation, two sets of quality control samples are analyzed in duplicate from different sample preparations at three different concentrations. The data may be presented as shown in Table 5.

Table 5. Interpolated quality control results

| Theoretical Conc. (ng ml^{-1}) | Actual Concentration (ng ml^{-1}) | | | RSD (%) |
| | Rep. 1 | Rep. 2 | Mean | |
	(% accuracy)*			
Day 1				
QC Set I 3	2.8 (93.3)	2.9 (96.6)	2.85 (95.0)	2.5
15	14.8 (98.6)	14.9 (99.3)	14.85 (99.0)	0.5
50	49.8 (99.6)	49.4 (98.8)	49.6 (99.2)	0.6
QC Set II 3	3.2 (106.6)	3.3 (110.0)	3.25 (108.0)	2.2
15	15.3 (102.0)	14.5 (96.6)	14.9 (99.3)	3.8
50	50.9 (101.8)	49.9 (99.8)	50.4 (100.8)	1.4

*% accuracy is determined by taking a ratio of the calculated to
theoretical concentration and expressing it as a percentage.

6.4.1. Determination of accuracy

An ANOVA is performed in the QC data using percent accuracy
with "days," "QC set," and "concentration" as factors (Table 6).

If no significant results other than "day" are apparent in the
ANOVA, a t-test is constructed to compare the overall mean accuracy
to 100% as shown below:

$$t = \frac{|\text{ Overall Average } - \ 100 \ |}{\sqrt{\text{Day MS} / N}} \tag{13}$$

Table 6. ANOVA table for response variables: QC accuracy

Source	DF	Sum of Squares	Mean Square	F-ratio	Prob>F
A (Day)	3	0.0112500	0.0037510	4.03	0.0187
B (QC set)	1	0.0093000	0.0093000	5.24	0.1060
C (Conc)	2	0.0003963	0.0001981	0.15	0.8602
AB	3	0.0053220	0.0017740	1.90	0.1558
AC	6	0.0077000	0.0012830	1.38	0.2636
BC	2	0.0033010	0.0016500	0.72	0.5240
ABC	6	0.0137300	0.0022890	2.46	0.0538
Error	24	0.0223500	0.0009314		
Total	47	0.0733600			

A significant effect means a consistent bias exists in the quality control sample results. If the t-test shows significance, examine the data to see if transcription errors or outlying data are responsible. For clarification, an ANOVA is performed for each quality control concentration separately with factors "days" and "QC set." Test the predicted versus theoretical (100%) for each concentration [eqn. (13)].

A significant effect for "QC set" indicates that one of the QC sample sets gives consistently biased results. If "QC set" in the overall ANOVA (Table 6) is significant, examine the raw data for errors and outliers. Remove the suspect data and re-analyze the remaining data. If no outliers or errors are apparent and "QC set" is significantly different, perform a t-test for each QC set at each concentration separately, comparing the mean result from the four runs to the theoretical concentration.

If any of the results of these outlying tests are significant, the data should be evaluated and outlying values rejected or the validation repeated according to the judgment of the analyst. If no significant difference is observed, the sample results are retained.

6.5. Control Charts for QC Samples

Control charts will be constructed using the quality control sample results from the validation. Two control charts will be constructed for each quality control concentration, an average chart which will monitor within- and between-day accuracy and a range chart to monitor assay reproducibility.

6.5.1. *Control chart limits*

Average chart limits are determined using the principles based on control charts for individuals (Duncan, 1986). The chart is constructed using the daily quality control averages and an average range based on the moving range of size 1. The average range (R) is calculated individually for each quality control sample. The average range for determining limits for the average chart, R, is the average of R_1 and R_2, respectively. The limits for the average chart are generated using $\bar{X} \pm 3R/1.28$. Sample data and calculations are shown in Table 7.

Table 7.   Control chart calculations: medium QC (15 ng ml^{-1}), accuracy (%)

SET I

Validation day	Rep. 1	Rep. 2	Average	Range	Moving Range
1	98.6	99.3	99.0	0.7	
2	100.0	102.0	101.0	2.0	2
3	100.0	102.0	101.0	2.0	0
4	100.6	99.3	100.0	1.3	1

SET II

Validation day	Rep. 1	Rep. 2	Average	Range	Moving Range
1	102.0	96.6	99.3	5.4	
2	100.0	110.0	105.0	10.0	5.7
3	100.6	102.0	101.3	1.4	3.7
4	100.6	98.6	99.6	2.0	1.7

In this example, the average moving range is 14.1/6 = 2.35. The overall average result is 100.8. The limits are $100.8 \pm 3(2.35)/1.128 =$

100.8 ± 6.3. The initial average control chart will have a mean of 100.8, with lower and upper limits of 94.5 and 107.1, respectively (Fig. 4). The average range from the eight sets of duplicates above is 3.1. The lower limit is 0 and the upper limit is 10.1 (3.10 x 3.27) (Fig. 5) (Bolton, 1984).

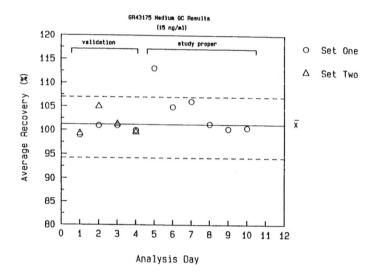

Fig. 4. Average chart

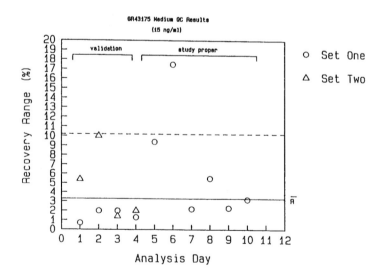

Fig. 5. Range chart

With these charts, there is a basis to reject a quality control sample after completing four runs. If duplicate samples fall within both the average and range limits, the data are considered acceptable. If the control charts show outliers, these values will be eliminated from the analysis and new charts prepared. For the validation run to be accepted, there must be at least one quality control value reported at each concentration with no more than one outlier per QC data set. After removing an outlier, the data set is unbalanced. This requires more caution in analyzing the remaining data. Note that these estimates of the standard deviation and range are approximate but may be used to initiate the control charts.

6.5.2. Modifying control charts

The initial control chart limits generated during the method validation should be used to determine whether the quality control results from the first study proper are acceptable. The average and range control chart limits are to be modified after the analytical runs and quality control results from subsequent clinical studies have been accepted.

6.5.3. Preparation of new QC sample

If the two QC sample sets prepared for the method validation show no significant differences, samples from either of the two sets can be used during the study proper. If the two QC sets show significant differences and both are used during the study proper, the duplicates during the study proper will consist of one from each of the two QC sample sets. If new QC samples are prepared, they must be analyzed prior to use to determine if they are statistically equivalent to the AC set being used in the study proper. If a single QC set has been used during the study proper, the new QC samples may be prepared by a single analyst. If two QC sample sets have been used during the study proper, the new QC samples are to be prepared by two different analysts. A two-tailed t-test will be performed to show the equivalence of the old and new QC sample sets. If the test shows a significant difference between QC sample sets, prepare a new set of QC samples and repeat the analysis.

6.6. Outliers

After completing the method validation, outliers can be identified either by significant effects in the ANOVA or by the control charts. If aberrant data are observed, the analyst should examine the calibration curve and quality control results to determine if the anomaly can be explained. The analyst will use judgment in deciding the seriousness of the problem and if the method validation or specific runs must be repeated. During the study proper, the control charts generated during the method validation will be the basis for rejecting quality control results.

7. Conclusions

The method validation presented provides a scheme in which statistical analyses of the bioanalytical data are used to determine the reproducibility and reliability of the method. Consideration has been given to emulate proper analysis conditions to understand the limitations and performance expectations of the method.

Acceptable criteria for the quality control samples and the assay are established based on the validation results and are used during the study proper. After accepting the analytical runs from a clinical study, the quality control results are incorporated into databases to modify their acceptance limits. This continuous validation process enables the analyst to monitor the performance of the method over time and to be confident that valid sample concentrations are being generated.

It is important to emphasize that this statistical approach in evaluating validation data is a more tangible concept than using arbitrary acceptance criteria. Our statistical design allows the bioanalytical data to reveal how reproducible the method should be on a routine basis. The analyst should interpret the statistical data using sound judgment in determining the reliability of the method and deciding if it should be used to support clinical studies.

Acknowledgments

The administrative assistance provided by Elsbeth van Tongeren, Lizbeth Liefer, and Amy Dobbins is gratefully acknowledged. Material presented in this chapter has been published previously in J.R. Lang and S. Bolton, (1991) *J. Pharm. Biomed. Anal.*, 9, 357– 361 and 435–442.

8. References

Bolton, S. (1984). In *Pharmaceutical Statistics*, Table V.10, p. 486. Marcel Dekker, New York.

Cardone, M. J. (1983). *J. Assoc. Anal. Chem.*, *66*, 1257.

Darioush, D. and M. R. Smyth (1986). *Trends Anal. Chem.*, *5*, 115-117.

Duncan, A. (1986). *Quality Control and Industrial Statistics*, 5th Ed. Irwin, Homewood, IL.

FDA (1985). *Draft Guideline for the Format and Content of the Human Pharmacokinetics and Bioavailability Section of an Application. (Docket No. 85 D-0275).*

Fisher, B. V. (1984). *Statistics in Chemistry*, (November) 443.

NCCLS Document EP10-T (1989). *Preliminary Evaluation of Clinical Chemistry Methods*.

Purich, E. (1980). *Bioavailability/Bioequivalency Regulations: An FDA Perspective in Drug Absorption and Disposition* (K.S. Albert, ed.). American Pharmaceutical Association, Washington, DC.

Shah, V. P. (1987). *Clin. Res. Pract. Drug. Regul. Affairs*, *5*, 51.

Vanderwielen, A. J. and E. A. Hardwidge (1982). *Pharm. Tech.*, 66-76.

Williams, D. R. (1987). *Biopharm.*, 34-36.

INDEX